MARINE, MUNICIPAL AND INDUSTRIAL
WASTE WATER DISPOSAL

MARINE, MUNICIPAL AND INDUSTRIAL WASTE WATER DISPOSAL

Proceedings of a conference held in Sorrento, Italy
23-27 June 1975

Sponsored by

Associazione Nazionale de Ingegneria Sanitaria
and
The International Association on Water Pollution Research

EXECUTIVE EDITOR — S. H. JENKINS

PERGAMON PRESS

OXFORD · NEW YORK · TORONTO · SYDNEY · PARIS · FRANKFURT

U.K.	Pergamon Press Ltd., Headington Hill Hall, Oxford OX3 0BW, England
U.S.A.	Pergamon Press Inc., Maxwell House, Fairview Park, Elmsford, New York 10523, U.S.A.
CANADA	Pergamon of Canada, Suite 104,150 Consumers Road, Willowdale, Ontario M2J 1P9, Canada
AUSTRALIA	Pergamon Press (Aust.) Pty. Ltd., P.O. Box 544, Potts Point, N.S.W. 2011, Australia
FRANCE	Pergamon Press SARL, 24 rue des Ecoles, 75240 Paris, Cedex 05, France
FEDERAL REPUBLIC OF GERMANY	Pergamon Press GmbH, 6242 Kronberg-Taunus, Pferdstrasse 1, Federal Republic of Germany

First edition 1979

Library of Congress Catalog Card No. 73-1162

ISBN 0 08 018070 1

PERGAMON PRESS

OXFORD · NEW YORK · TORONTO · SYDNEY · PARIS · FRANKFURT

Printed in Great Britain by A. Wheaton & Co. Ltd., Exeter

CONTENTS

SCIENTIFIC ORGANIZING INSTITUTES AND ASSOCIATIONS

Politecnico di Milano — Instituto di Ingegneria Sanitaria (Italia)

University of California — Berkeley (U.S.A.)

Università di Napoli — Istituto di Acquedotti e Fognature (Italia)

International Association on Water Pollution Research — I.A.W.P.R.

Associazione Nazionale di Ingegneria Sanitaria — A.N.D.I.S. (Italia)

SCIENTIFIC STAFF OF THE CONGRESS

President

Eugenio de Fraja Frangipane (Italy)

Scientific coordinator

Erman A. Pearson (U.S.A.)

Rapporteur

Harvey F. Ludwig (Thailand)

Scientific Secretary

Costantino Nurizzo (Italy)

SPEAKERS

Aubert, Maurice

Research Director, Institut National de la Santé et de la Recherche Médicale
(I.N.S.E.R.M.); Director, Centre d'Etudes et de Recherches de Biologie et d'Océanographie
Médicale (C.E.R.B.O.M.), Nice (France)

Bascom, Willard

Director, Southern California Coastal Water Research Project; El Segundo, California
(U.S.A.)

de Fraja Frangipane, Eugenio

Director, Istituto di Ingegneria Sanitaria del Politecnico di Milano; Vice-President,
Associazione Nazionale di Ingegneria Sanitaria (A.N.D.I.S.) (Italy)

Gameson, A.L.H.

Water Pollution Research Laboratory; Stevenage (U.K.)

Giaccone, Giuseppe

Professor of Algology, Facoltà di Scienze dell'Università degli Studi di Trieste (Italy)

Gilad, Alexander

Project Manager, Environmental Pollution Control Project, World Health Organization; Athens (Greece)

Isaacs, John D.

Director, Institute of Marine Resources, Scripps Institution of Oceanography; University of California, La Jolla (U.S.A.)

Jenkins, Samuel H.

International Association on Water Pollution Research I.A.W.P.R.; Executive Editor of *Water Research* and *Progress in Water Technology*; Birmingham (U.K.)

Lacy, William

Director, Industrial Pollution Control Division, Office of Research and Development, United States Environmental Protection Agency; Washington (U.S.A.)

Lewis, Robert

Chief, Division of Research and Planning, State Water Research Control Board; Sacramento, California (U.S.A.)

Lucas, Robert

Chief, Naval Engineering Division, U.S. Coast Guard; Washington (U.S.A.)

Ludwig, Harvey F.

Consulting Engineer, Southeast Asia Technology Co. (S.E.A.T.E.C.); Bangkok (Thailand)

Markantonatos, Gregory

Director, Sanitary Division, Ministry of Social Services; Co-manager Environmental Pollution Control Project; Athens (Greece)

Mearns, Alan

Biologist, Southern California Coastal Water Research Project; El Segundo, California (U.S.A.)

Mendia, Luigi

Director, Istituto di Acquedotti e Fognature, Centro Studi e Ricerche d'Ingegneria Sanitaria della Università di Napoli; Vice-President Associazione Nazionale di Ingegneria Sanitaria (A.N.D.I.S.) (Italy)

Oakley, Horace Roy

J.D. & D.M. Watson; Bucks (U.K.)

Olivotti, Raffaello

> Professor, Istituto di Idraulica e Costruzioni Idrauliche dell'Università di Trieste;
> Trieste (Italy)

Paoletti, Alfredo

> Director, Istituto di Igiene della Facoltà di Scienze — Università degli Studi; Napoli (Italy)

Pearson, Erman A.

> Professor of Sanitary Engineering, University of California; Berkeley (U.S.A.)

Portmann, John E.

> Ministry of Agriculture, Fisheries and Food, Fisheries Laboratory, Essex (U.K.)

Shelef, Gedaliah

> Associate professor of Environmental Engineering, Technion-Israel Institute of
> Technology; Haifa (Israel)

Snook, Gerald

> Tarry, Froude, Snook & Akerman; Clevedon (U.K.)

Storrs, Philip N.

> Senior Vice-President, Engineering-Science Inc.; Lafayette, California (U.S.A.)

FINAL REPORT OF THE SORRENTO CONFERENCE

Harvey F. Ludwig

Southeast Asia Technology Co. (S E A T E C), Bangkok, Thailand

1. PURPOSE AND THEME OF MEETING

The opening session dealt with the purpose, aims and objectives of the Third Congress, and included presentations from each of the sponsoring and collaborating agencies. These included the local governmental hosts at Sorrento, the Universities of Milan, Naples and California, the IAWPR, the Italian national sanitary engineering association, the WHO and the ISWA.

It was noted that the choice of Sorrento as the site for the Third Congress (following those of Trieste in 1972 and of Sanremo in 1973) was most appreciated in that Sorrento typifies a natural coastal setting of great beauty and tourist value which must be protected from pollution. Also Sorrento is the site of Italy's first waste treatment studies started in the 1950's by the University of Naples.

Prof. Eugenio de Fraja Frangipane, Chairman of the Congress, noted that the series of Congresses stemmed from establishment in 1971 by the Ministry of Public Health in Italy of rigorous standards for marine waste disposal, which, because they are more severe than in most other countries, had raised the question of their validity as related to rational and economical design of marine waste disposal systems. He further noted a worldwide trend toward establishment of marine waste disposal standards based on "do not harm ecology" concepts, and moreover requiring uniform high levels of treatment regardless of the natural waste-absorbing capacities of the ocean receiving waters, without evidence of scientific bases to justify the requirements. The result has been widespread design and construction of unrealistic treatment and disposal systems with gross loss and wastage of scarce public and private funds. It was the primary objective of the Sorrento Congress to try to clear up this confusion, and help develop guidelines for assisting both the regulatory authorities and the engineers with responsibility for designing treatment plants to reach agreement on standards that are rational, realistic and practical – which assure adequate protection without waste of energy and money.

Dr. A. Gilad, representing WHO at the Congress, reviewed recent developments on water pollution control in the Mediterranean region and noted this region historically has made great contributions to the progress of civilization, and that the solutions being worked out for protecting the environmental resources of the Mediterranean from pollution could similarly contribute to advancement of marine waste disposal and environmental protection technology throughout the world.

Dr. E.A. Pearson, representing the University of California, emphasized the importance of this type of conference in view of the billions of dollars to be spent around the world on marine waste disposal. To the extent that such meetings can contribute to design of treatment systems based on scientific fact rather than on emotion and other non-technical factors, great savings in public money will accrue.

The representative of the IAWPR and ISWA stressed the international aspects of marine waste disposal in the region. Lack of adequate attention to the problem in one part of the Mediterranean could lead to detrimental effects in other countries, hence a comprehensive coordinated approach needs to be developed. They stressed the importance of the Congress in furthering technology transfer and mutual under-

1

standing between the participants so essential in achieving international collaboration.

Representatives of the University of Naples, of the city of Sorrento, of the regional tourist agency and of ANDIS noted, despite concerns with local pollution problems in the Bay of Naples and studies and efforts to control this pollution over the past two decades, there has been a gradual depreciation in the environmental values of the region. It is hoped this Congress will come up with some new concepts of solutions which can reverse this trend. Attention should be given to the possibilities of waste reclamation and reuse together with discharge of effluents to the sea as a balanced approach for achieving both conservation and protection of natural resources.

2, ECOLOGICAL EFFECTS OF MARINE WASTEWATER DISPOSAL

This session was oriented to the need for bringing rationality and balance to the subject of influence of wastes on marine creatures, to place the role of pollutants in proper perspective among all the factors exercising influences. These include long-range cyclic and episodic effects, by which species incidents have been known to vary greatly over the past 200 years and more, natural phenomena (such as the explosion of Krakatoa) which disturbed the planetary environment far more profoundly than total waste discharge today, and many other activities of man which have been long accepted but which are ecologically very disruptive such as the farmer's ploughing of fields.

The phenomena involved in the relationship between pollution and the biosphere are most complex and difficult to quantify but from the mass of accumulating data it appears the most important effect of present day waste discharges to open coastal waters results from trace pollutants (such as chlorinated hydrocarbons) in the food web affecting certain top predators. Populations of coastal marine creatures exhibit great adaptability to environmental changes in the water environment, whereas marine birds and marine mammals appear to be the most seriously affected, possibly because of the long nongrowing period of adults which makes them vulnerable to trace material concentration.

Studies of the effects of waste discharge off the coast of Southern California, where a comprehensive programme of environmental research has been carried out over the past 5 years, have shown that many species of fish and invertebrates are abundant both near and away from the major outfall discharge sites. Near these sites the organism show changes in community structure, disease frequency, and rates of growth. There is greater biomass and lesser diversity near these sites. However it appears such discharges have little impact on the total productivity and fisheries of the region. However, DDT and other persistent synthetic organics display a wide coastal distribution and possible wide-ranging sub-lethal effects. Trace elements from the wastes and bottom sediments are not concentrating in fishes perhaps because these metals are not being discharged in biologically available forms, or because of compensatory adjustments in the organisms or communities. One of the outfalls has resulted in an increase in fin erosion disease and in the suppression of slime production in local resident benthic fish for reasons not yet identified; however, it was noted that the actual incidence rate of other diseases in outfall vicinities is not statistically different from areas outside the influence of the outfalls.

Continuing studies in France on the impact of wastes on marine organisms show that toxicological effects may be both natural in origin such as from red tides, or due to discharges to the sea from wastes, land runoff, rainfall, etc. The action of these toxic products may be exerted directly on individual species, or indirectly in affecting consumers of marine products. Such processes are now being studied in the laboratory by simulating trophic marine chains, and by this means it is anticipated it will be possible to organize controlled experiments by which toxic substances are transferred progressively up the food web, thus indicating the maximum limits which could result from such transfers which might occur in nature. Thus far it has not been possible to observe such series transfer from *in situ* studies in the field. The laboratory models used for simulating the trophic marine chains include pelagic and benthic organisms as well as crustacea and molluscs. It was pointed out that such linear transfers in a food chain may exist *nowhere* in the sea, however, and that the strong "downward" flux of detritus and reproductive products, together with a marine economy dominated by young, growing organisms and opportunistic forms may effectively preclude such bioaccumulation.

Another finding from the toxicological studies is that exogenous chemical like chlorinated hydrocarbons appear to disturb the naturally occurring chemicals which serve as chemical telemediators in the water medium, resulting in another type of toxicological effect called "ecological drift". Other aspects of the studies, dealing with the biostimulatory effects of wastes, note that discharge of nutrients to open ocean waters can be beneficial in increasing biological productivity.

Studies in Italy at Trieste have evaluated effects of sanitary wastes on marine phytobenthic organisms. It was found that the phytobenthonic communities are a much better indicator of waste impact than plankton or nekton because of their close connection to the substrate, their sedentary habits and their sensitivity to many water column parameters such as transparency, and that the effects of wastes on these organisms can be evaluated only by *in situ* field survey. Also, laboratory bioassay procedures were developed, using ten species of benthic algae, for determining their tolerances to various substances including MBAS and selected heavy metals. Within test periods of 3–5 weeks results can be obtained that are useful for guiding treatment plant design.

Studies in England on persistent organics in municipal and industrial wastes, including PCB's organo-chlorine pesticides, halogenated substances, organo-metals, phenols, detergents, and persistent oils, show that persistence does not necessarily mean such substances will be harmful in the marine environment unless their levels exceed tolerable limits which differ according to the ecological structure and other features of the area. Experience has shown that with proper design of waste treatment and disposal systems together with source controls it is possible to reduce hazards from these substances to negligible or tolerable levels and that the concept of total prohibition of such discharges is unnecessarily restrictive and costly. The main need is for research to establish definite safe standards for each type of persistent substance so that regulatory authorities will be in a position to approve of treatment systems based on rational design.

Studies in Denmark evaluated effects of waste discharges from sugar factories in stimulating growth of bacteria of types pathogenic to finfish and shellfish. The investigations show that *Vibrio anguillarum*, which is pathogenic to fish and shellfish, is found in both the water columns and bottom sediments in numbers exceeding 1000/ml and 100,000/g respectively. The data indicate that such "*Vibrio* eutrophication" results from discharge of wastes containing nitrogen together with cellulose and starch. This phenomena is significant for their reproduction. It was also noted, with respect to fishes exhibiting fin erosion diseases in the vicinity of marine outfalls in Southern California, that the diseased fish contained fewer pathogenic bacteria than did non-affected fish from coastal areas.

Studies in France evaluated the use of gamma spectroscopy for monitoring loss of radioactive materials into the marine environment from plants processing nuclear fuels, which perhaps represent a greater hazard to the environment than nuclear power plants because part of the radioactive residues are discharged to the sea. The selected tracers are ^{137}Cs and ^{106}Ru, and the monitoring may use both field surveys and laboratory methods, which serve to complement each other. Each plant is a special case and needs independent investigation. The long-term objective is monitoring of levels in bottom sediments rather than in the sea itself. It was noted that gamma spectroscopy, while a valuable monitoring tool, is not applicable for detecting the transuranic elements, which are the most hazardous of the radioactive materials which may be discharged.

In the question and answer period following the presentation, the discussion served to bring out the main theme of the session, namely (1) the effects of wastes on marine ecology are most complex and continuing research will be needed to quantify these relationships sufficiently to serve as the basis for establishing standards of treatment levels to be achieved for particular substances; and (2) the experience to date in design and operation of waste disposal systems indicates that present technology properly applied can result in treatment and disposal systems that both make use of the absorbing capabilities of the sea and provide good protection of marine ecological values.

3. PRELIMINARY SURVEYS AND STUDIES

This session was devoted to the technology of preliminary investigations that furnish the basic inform-ation for determining the best treatment and disposal system for solving a pollution problem, and for furnishing background data needed for proceeding with detailed design and construction and also for establishing the picture of the existing environment against which changes in environment resulting from operation of the system can be assessed. Over the past two decades the scope and depth of such studies, preparatory to design, have steadily increased in response to increasing public concern that a reasonably good understanding of probable environmental impacts be established before the commitment is made to proceed with design and construction.

The technology of preliminary studies and surveys as now practiced in California represents probably the most advanced methods and procedures thus far developed, based on extensive evaluations of effects of waste discharges on marine environment dating back to 1950, mostly carried out under sponsorship of the State Water Resources Control Board. The California experience has led to development and "codifi-cation" of detailed procedures for assessing marine environmental values based on use of all known applicable physical, chemical and biological parameters.

At present the California State Board requires both reconnaissance surveys and intensive advance surveys. The reconnaissance surveys have the objective primarily of indicating the general pattern of treatment and disposal to be used — how many communities and/or industries should be served by a common system and the general location of the treatment plant and outfall. The intensive surveys are intended (1) to obtain a statistically reliable picture of the existing ecology; including both the water column and the bottom sediments and including essentially four habitat groups of biota, namely phytoplankton, zoo-plankton, fish, and benthic organisms; (2) to obtain physical data needed for structural design of the outfall system, and (3) to ascertain current, dispersion, and coliform dieaway patterns sufficient for developing a model for predicting the distribution of waste constituents within the receiving waters. The key parameters for relating treatment levels to effects on marine environment are usually the coliform decay rate, floatables, toxicity (as determined by bioassays), and transparency/transmissivity. Also, the use of artificial substrates (setting plates in the receiving water and observing accumulations of organisms) appear to be a promising tool, now being evaluated, for assessing effects of waste discharges. Aerial photo-graphy and ERTS satellite imagery are being used to supplement the oceanographic surveys.

As an example of the application of the above procedures, surveys made in 1971 at Santa Barbara, California, were described including use of results to design of the outfall system. In this case isobars were established fixing the location of the offshore boundary where initial dilutions of 100:1 (and 200:1) would be obtained, and also where coliform dieaway would be sufficient for protection of the local beaches. The outfall discharge point was then selected to be outside both these boundaries and also to avoid proximity to local kelp beds and bottom rocky areas rich in benthic fauna.

Another example of preliminary oceanographic surveys is current work at Rio de Janeiro, in the Barra da Tijuca area. A series of curves were developed which permit selection of the necessary outfall length to meet any specified coliform standard. The initial fecal coliform concentration, following treatment limited to removal of floatables, was estimated at 7×10^7 MPN/100 ml, and the T_{90} dieaway rate, based on field observations, is approximately 1.0 h. The data show the onshore current velocity to have considerable effect, e.g. an increase in velocity from 0.37 to 0.51 knots (+ 38%) results in approximate doubling of the beach coliform concentration.

As an illustration of the types of sophisticated methods now available for use in marine ecological monitoring, the equipment developed and used by the Southern California Coastal Water Research Project was described in detail including various kinds of benthic samplers, underwater television systems, current meters, samplers for collecting floatables, microbe samplers, mussel buoys (for measuring pollutants taken up by filter-feeding animals), fish trawls, baited cine cameras, and atmospheric fallout samplers. This project, representing the most intensive marine ecological field studies ever made, literally has had to upgrade the entire line of existing sampling equipment to meet the requirements for advanced precision work.

With respect to current measurements, new techniques involving radar scanning of wave patterns indicate great promise for describing synoptic current patterns over entire bays or zones. This will provide an analytical tool, which is vastly superior to present methods which obtain data limited to a relatively few points.

A detailed presentation was made of the basic principles and of methodology for planning and executing comprehensive environmental pollution control studies, including a chart showing the interrelationships of technical, economic, and political factors involved in arriving at decisions on required treatment levels. Because costs of pollution cannot be precisely ascertained, decisions on treatment requirements cannot be entirely rational and judgements will be involved. An application of this methodology now under way at Athens, Greece, was described in detail. Special features of this study include evaluation of (1) movements of water masses in the Saroniko Gulf, involving a complex mixing of essentially four different basic water masses (outer gulf, inner gulf, central water and Elefsis Bay water) including salinity and temperature observations, (2) circulation patterns as affected by winds, and (3) nutrient circulation and primary biological production. The water mass movement studies show two freshening-salting cycles each year, a fall and a spring freshening, with salting in mid-winter and summer. These data are intended to permit design of a regional treatment and disposal system which not only conserves existing ecology but may help reduce problems of eutrophication and enhance aesthetic clarity.

In summary, much progress has been made in monitoring marine ecology in the vicinity of proposed marine waste disposal systems, so that the existing ecology can be described in statistically reliable terms and hence used for assessing changes due to operation of the disposal system following construction. However, the actual design of the treatment and disposal systems is still based on use of only a few parameters, such as coliforms, floatables, and transparency, hence much of the information from monitoring does not yet contribute to design. The primary reason for this is lack to definitive concepts on the proper role of marine water resources both as to conservation and utilization. Most regulatory concepts for marine waste disposal are still copied from fresh water pollution control technology, but the nature of ocean receiving waters is vastly different. The critical need for clearing up the existing confusion is to obtain concurrence on the proper role of ocean waters, after which both regulatory standards and design solutions can be established in harmony on rational scientific bases.

4. ENGINEERING DESIGN OF MARINE WASTE DISPOSAL SYSTEMS

This session focused on how to design marine waste disposal systems based on the information developed by the preliminary studies discussed at an earlier session. The ocean waters possess great purifying capacity, hence for most systems it will be desirable to use an outfall extending a sufficient distance offshore to reach deeper waters and currents where the waste will be rapidly mixed and dispersed. The engineer's choice varies between two extremes, a long outfall with no treatment and a treatment plant with little or no outfall.

Generally it will be much more economical to place primary dependence on the outfall, complemented by the necessary level of treatment. In many instances where discharge is to the open sea the most efficient plan is a long outfall together with removal of floatables. However, in some cases the regulatory agencies may require treatment to certain levels (sometimes secondary treatment) regardless of whether an outfall is used, and such regulations may then become the controlling factor in design. The session included detailed reviews of engineering design practices in England, in Italy and in Israel.

The great purifying capacity of the sea was illustrated by presentation of findings of studies of the fate of coliform bacteria discharged in wastes to the Bay of Naples. These data show that bathing in coastal waters poses very little health hazards except where gross pollution may occur, because the sea environment causes rapid dieaway of most pathogenic bacteria as well as the coliform indicator organisms. Transmission of disease requires that a sufficient number of the pathogens be ingested and moreover that these be virulent, but the adverse effects of the sea environment decreases the numbers very rapidly and attenuates the remaining organisms. In addition to pathogens contained in sanitary wastes, toxic chemical substances and radioactive materials, especially in industrial wastes, pose hazards to public health hence require monitoring and control. Also, most viruses are believed to have slower dieaway rates than pathogenic bacteria, but the significance of this is not known because the coliform standards now used are empirical indicators which may provide protection from viruses equally as well as for pathogenic bacteria.

From experience in England, useful tabulations were presented showing: (1) the performance of several alternative types of treatment in removing organic matter, major nutrients, detergents, and coliforms, and schemes including both capital and operating costs and including both energy requirements and labor requirements. The alternatives considered include primary treatment, chemical precipitation, floatation, partial biological, full biological (secondary treatment), and electrolytic treatment. It was noted that sludge disposal costs in municipal treatment systems may amount to 40% of the total cost. Physical-chemical treatment methods not included in the tables can also be used but usually would be more costly for treating most municipal or sanitary wastes.

Experience with marine waste disposal in England has led to development of an efficient system of treatment which dispenses with any need for separate sludge disposal. This design features use of a long outfall (sufficient to meet California Beach coliform standards) together with removal of solids by very fine screening. The screenings are then comminuted or disintegrated to a state of near liquefaction, then returned to the sewage flow for sea discharge. By means of a balancing tank, which receives the flow after screening, and which has a capacity at least half the volume of the submarine pipeline, the hydraulic flow into the outfall is maintained at a fairly steady state. Also, the liquified solids are discharged to the outlet chamber of the balancing tank so that the suspended solids content of the flow from the balancing tank is about the same as into the tank (about 250 ppm). Several such schemes have been installed in England over the past 6 years and monitoring of the bathing waters has shown the coliform concentrations to be well within the California standards (10/ml), even during periods of severe on-shore gales. Problems of septicity within the system, if they tend to occur in warm weather, can be managed by air injection. Also, provision is made for periodic removal of grease which may accumulate in the balancing tanks. The design layout also provides space for additional treatment steps should such ever be shown to be required. An interesting note is that this type of design can provide levels of protection 10 times higher than would result from a system of the same cost using secondary treatment and a short outfall meeting the regular British Government (Royal Commission) standards.

In Israel it is proposed to use a combination of rotor straining and chlorination of raw sewage as an interim treatment scheme for protecting the beaches of Greater Tel Aviv. Using a stainless steel rotor strainer with 0.75 mm screen openings, followed by chlorination at a dosage of 20 mg/l with 15 min contact time, settleable solids are reduced by 41%, suspended solids by 38%, floatables to a very low level, and fecal coliforms also substantially reduced. The rotor straining residue contains 15% solids and is amenable to further dewatering preparatory to disposal by a variety of methods. These improvements are expected to serve reasonably well for beach protection until a full-scale treatment plant can be built which will permit reclamation and reuse of the waste as a supplementary water supply. A tentative design under consideration would utilize the activated sludge process (10 h aeration) together with denitrification and chemical precipitation with alum or iron salts for removal of phosphates, and with the excess activated sludge returned to the outfall for discharge to the sea. Preliminary evaluations indicate the excess activated sludge, being free of floatables, should enhance the marine environment in that the excess sludge may be considered as valuable nutrients entering a nutrient-deficient sea. In addition, no problem is anticipated with fish toxicity in the receiving waters due to the aluminium or iron from the coagulating process because it is expected that these metals will be removed from the water medium through precipitation of phosphates. In this system denitrification is considered necessary because the treated effluents will be used for recharge of ground-water where it is desired to avoid buildup of excessive nitrates.

A presentation on methods for constructing ocean outfalls noted that intensification of construction of offshore oil and gas pipelines over the past decade has resulted in improved techniques of design and construction which are equally useful for submarine outfalls. Details were given on the presently-used methods for constructing outfalls including (1) the bottom-tow method, in which "strings" of pipe are assembled on land and successively winched to sea by offshore barges; (2) the pipe-by-pipe method, including use of the floating barge or "horse" for positioning the pipe, with divers joining the pipe sections under water using mechanical joints; (3) the float-and-drop method, which involves floating a pipeline (complete in sections) to the appropriate location, then sinking it; (4) the lay-barge method, used for laying oil or gas lines up to 1,250 mm, with all jointing being done on the barge, which is feasible if a lay-barge is available between major oil/gas pipeline contracts; (5) the reel-off-barge method, limited to 250 mm pipe, where the whole pipe is rolled off a large drum on the barge, especially applicable for plastic pipe; and (6) tunneling, such as used recently on the west coast of Scotland.

Trenching is usually essential for protecting the pipeline near the shore, and the main problem is in maintaining the trench long enough to permit installation of the pipe. Methods for trenching include submarine ploughs, high-pressure jetting, grab excavation, and dredging. Steel is the traditional pipeline material, but needs to be lined and coated to resist corrosion. Aluminium is being increasingly used, and ductile iron pipe may well prove to be more economical and satisfactory than steel and other materials including plastics. Plastics have come into use in recent years but must be weighted to be held in position. An interesting combination of materials is a high-density polyethelene liner reinforced with steel pipe and reinforced wiring.

Additional experience reported at the meeting may be summarized as follows: (1) sludge disposal by barging to the open Atlantic ocean has been used by the City of Philadelphia for many years, without any evidence of accumulations of bottom sludge, nor of heavy metals, nor of any damage to ecology. Nevertheless the city has been required by the U.S. EPA to move its disposal site farther offshore, and by 1981 to discontinue this method of disposal, despite the scarcity of land available for other disposal methods and the relatively very low cost of sea dumping (about $ 11 per 1,000 gallons); (2) current practice at New York City, for discharge of sewage effluents to confined marine waters, is to achieve at least 85% removal of BOD and suspended solids thus reducing suspended material to levels where disinfection with chlorine can more readily destroy pathogenic bacteria and viruses; (3) studies of receiving water conditions off the Belgian coast indicate municipal wastes may be economically disposed of by discharge to the open sea without adverse effects; and (4) a system of outfalls discharging wastes around the periphery of the island of Puerto Rico has been planned to give comprehensive waste disposal service for the entire island; recent oceanographic surveys show by far the most economical treatment and disposal scheme for these outfalls is to use the combination of long outfalls with limited treatment.

5. INDUSTRIAL WASTES

Design of treatment and disposal systems for industrial wastes is often much more difficult than for municipal or sanitary wastes because of the great variations between the waste characteristics of different industries and the complexity of characteristics of many wastes. The subject has been so complex that only in the past few years, as a result of an intensive program of study by the U.S. EPA, is definitive information being developed in sufficient detail to assist the engineer to prepare rational designs. The "point source" manuals being issued by the EPA, for each category of industry, represent virtual "bibles" of information on the various industries including identification of the significant parameters applicable for each category.

A presentation on current regulatory concepts for control of industrial wastes in the U.S.A., now being developed by the EPA, shows that requirements for in-plant recycling and reuse are generally less demanding than for municipal water reuse systems. Hence direct industrial water reuse through closed-cycle systems should be technically and economically achievable much sooner than for municipal systems. While the impact of water pollution control on industrial plants will involve only small increases in total manufacturing costs, from 1 to 5%, nevertheless these costs will be sufficient to encourage industry (1) to reduce both the quantities and strengths of wastes discharges, and (2) to progress towards more and more use of closed cycle systems (energy cost for environmental quality is less than 0.25% of total energy cost). This will emphasize recovery and reuse both of water and of the suspended and dissolved materials normally discharged as waste impurities.

A review of current practice in disposing of industrial wastes to the marine environment noted the importance of using an outfall to gain access to the unconfined deeper offshore waters, beyond the zone of nearshore water to be protected, where the dilution, dispersion, and absorption capacities are high compared to discharge into shallow confined waters. For discharge into confined marine waters, the treatment requirements will be virtually the same as for discharge to inland fresh waters, hence secondary and even tertiary treatment will often be required. For discharge to open coastal waters only a few parameters will usually be applicable such as floatables, colour, and toxicity, hence the treatment requirements will be greatly simplified. Usually laboratory bench scale or pilot scale testing will be needed for evaluating a particular waste's response to various treatment measures. Following such testing and field oceanographic surveys of the receiving water situation, a treatment and disposal plan can be formulated based both on the technical facts and the regulatory constraints. Monitoring of the affected receiving water is very important to "prove out" the validity of the design, especially because of the present confusion from the regulatory point of view on the extent to which marine waters may properly be utilized for waste disposal.

A presentation on control of oil waste discharged from ships reviewed regulations adopted by the International Conference on Marine Pollution (London, 1973). One of these regulations specifies that all oil tankers of 150 gross tons or more and all other ships of 400 gross tons or more have oil-water separating equipment to produce an effluent with not more than 100 ppm oil, plus an oil filtering system which will further reduce the oil content to not more than 15 ppm. In response to this policy the U.S. Coast Guard has been conducting research since 1970 to develop treatment units of various sizes suitable for removing oils from its own ships. The most serious treatment problem on Coast Guard ships is in processing cold fresh water contaminated with used diesel engine lubricating oil. The resulting separator consists of three pressure vessels in series. The first filters out dirt and removes entrained air; the second and third use filter-coalescing cartridges which demulsify the oil droplets with the rise to the top where oil accumulations are removed automatically by means of a capacitance probe and sent to the ship's oil tanks. A monitoring device continually analyzes the effluents for oil and automatically recirculates unsuitably-cleaned water. This system represents a satisfactory answer to the problem except for its inability to handle heavy fuel oil discharges characteristic of some types of ships.

In Israel attention is now being given to controlling pollution from industries, virtually all of which are located along the Israeli coast. Most of these have discharged untreated wastes to the sea via local stream. The regulatory approach has been (1) to require the industries to reduce water consumption an average of 80% thus both conserving water and concentrating the wastes, and (2) establishment of regional industrial waste treatment systems serving groups of plants, featuring in-plant recycling and reuse, use of some of the treated waste as irrigation water, isolation of toxic wastes for special handling (e.g. discharge to abandoned oil wells), and discharge of residual wastes via outfalls to the sea. This results in very flexible systems capable of continuing adjustments to suit changing conditions.

Other finding on the subject of industrial wastes included: (1) biological treatment can be adapted to wastes having high salinities including both polyalcohol wastes and municipal sewages; (2) a system in southwest Scotland successfully disposes of highly acid wastes to an estuarine environment which readily neutralizes the waste; the discharge is limited each day to a 2-h period following high tide to insure against upflow in the channel, and a comprehensive system of instrumentation with a mini-computer is used to control the discharge including alarms in event of malfunctionings; (3) in addition to the oil-water shipboard separators described above, other studies in the U.S.A. have under development various methods of oil-water separation and oil spill recovery using absorbing foams which can be regenerated by squeezing and microfiltration systems for treating all types of shipboard wastes including sanitary wastes; and (4) installation of mercury removal at caustic soda plants in northwest Italy resulting in significant decreases in mercury accumulations in marine organisms as determined by a program of continuing sampling and analysis, indicating a return to normal concentrations in tussues of the organisms in about 2–3 years.

6.   REGULATORY ASPECTS

Because of pressures in California and elsewhere in the U.S.A. from the EPA, the decision has been made to require secondary treatment for all municipal waste systems discharging to open coastal waters, regardless of (1) the record of many major systems which have given good service over the years using primary treatment together with long outfalls, (2) the lack of scientific data to justify the decision, and (3) the very high extra costs which will be involved. One reason for this is the lack of sufficient records of performance of the existing systems to prove definitely that secondary treatment should not be required. In the industrial waste field a different pattern is evolving with respect to attempts by EPA to impose as standards for all discharges (whether to fresh or marine waters) the limiting emission levels set forth in the EPA industrial waste manuals. It was stated that these proposals will be the subject of legal actions in the U.S.A., brought either from conservationist who regard the limits as too lax, or dischargers who regard the proposed limits as too severe. If these differences have to be resolved in the courts, it should eventually lead to official and widespread recognition of the importance of monitoring to assess the impact of wastes on ecology as the only means for safeguarding public health and delineating the marine ecological values to be protected and the necessary limits of controls.

A comprehensive review of the legal aspects of marine waste disposal was presented which noted that most regulatory agencies depend upon use of effluent and/or receiving water standards, and monitoring to insure compliance, as the only feasible means for applying legal control methods. Regulatory agencies in most countries are independent agencies with powers limited to setting and enforcing standards. They do not also have responsibilites for plant siting, construction, and financing, hence can presumably weigh the impact of these influences from an objective point of view. Most standards for municipal wastes utilize the coliform index as the key parameter, with the numerical limits allowed usually being expressed on a statistical basis. Recently attempts have been made to relate coliform incidence in bathing waters to the risk of infection from viruses. A committee of the Council of European Communities has recently recommended adoption of uniform standards for receiving waters including limits for other parameters. Where coliform counts have been too high, the legal response has been either quarantine of bathing or shellfishing areas or prohibition of shellfish sale. Another legal approach to controlling pollution is to use the power of permits and/or the power of financial aids or grants to hold up construction and development. With respect to sludge dumping at sea, which has lately been the subject of much legal controversy, the scientific data indicate that this practice, when properly monitored, is a satisfactory and very economical method of disposal at most locations.

7. SUMMARY

The Third Congress at Sorrento has been most timely in being able (1) to focus on the current controversial situation throughout the world on the extent to which the marine environment can properly be used for disposal of wastes, and (2) to contribute significantly in defining the current situation and in delineating guidelines for use by waste dischargers in designing waste treatment and disposal systems and by regulatory agencies in managing water pollution control programs.

Over the past two decades there has been a vast increase in research and development on the technology of waste treatment and disposal, including marine waste disposal; hence a great deal of scientific information is now available to guide the engineer in designing rational waste disposal systems. However during this same period there has been a comparable escalation of the bureaucracy of regulating water pollution control. A key aspect of this development in regulatory procedures has been a tendency by the regulatory agencies to require uniform levels of treatment regardless of the receiving water capabilities and moreover to set the requirements at high treatment levels. While this system of regulation sounds simple as well as simplistic, it is also very costly, very wasteful, and probably will not achieve the desired protection of environment.

The consensus of the Third Congress is that waste disposal systems should be designed on the basis of the scientific facts, as determined from field surveys to assess the existing ecology, so that the selected design profits from use of the valuable assimilative capacities of the ocean while at the same time assuring protection of all essential marine environmental values.

Another important contribution of the Third Congress is recognition that waste disposal to open coastal waters is a new technology, quite different from the technology of treatment and disposal relating to discharges to inland fresh waters. The present disparity of opinions on the extent to which the ocean should be utilized for absorbing wastes stems mostly from continued adherence to questionable or anachronistic theory (e.g. food chain entrainment of toxic substances) and from lack of concurrence on the proper role of marine waters as compared to other disposal recipients including fresh waters and land. At this time it appears that the main parameters for assessing effects of wastes on the marine environment are those relating to aesthetic values (floatables, colour, turbidity), to particular toxic effects (such as chlorinated hydrocarbons), and to public health (pathogenic bacteria and viruses). Another emerging concept is that wastewaters should be reclaimed and reused to the extent feasible, with the residual waste, properly controlled and treated, discharged to the sea. The various presentations at the Third Congress included description of numerous treatment and disposal systems, including municipal and industrial and ship-board wastes, mostly in Italy, the United Kingdom, Western Europe, and the U.S.A.

The excellent attendance at the Sorrento Congress (more than the second, which was more than the first) attests to the success of the Congress as a means for promulgating the technology of marine waste disposal. Also, attendance by consulting engineers and representative of manufacturers was much higher than before. In fact, the Congress was well attended by representatives of all sectors concerned with the technology of water pollution control excepting, only and unfortunately, representatives of the legal and lay group, that are so busy promulgating new water pollution control regulations these days.

PRESENT STATE OF KNOWLEDGE ON MARINE WASTE WATER DISPOSAL

E. de Fraja Frangipane

Istituto di Ingegneria Sanitaria, Polytechnic Institute of Milan, Italy

Different problems concerning the design of marine waste water disposal systems are being discussed all over the world and constitute the object of studies and investigations intended to attain their definition in scientifically correct terms.

An examination of such topics seems quite convenient now that the whole problem concerning the development of a methodology and of efficient and rational standards of pollution control is in such a state of uncertainty. As a matter of fact, depending on the terms of such a control, it will be possible to single out the mechanisms of intervention; such a control, in fact, involves a prior definition, in scientific terms, of what is generally meant by pollution.

Now that everybody speaks of pollution, it might seem superfluous to look for a definition or for an evaluation of such a condition. By considering the extent of the rumour about it and especially of what is done, at least in some areas, most people would think that everything is known about pollution.

Instead, man has only lately begun to care about the possible consequences of his own activities on the natural environment; all attempts to scientifically determine such effects are even more recent and generally incomplete. At present it is absolutely impossible to draw any certain conclusion about the behaviour of the ecosystem against the impact of pollution and about its capacity to absorb all wastes produced by human activity.

The restrictions imposed, mostly by standards of effluents and much less frequently on the receiving waters, still largely involve that "no negative effect must occur". That would actually mean that every discharge should be forbidden.

As a matter of fact, no significant standard can be established so far on rational bases, owing to the lack of scientific knowledge that might allow the definition of adquate parameters and of such acceptability limits to guarantee a valid ecological protection. If, on the one hand, the standards based on the principle that "no negative effect must occur" would often lead to unnecessary and vainly expensive interventions, on the other hand the natural delays occurring in the ecological processes increase the risk of underestimating that measures of control are necessary.

Certainly, up to a given limit of saturation, the wastes produced by human activity may be safely absorbed by the natural ecosystem; beyond it, they start accumulating at noxious levels. Such a limit is not known and its determination constitutes a very important goal, essential for a correct characterization of the interventions.

Hence, the priority task of scientific research must be that of giving a better scientific basis for both design and operation of marine disposal systems.

Such a scientific basis being lacking or inadequate, the trend has lately developed to establish government policies tending to actions that, independently of costs, would offer the advantage of immediacy, spurred as they are by public opinion, i.e. by emotional feelings without any scientific ground. Therefore, the Governments' tendency is to adopt and impose uniform standards that require high treatment levels, independently of correct environment protectional requirements. A policy of this kind may be hardly justified on scientific bases and may be responsible for considerable economic loss; works may be installed that may prove to be useless and of uncertain adequacy.

It must be stressed that enormous funds are required for works for the reclamation of waste waters and their prompt availability is hardly possible. Hence, expenses must be thoroughly justified and capable of attainment by public works carried out under conditions that produce the best possible results from the very beginning.

In this regard, there is a genuine risk of spending all available funds for quite expensive, though little useful works, and later, if failure of such works occurs, to lack the funds necessary to achieve the right solution by means of better works.

Under such conditions and from the practical point of view, since wastes must somehow be dumped, the designer of marine waste water disposal systems must supply them with the highest flexibility, i.e. they should be such as to be easily modified and technically improved whenever necessary. Furthermore, he has also to consider the possibility that the systems initially foreseen may adversely affect the receiving environment and may even be inadequate from the technical point of view.

Marine pollution and relevant remedies have certainly been dealt with on several occasions: conferences, seminars, numberless theoretical and experimental investigations have been devoted to the problem.

What has been done so far supplies researchers and designers with a large amount of information with regard to both the peculiar characteristics of the different types of marine pollution and the technical-scientific criteria that should pervade the general design and dimensioning of systems.

However, as happens in other fields too, in the development of scientific research, full advantage is not taken of the work done so far, and this is even sometimes ignored in its application.

In some respects several scientific aspects of the problem of pollution have already been satisfactorily resolved — as may be seen by a careful examination of the literature. In other respects, it is enough to remember that in a number of practical circumstances, the problem of marine pollution and of its technical remedies is dealt with exactly like that concerning reclamation of freshwaters (rivers, lakes). However, the fact that several scientific conferences have been expressly devoted to this topic and that research groups and scientific associations have been established with the main purpose of acquiring further knowledge on it suggests that the problems mentioned (marine pollution and pollution of freshwaters) are quite different in many respects, with regard to both the spread and sequence of different types of pollution, and the criteria that have to be used to measure and control it.

The purpose of this conference is essentially to deal with some features of marine pollution with particular reference to the design of treatment and disposal systems and to the execution of *in situ* investigations, which, from case to case, should direct such planning. Futhermore, some biological and hygienic questions strictly pertaining to the mentioned topics are tentatively tackled.

Before highlighting the features that will be particularly probed during this conference, we wish to recall some of the most significant conferences and seminars held on the topic.

On a national level, mention must be made of the first ANDIS meeting, Trieste, 1955, because of the wealth of information contained in the communications delivered and the value and relevance of the main lectures by De Chigi and Rio, and by Ippolito. We also have in mind the meeting organized by the "Associazione Ingegneri della Provincia di Bologna" in 1966, the International Conference of FAO, Rome 1970 and the meetings on pollution of the Tyrrhenian and Adriatic Seas organized by the Chamber of

Deputies. Furthermore, a number of symposia, panel discussions seminars have been devoted to the problem, on the occasion of exhibitions of equipments and materials for sewers and purification plants: among them, we would mention those periodically held at Padua and — with the cooperation of FAST — at Milan.

On an international level, among the many meetings on the topic, some were particularly successful for the soundness of the contributions and for their relevance to the application of existing knowledge: e.g. the 1st International Conference on Waste Disposal in the Marine Environment, Berkeley, 1959, the two recent conferences (Bournemouth and London) organized in Great Britain by the Institute for Water Pollution Control and by the Water Pollution Research Laboratory of Stevenage and finally that held in Arhus (Denmark) organized by the International Council for Exploration of the Sea.

Furthermore scientific conferences are organized every other year by CERBOM, on a variety of chemical, physical, biological, microbiological problems concerning pollution of the marine environment and in particular of the Mediterranean, the proceedings which are regularly published in the *Revue Internationale d'Oceanographie Médicale*. These are of a particular interest for the practical purposes of design, in view of the attention that is given to the application of information obtained by scientific investigation.

Outstanding scientific contributions may be found in the proceedings of the meeting organized every 2 years by IAWPR and in the journal of the Association *Water Research*. The proceedings of this Conference are published as a one volume supplement to its sister publication *Progress in Water Technology*, Vol. 4.

In addition to the several meetings, some training courses have been devoted to this topic all over the world; among these we mention that on "The Control of Pollution in Coastal Waters" sponsored by OMS, Arhus (Denmark, 1970) and that on "Pollution of Coastal and Estuarine Waters", University of Berkeley; the summaries of the lectures, which are available at our Institute like the other papers mentioned, are of particular interest for the design of sewers, diffusers and reclamation systems.

Marine pollution and related remedies constitute an important section of the training courses that are regularly held at Delft; with regard to Italy, we may mention that some lectures delivered at the updating courses on sanitary engineering that are periodically held at the Polytechnic Institute of Milan, concern this topic; their texts have been published in the course notes.

The fundamental paper by Pearson, which — though written 20 years ago — is still up-to-date, is nearly all what we have on the design features of dumping systems. Although several articles may be found in technical literature, we actually lack a comprehensive work, like a handbook, orderly collecting the methods of calculation and the data in common use in the design phase.

Instead, a careful examination of the present state of knowledge and a summary of the problems that must be urgently solved, are reported in a recent paper edited by Pritchard and Pearson, entitled "Wastes Management Concepts for the Coastal Zone — Requirements for Research and Investigations". It consists of 125 pages, summarizing the contents of 16 reports written by specialists, under the patronage of the National Academy of Sciences and of the National Academy of Engineering.

We may conclude by reference to the reports on pre-design and control investigations that have been carried out in several countries; though such data are not general in character (as a matter of fact they were accomplished in well definite places and conditions), to the specialists their interest is beyond the particular case with regard to investigation methods and results. It is enough to recall the investigation done in the San Francisco Bay and the studies and investigations entitled "Southern California Coastal Water Research Project", which are of exceptional interest and have already been issued.

As to Italy, pre-design investigations have been carried out in the Gulf of Trieste and of Naples.

This short review of the available literature leads to some general observations.

First of all, the determination of the present situation of knowledge about marine pollution would require a number of treatises by the respective specialists (in this regard we have already the report "Wastes Management Concepts for the Coastal Zone").

Secondly, little is done to encourage practical use of the large body of information already available. As already mentioned, in some circumstances, the existence of the problem of marine pollution and of its peculiarities is even ignored.

Hence, one may wonder whether it is more profitable for practical purposes to direct efforts toward research in order to increase knowledge or toward the application of what is already known.

An obvious answer to such a question may be that both paths must be followed. However, if the dissemination of information and its application is not pursued, unsatisfactory consequences will occur more frequently.

As a matter of fact, the above considerations seem consistent not only with the Italian situation, but also to a lesser or greater extent with situations in other countries.

A further question worthy of consideration and partly connected with the above statements concerns the area where the planned disposal systems must be situated and predesign investigations. It is generally recognized that reclamation of coastal areas involves the construction of sewers, the siting and building of purification plants, the examination of the possible re-utilization of treated effluents, the study of points of discharge disposal.

Hence, instead of tackling the problem by considering every single discharge — as is sometimes done for reasons of urgency, e.g. when the bathing season is drawing near — coastal resorts are recommended to arrange general projects that make best use of the various technical possibilities to deal with all discharges.

We are pleased to note that the "special projects" of the Cassa per il Mezzogiorno for the reclamation of the Gulf of Naples, which are already on the straight path as far as design is concerned, will follow such a trend.

Obviously, both pre-design investigations and the realization of such general designs require adequate funds and time.

Hence local administrations are often prone to act in a different way for a number of reasons: urgency (as already mentioned), the wish to exhibit some completed works to satisfy critical public opinion, or the lack of necessary funds. Therefore, it is desirable that local administrations should organize adequate, comprehensive projects that are backed by funds necessary to carry out an investigation and to elaborate a suitable project. Furthermore, such "intervention projects" would involve identifying the more precise responsibilities at the scientific and technical levels and probably also indicate the various professional efforts that will be required.

As to investigations *in situ*, greater attention should be drawn to checking the efficiency of works already built; as a matter of fact, such investigations are even more sporadic than the pre-design ones, being less well financed since the need for such an exigency is rarely anticipated by the client.

In Italy, as in several other countries, the microbiological quality of coastal waters is conventionally defined as a function of coliform numbers. Such a criterion suffers from various defects: on one hand, bathing may be forbidden in a given area without any adequate scientific justification; on the other hand, it does not seem to guarantee mollusc eaters from risks connected with the ingestion of viruses and toxic substances accumulated by microorganisms. However, that parameter may be considered as a sensitive indicator of faecal contamination; furthermore, rational works designed for the re-establishment of aesthetically satisfactory conditions and for keeping the coliform levels in coastal waters below a given standard, are generally beneficial for other purposes (e.g. toxicity concentration in sea waters, disposal of the degradable load, distribution of nutrients, etc.). While this may be true wherever the coliform limit is attained without chlorination it is not necessarily the case if the limits are achieved by waste water chlorination, since this may be accompanied by toxicity and it does not permit natural purification to effect bacteriological load abatement. Nor does it allow natural degradation of the organic substances and assimilation of nutrients.

Hence, a more precise picture of a given environmental situation (in this case, a more satisfactory control of the efficiency of the works already built) might be obtained by simultaneously performing periodical coliform counts on waters used for bathing, mollusc breeding and free harvesting, as well as systematic controls of the effluent toxicity and of some other easily-measurable parameters (e.g. floatables, macroscopic solids, settleable solids or turbidity, etc.). Obviously, integration of personnel and of laboratory and field equipments is necessary to fulfil these new tasks if staff and resources are not to be overburdened.

Under such conditions it seems that the seriousness of a given situation as well as its gradual progress as a function of the real efficiency of the works built, might be better demonstrated.

Controls of this type, systematic and applied to the whole coastal line subjected to sampling for microbiological analyses, might supply an approximate, though wide, picture of the situation. In the most important cases, they should be accomplished jointly with detailed hydrobiological and biological investigations for the purpose of quantitatively ascertaining the ecosystem response to changing situations.

Such investigations are dealt with in some of the communications to this Conference. It would be interesting to gather proposals about systematic surveys mentioned above and the pattern-schemes for biological investigations (e.g. on phytobenthos) which it is suggested should be considered as compulsory in all important cases. The performance of such investigations is in one sense anticipated in Italy from the laws concerning sea fishing; according to such a criterion, some satisfactory results have lareday been obtained, e.g. at Trieste.

With regard to the design of treatment and disposal plants, many of the ideas put forward are up-to-date and worthy of being considered in detail, and they will be extensively dealt with by various speakers. Above all, it must be emphasized that treatment and disposal plants must be conceived within a comprehensive overall system which is adequate for all purchases.

With regard to outfall dimensioning and lay-out, the existing knowledge acquired is satisfactory and widely supported experimentally. Furthermore, the not infrequent outfall breakdowns or stoppages are due more to the fact that builders do not apply known techniques rather than to a scarcity of technical progress in the area.

With regard to the functional requirements according to which outfalls are generally designed (the aim in general is the attainment of a certain degree of dilution obtainable by suitable diffusion and the observance of a given coliform standard in waters used for bathing or mollusc breeding), it may be stated that, after completing the required design investigations, the present state of knowledge is adequate to reach the required goals.

In this case the conclusion may be reached that, if the required design objectives are not met it is because designers have not carried out sufficient preparatory study and design and funds are not enough.

Indeed, it sometimes happens that outfalls are dimensioned according to the available funds and not to the actual environmental needs.

Pinch penny policies are sometimes adopted in the field of purification plants. In this case, efforts have been made to remedy the situation resulting from underdimensioned plants, by imposing rules to all designs; so as to avoid chronic overloading of plants an interesting example is supplied by a set of technical directions formulated by a qualified office of the "Cassa per il Mezzogiorno".

At this point we wonder whether technical directions could not be elaborated even with regard to outfall designs and construction, that might be general enough to avoid roughly inadequate dimensioning and construction, but allowing sufficient flexibility to meet the variations that occur from place to place.

With regard to the functional feature of outfall designs, two major questions concerning marine environmental protection elude the methods of calculation and the forecasting patterns commonly adopted. Indeed, it seems that as yet no satisfactory correlation can be found among the effects on primary productivity in a given sea area caused by a discharge, whose position, type of outlet, yield, biostimulating nutrient potential are known. Furthermore, the long-term distribution of persistent substances discharged into the sea from a given outfall cannot be foreseen to a satisfactory extent on the basis of the methods of calculation and oceanographic investigations commonly performed.

It may be stated that the problem of a too high enrichment of sea waters usually is far less worrying than that concerning inland waters and especially lakes. With regard to the Mediterranean, instead, the problem is the opposite, in view of the well-known nutrients scarcity characterizing it. On the other hand, introduction of persistent toxicants must be suitably controlled by limitation and control at the point of origin so as to leave an amount of substances that can be removed by subsequent purification of the waste water.

Even if such considerations are valid, urgent problems still remain: in fact, the problem of nutrients exists even along the Mediterranean coastline directly affected by outfalls: there, nutrient concentrations may be so high and their residence times so long that sea water transparency and submerged vegetation may suffer. Furthermore, under particularly unfavourable physiographic circumstances − e.g. when wastes are dumped into bays, inlets, coastal regions where water exchange is low in respect of the discharged amount of nutrients and biostimulants − the consequences on benthonic communities may become serious. It is known that by using treatments commonly adopted, an appreciable amount of persistent toxic substances remain; consequently, the residual toxic effects on the marine environment largely depend on both the effectiveness of inland removal systems and on the level of dilution and dispersion to which the effluent is subjected.

The characteristics of waste waters should be better represented not only by the parameters already known (suspended solids, oils, greases, etc.), but also by a toxicity factor and by an expression of their nutrient and biostimulant potential (e.g. the former may be detected by fish toxicity tests and the latter by testing with algae cultures).

Obviously, such investigations should preferably refer to named organisms and be performed according to well defined methods. For example, toxicity may be expressed by toxic units and the overall amount of discharged substances by "relative toxicity".

Therefore, calculations should be carried out that allow a forecast to be made − on the basis of precise oceanographic measurements − of the results of exposure of the various marine environments affected by discharges, to their toxic and/or eutrophic elements.

With suitable reservations on the fitness of a procedure of this type to satisfactorily represent the effects of a discharge on the ecosystem, the interest of this procedure — which has already been experimentally tested with some success in California — must be evaluated in connection with the possibility offered of comparing different solutions for aquatic environment protection on a uniform basis and in quantitative terms. By continuing *in situ* investigations on the actual response of aquatic biota to the impact considered (i.e. exposure to the inhibitory and "eutrophicating" action of the discharge expressed by established indices), it should be possible to further improve this procedure.

In this short survey of the basic elements of waste water disposal in the marine environment consequences, some highly significant facts emerge: the available scientific information is still insufficient to the identification of the ideal solution to a problem.

In such circumstances, the engineer must therefore select a realistic policy and act without having all the desirable information data. Consequently his design systems must be capable of modification and operation with the maximum of flexibility.

This is not always easy or possible, especially in countries where traditional design concepts are deeply rooted. Such conservatism occurs rather frequently because the technology concerning waste water disposal tends to follow empirical processes which go back many years.

Only in the last few years have efforts been directed toward the study, understanding, control and codification of all the factors concerned with the environment response to pollution, for the purpose of enabling the technologist to establish the significant parameters that must be used as a basis for engineering comprehensive pollution abatement policy. This topic is being widely studied and discussed, but it will take years to obtain all the information that is required. Criteria are having to be used that are of no help because they suffer from a lack of scientific basis. Being arbitrary, the criteria are sometimes too severe, although they may be administratively attractive. However, criteria may also be too lenient and ignore the actual risks. It follows that the simple observance of criteria — which may be legally admissible — in general does not give the optimal result, i.e. environment protection at the lowest cost.

It may take years to establish criteria based on scientific grounds. Hence, when the problem under examination is of great importance and the solution costly, the only satisfactory solution must be based on a tailor-made pre-design investigation, that is an investigation leading to the identification of the essential parameters in the minimum time possible.

If discrepancies arise between scientific results and numerical criteria, no conflict should arise provided the criteria are adjusted so as to conform to the scientific results and, where necessary, fit the parameters. Faults in design can then only arise by ignoring scientific fact for the sake of conformity with preconceived criteria.

RESPONSES OF COASTAL FISHES AND INVERTEBRATES TO WASTEWATER DISCHARGES*

Alan J. Mearns

Southern California Coastal Water Research Project, El Segundo, California, U.S.A.

Summary — Populations of nearshore fishes and invertebrates respond to municipal wastewater discharges through observable changes in abundance, diversity, community structure and diseases. Both groups of organisms show some enhancement in numbers with a consequent decrease in diversity in the immediate vicinity of large (12—360 MGD) coastal outfalls. A disease caused by a still unidentified component from one outfall affects a variety of resident benthic fish. Effects of persistant synthetic materials, such as DDT isomers, extend well beyond dilution zones and are clearly undesirable. However, there appears to be little or no bioaccumulation of trace elements (metals) originating from primary effluent. Effects of the discharges on major fisheries resources are small relative to other factors and are thus difficult to assess.

1. INTRODUCTION

The complexity of life in the sea seems enormous to comprehend. Yet as trustees of water pollution control facilities, we are obligated to protect the valuable as well as the not so obvious marine resources.

The complexity of marine biological activity is nowhere more apparent than in coastal waters, especially those utilized by man. Fortunately, the process of understanding coastal ecosystems has already begun in California and other areas of the United States where comprehensive regional coastal assessment programs are underway.

This paper reviews some findings which have resulted from a 5-year field investigation in the heavily populated coastal zone of the Southern California Bight. The purpose of this paper is to point out some important biological responses of animal communities which occur in our waters and which should provide some guidance toward defining realistic criteria for the wise use of this and other open coastal areas.

The methods of our field investigations have been described in other reports [1,2] and include use of trawls, cine cameras and closed circuit television to sample fish populations and grab and core devices to sample bottom sediments and benthic invertebrates. Our work involves close collaboration with many agencies and individuals, to whom we are particularly grateful, and is sponsored by a consortium of public agencies** and grants from the United States Environmental Protection Agency (EPA) and the California Cooperative Oceanic Fisheries Investigations (CalCOFI).

---

* This work is supported in part, by Grant R801-152 from the United States Environmental Protection Agency.

** Sanitation districts of Los Angeles County, Los Angeles City, Orange County, City of San Diego, and Ventura County. The Project is under the control of a Board of Commissioners and receives scientific guidance from a Board of Consultants composed of nationally known Scientists.

2. PHYSICAL, CHEMICAL AND BIOLOGICAL BACKGROUND

One billion gallons (3.7 million cu m) of treated municipal wastewater is discharged into the southern California coastal zone each day. Over 95% of this is discharged by five large municipalities via deep water (60–100 m) diffusers with flow rates ranging from 12 to 360 MGD (million gallons per day). Approximately 90% of this wastewater receives primary treatment only; the remaining 100 MGD is given biological secondary treatment.

2.1. Characteristics of offshore discharge sites

These wastes are discharged offshore of major cities into coastal waters of the Southern California Bight, a large indentation of the coast with a seaward boundary formed by the large southerly flowing California Current (Fig. 1). The coastal shelf is narrow (2–13 km wide) and descends into several series

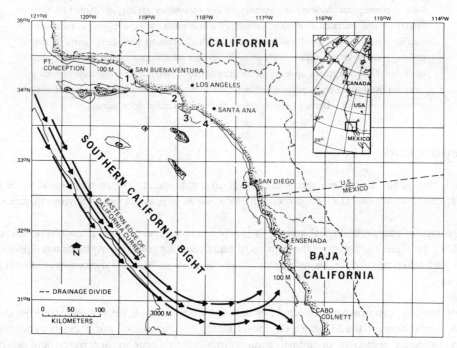

Fig. 1. The Southern California Bight and the adjacent coastal basin.
The major municipal wastewater discharge sites are: (1) Ventura County; (2) The City of Los Angeles;
(3) County Sanitation Districts of Los Angeles County; (4) County Sanitation Districts of
Orange County; (5) The city of San Diego.

of deep basins characterized by anaerobic cold water (4–6°C). A strong summer thermocline forms to 20 m and effectively bisects the coastal shelf into a warm shallow zone (16–20°C) and a deeper cold zone (9–14°C). Although the thermocline prevents wastewaters from surfacing during much of the year, it can be penetrated by deeper waters during periods of upwelling; it also causes a major discontinuity of the benthic fish and invertebrate fauna as one moves offshore.

The large sewage outfalls discharge above a benthic environment composed of silt and clay with little sand and little or no relief from rocks or reefs. This fact is of major importance because the fauna most influenced by these discharges is characteristically dominated by soft bottom fishes and invertebrates rather than by reef or pelagic organisms. This means that the mere presence of a large outfall structure can effectively attract a reef fauna and cause attendant ecological changes such as increased predation on nearby benthic resources.

Since most of the discharged waste is composed of primary effluent, fine particulate matter tends to settle on the bottom near the diffusers and to bring down measurable quantities of attached materials such as trace elements, chlorinated hydrocarbons and material of high BOD content. Thus depending on subsurface currents and on the volume and duration of discharge, large concentration gradients of these materials occur in bottom sediments as one approaches the sites of heaviest fallout [3] (Fig. 2). Concentrations of trace elements and hydrocarbons in the water column are low and difficult to measure

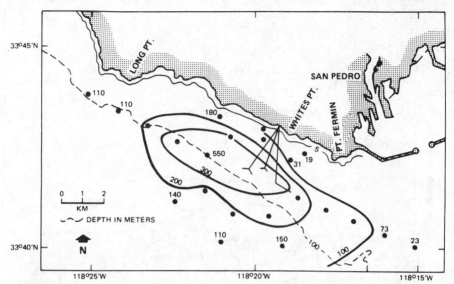

Fig. 2. Copper concentrations (mg/dry kg) in surface sediments around the outfall system
of the County Sanitation Districts of Los Angeles County, Palos Verdes, May 1970.

although some such as DDT can be detected in concentrating organisms living many miles from the discharge sites (Fig. 3).

2.2. Other sources of influence

In addition to the wastewater discharges, other sources of trace materials include dry and wet river runoff, aerial fallout and ships and harbors, ocean dumping and advection from the California Current. In our area, a major source of lead in coastal waters is aerial fallout (automobile emission). Biologically available forms of copper and zinc appear to be contributed by harbors and marinas [1].

Evaluating the responses of coastal marine resources to wastewaters must also include an account of other factors including fishing, flood control projects, construction of marinas in lagoons and estuaries and intermittent factors such as oil spills. Fishing, in fact, may have both direct and indirect effects; for example, there is no commercial trawl (bottom fish) fishery in southern California as there is to the north. But the sportfishery, which takes large numbers of predatory fishes, is sizeable as is the pelagic wet fishery [1,4]. Changes in a fishery caused either by nature or overfishing result in shifts in effort to less desirable forms while the removal of top predators may contribute to major population changes in prey species and forage fishes.

2.3. Natural biological variability

Long-term and short-term natural environmental changes occur continuously. These cause major changes in fishery stocks and in unfished populations of fishes and invertebrates. A notable catastrophe of economic significance was the decline in the sardine fishery which occurred in the early 1950's (Fig. 4).

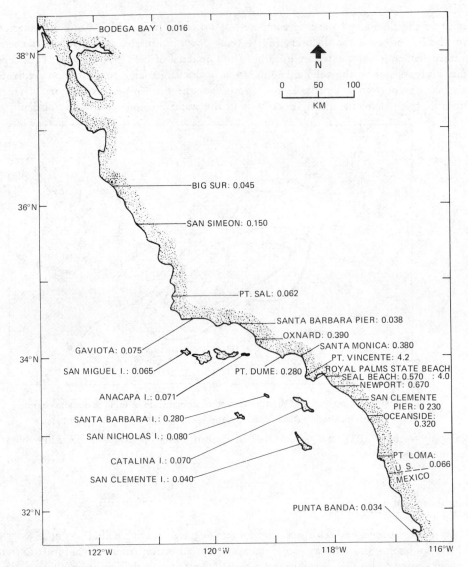

Fig. 3. Concentrations (ppm, wet weight) of total DDT in whole soft tissue of *Mytilus californianus* in southern and central California and Mexico, 1971.

Overfishing was a primary factor [5], although longterm, large-scale environmental changes were also influential. Soutar and Isaacs [6], in a unique study of dated core samples from an anaerobic basin, found that the sardine and other pelagic species have undergone several major population explosions and declines over the past 200 years.

Shorter term fluctuations of biological populations (months and years) are also apparent. Aperiodic red tide occurrences and apparently unpredictable invasions and mass strandings of squid, pelagic red crabs, jellyfish, and other organisms suddenly appear in abundance in a matter of days and disappear just as quickly. Equally significant is the destruction of kelp beds by extended periods of warm water intrusion and the effects of over grazing by enhanced populations of sea urchins [7].

Natural gradients in bottom water temperatures and dissolved oxygen appear to play a major role in determining the relative abundance and distribution of fishes on the coastal shelf (Fig. 5). Cold water of low oxygen content from the offshore basins can mix with warmer water over the shelf creating large gradients in bottom water isotherms. In addition to causing a natural depression of dissolved oxygen

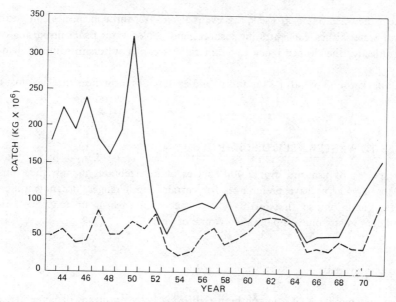

Fig. 4. Commercial fish and shellfish landings into the Los Angeles and San Diego regions from California waters, 1943–71. The solid line is total catch; the dashed line is total catch minus sardine and anchovy landings.

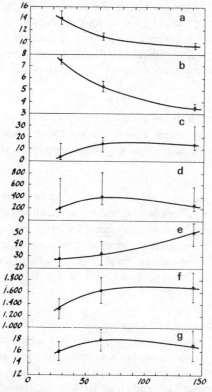

Fig. 5. Depth distribution of 7 physical and biological variables measured during surveys around the outfalls on the coastal shelf of Los Angeles and Orange counties; September 1973.
The plots are: (a) Mean temperature ($^{\circ}$C); (b) Mean dissolved oxygen (mg/l); (c) Median fish biomass (kg/haul); (d) Median fish abundance per haul (No. fish/haul); (e) Mean fish size (g); (f) Mean Brillonin diversity; (g) Mean species (No. species/haul). The abscissa is depth in meters.

(to 3.5 mg/l) near the wastewater diffusers, this event also causes shifts in species composition and abundance, with warm water fishes concentrating inshore, and cooler-water fishes dispersing over larger areas of the shelf. Ironically, the deeper fauna also experiences cooler waters in summer than in winter [2].

It is against this background of natural variability and extraneous influences that we must measure effects of waste discharge.

3.   RESPONSES TO WASTEWATER DISCHARGES

Well over 200 species of fish and over 2,000 species of invertebrates inhabit the southern California coastal shelf. Most appear to have preferences for certain depth ranges, oceanographic conditions and bottom substrate conditions so that relatively few species of each group (e.g. 10—50) are dominant, common or abundant in any one sampling site or time of year.

3.1. Abundance and diversity

The fauna at coastal discharge sites in southern California is by no means depauperate. However at the largest sites there are measurable increases in abundance of some species and a loss of others.

3.1.1. *Demersal fishes.* The use of 25 ft (headrope length) otter trawls towed at 2—2.5 knots (3.6—4.5 km/h) for 10 min (on-bottom time) has been generally accepted as a standard unit of effort in southern California demersal fish monitoring surveys [8]. Using this unit of effort, average catches from individual coastal fish surveys surrounding wastewater discharge sites range from 50 to nearly 800 fish, from 2.0 to 36 kg and from 9 to 20 species per haul [1,10]. Based on some knowledge of gear performance, we estimate that these statistics correspond to about $7.5 - 120 \times 10^4$ fish per sq. km and about 3—50 t of fish per sq. km.

Catches show distinct seasonality, with low abundance and diversity in winter and higher values in late spring or summer. And, during any single survey, catches appear to increase with depth (from 20 to 100 m [9]; a feature which may be due, in part, to visual avoidance of some species in shallower, well illuminated areas.

Because of the great variability among individual catches it has been difficult to detect effects of wastewater discharges on abundance and diversity of demersal fishes. However, there does appear to be a trend for a few species to dominate the demersal fauna near discharge sites where catches appear to be large and diversity lower than at the few localities surveyed where no major discharges occur (Table 1). In fact, at only one or two stations at one of the three largest discharge sites (Palos Verdes) do we consistently retrieve small catches of low diversity. It is also here that fin erosion disease caused by the waste discharge prevails (described below, 3.3.3).

Fishes which dominate waste discharge sites are characteristically benthic-feeding fishes and include both free swimming (perch, rockfish, croakers) and truly benthic forms (sculpins, sanddabs, flounder, sole). Near the areas of heaviest contamination, catches are dominated by infaunal feeding fish (sole) indicating that the infauna of the benthos (polychates, clams) is enhanced as a result of accumulation of waste particulates in the sediment. Conversely, epifaunal feeders are reduced in numbers, a condition due possibly to reduced population of amphipods and other small invertebrates normally active just above the bottom.

Although only a few of the 40—60 species encountered in each survey are presently important (recreational or commercial) resources, many provide forage for larger fishes and are effectively contributing

Table 1. Comparison of standard trawl catch statistics from four coastal areas near large and small
wastewater outfalls, August and September, 1973

	Santa Monica Bay	Palos Verdes	San Pedro Bay	Dana Point
Month	September	September	September	August
No. trawls	9	9	9	8
Depth range (m)	27–150	27–150	27–150	28–88
$\bar{x}$ species/sample	17.6	16.7	15.7	20.7
$\bar{x}$ specimens/sample	642	648	336	402
$\bar{x}$ diversity (H)/sample	1.70	1.50	1.75	1.96
Total No. species	51	52	41	50
No. species with fin erosion	3	5	3	1
No. Dover sole	239	651	256	112
% fin erosion in Dover sole	2.1	37.1	1.5	0
Outfall characteristics				
Flow (MGD)	335	371	130	3.5
Diffuser depth (m)	60	60	60	12–18

to the overall productivity of the coastal shelf. They also coexist with populations of epibenthic shrimp, and large species of echinoderms and mollusks which must not be overlooked.

3.1.2. *Benthic invertebrates*. Although requiring considerable laboratory effort and expense, benthic infaunal surveys provide a great deal of information on effects of waste discharge [10]. Decrease in variety and increase in abundance of benthic and fouling communities have provided useful indicators of effects, lack of effects, or recovery from effects of waste discharge in harbors [11], and coastal areas [12,13]. In southern California, these kinds of changes have been measured at all major discharge sites and at several lower flow (less than 10 MGD) sites in shallow water [1].

General response of the benthos to coastal discharge of primary wastes appears to include enhanced biomass and reduced diversity due, primarily, to enhancement of a few abundant forms and loss of rarer species. These responses generally occur in areas where organic material begins to accumulate, along with increasing concentration of organic carbon and sediment sulfides.

Direct observation and enumeration of organisms by diver or deep submersibles is coming into increasing use in southern California and is providing data on the abundance and diversity of common epibenthic organisms (large clams, starfish, colonial coelenterates) not adequately sampled by either trawls, dredges, grabs or cores.

3.2. Community structure

Despite the apparent complexity, marine organisms are not randomly distributed, especially with respect to each other and to major environmental gradients in temperature and depth. In fact, both coastal fish populations and benthic infauna show distinct species assemblages which in turn associate with specific site characteristics (substrate, depth, temperature, etc.) on the coastal shelf.

On our coastal shelf, the demersal fish fauna shows three or more species assemblages arranged by depth as shown in the recurrent group analysis in Fig. 6 [14]. Within assemblages occur a few common fish species of very different morphology and structure, i.e. each is performing a specific function within the assemblage (e.g. feeding on infauna or epibenthic organisms or on pelagic organisms, etc). In deeper waters, there is a different assemblage, but feeding functions are similar and are divided among a different group of species which may resemble those inshore.

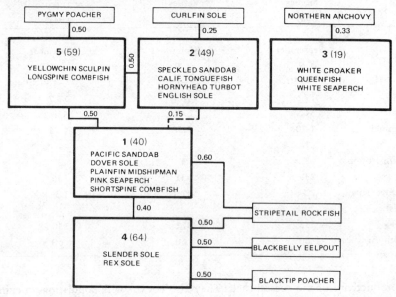

Fig. 6. Species associations of southern California nearshore demersal (10—360 m) fishes, 1969—72. The bold number is the group number; the number in parentheses is the number of samples out of 303 in which the group was found. The groups were identified by recurrent group analysis using an affinity index of 0.50. Other associations, indicated by connecting lines, were determined by Connex analysis. Groups 5, 2, and 3 are found in shallow waters; Group 1 occupies middepth waters, and Group 4 is found in deep waters.

These findings allow us to test an ecological model which predicts which benthic species ought to occur at a given depth and with what other species [2]. For example Fig. 7a—d shows the expected depth distributions of three species of fish (a) which are potential competitors and which can replace one another depending on fish size (or jaw size) and depth. Figure 7b and c represent data from two trawl transects several miles distant from the Hyperion wastewater outfalls in Santa Monica Bay. The patterns generally agree with the expected distribution. However, a transect of stations taken along the outfalls (Fig. 7d) show a distinctly different pattern: the speckled sanddab has extended its distribution well into the outfall depths (about 100 m) replacing the Pacific sanddab. It is possible that the Pacific sanddab is more sensitive to receiving water or sediment condition than the other two species.

Very distinct patterns in infaunal invertebrate communities occur as well. The patterns are particularly obvious in areas of extreme environmental gradients (sediment metals, DDT, low pH, percent silt). The analyses revealing these patterns are based on calculation of similarity coefficients, for samples and species, and application of clustering techniques to form groups and to determine relationships between groups [2]. Figures 8 and 9 summarize the results of an analysis from a biological and chemical survey of sediments surrounding the Los Angeles County Sanitation Districts outfalls (360 MGD) off Palos Verdes, an area inhabited by well over 300 species of invertebrates. Figure 8 shows similarities among sampling sites based on species composition and relative abundance while Fig. 9 shows similarities among the same sampling sites based on physical characteristics (depth, sediment texture) and chemical characteristics (organic nitrogen, vegetative matter, hydrogen sulfide, total DDT and mercury). The similarities between the biological and physical patterns are remarkable and greatly aid description of the responses of complex benthic communities to sediment contamination from discharged wastes. Moreover, these patterns are considerably more informative than patterns based only on abundance or diversity.

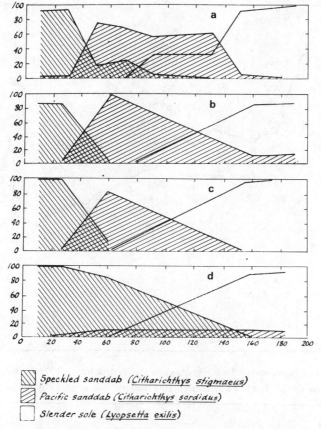

Speckled sanddab (*Citharichthys stigmaeus*)
Pacific sanddab (*Citharichthys sordidus*)
Slender sole (*Lyopsetta exilis*)

Fig. 7. Expected and actual depth distributions of three species of benthic fish filling the "sanddab" role.
The plots are: (a) The expected distribution; (b and c) The distribution found along two transects in
Santa Monica Bay several kilometers from the outfalls; (d) The distribution found along a transect
adjacent to the outfalls (discharge depth: 75—100 m). The ordinate is relative abundance in
percent of the total catch of the three species; the abscissa is depth in meters.

Our analyses suggest that the major biological effects of discharge of municipal primary waste is through
overt organic enrichment of bottom sediments, causing loss of filter feeding organisms and those requiring
coarser sediment and an enhancement of detrital forms and their predators (including invertebrates and
infaunal feeding fishes). The toxic effects of hydrogen sulfides are probably also important as are effects
of metals and chlorinated hydrocarbons, and work is in progress to separate their effects from those due
to silting and organic enrichment.

3.3. Diseased and abnormal fish populations

Diseased and abnormal fish occur in our coastal waters in measurable frequencies. Most of the disorders
are rare and do not appear to be associated with municipal waste discharge sites. However, several diseases
are indirectly caused and maintained by the continuous discharge of municipal wastes offshore.

Measuring and evaluating the actual health of marine fish populations is a complex, but intriguing and
rewarding problem and has provided insight into relating laboratory and field observations.

At least nine classes of diseases and disorders are recognized in our coastal fish populations. These include
abnormal pigmentation and coloration, parasite infections and exopthalmia [1], bone deformities and
asymmetry of meristic features [15], epidermal tumors in flatfish [16], lip oral papillomas in sciaenid

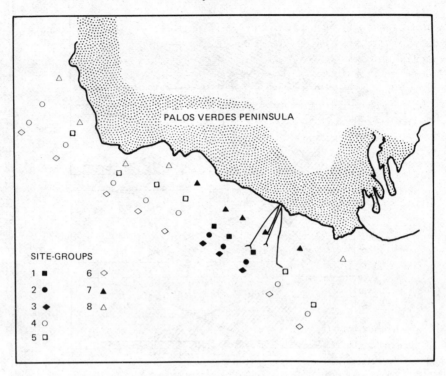

Fig. 8. Site-groups generated using species abundance data from a 1973 benthic survey off Palos Verdes. The black symbols define the area most affected by the outfalls.

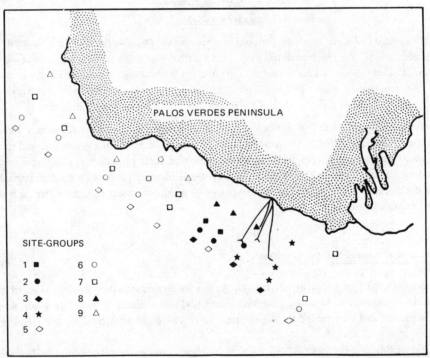

Fig. 9. Site-groups generated using physical/chemical data (sediment coarseness, hydrogen sulfide in a 1973 benthic survey of Palos Verdes. The black symbols define the area most affected by the outfalls.

fishes [17], infectious tail rot or tail erosion in free swimming fishes, especially sciaenids [1,18] and fin erosion or fin rot in offshore demersal fishes [16].

Assuming that sampling is sufficiently comprehensive in time and covers a region considerably larger than the migratory ability of the populations, then a disease originating from pollution ought to increase in frequency with proximity to the point source. As in human diseases, it may or may not be infectious, seasonal, species specific or specific to certain ages, sizes, sex or microbial or chemical influences, but these factors can be useful in the complete biological assessment. With these criteria in mind, we and our co-workers have developed a rather comprehensive epidemiological approach to assessing some of these diseases involving a combination of histochemical, chemical, microbiological, behavioral and population structure analyses applied to numerous samples from field surveys [15,16]. Although still in progress, these studies have shown three types of diseases that do have some direct or indirect relation to waste discharge (tail erosion, fin erosion and asymmetry). Two of these, tail erosion in sciaenids and asymmetry of meristics in sand bass, grunion and barred surf perch, occur in populations well beyond the immediate discharge sites (point sources).

The tail erosion disease appears to be microbial in nature and is particularly prominent in the Los Angeles Harbor area especially near cannery effluents, wastes very high in BOD. It also occurs in a family of fishes (Sciaenidae, croakers) which may congregate near freshwater sources during spawning. And, the disease does appear at sites removed from waste discharges. These and additonal data suggest a disease of natural microbial origin which can be enhanced by waste discharge.

Asymmetric growth of bilateral features and bone deformities occur in at least three kinds of fishes and have been shown to form a gradient extending hundreds of miles from the Los Angeles are [15]. Valentine et al. [15] postulate that the conditions are caused specifically by DDT, a persistent hydrocarbon which we know to be distributed and accumulated in other biota (Mytilus, Fig. 3) ranging from 50 to 150 km from a previously large point source in the Los Angeles area.

A fin erosion disease in offshore benthic fish is not found away from discharge sites, is chronic (not seasonal) and is not specific to species, sizes or sexes at its known point of origin (Palos Verdes) [16]. Diseased adult Dover sole appear to migrate to other localities including other outfall sites. And, although microorganisms are recovered from diseased fins, there are no systemic infections or internal pathogens in Dover sole [16]. The highest frequency is found in fish living over sediments containing high concentrations of metals, chlorinated hydrocarbons and sulphides, but only the chlorinated hydrocarbons are accumulated by the fish. The diseased fish have enlarged livers and are incapable of mucous production, two conditions which suggest some metabolic disturbance. Thus the discharge may be providing a toxic material (to which the fish must expend energy detoxifying) and sediments of obnoxious composition which affect the fins of compromised fishes. This is the kind of "environmental disease" model considered by Sniesko [19], involving not only the fish and disease agent but a stressor as well (i.e. possibly DDT).

A skin tumor condition of Dover sole previously reported by Young [18] appears to be an epizootic disease quite unrelated to waste discharge, but very specific to pleuroneclid flatfishes of the northeast Pacific Ocean [16]. Finally, oral tumors in croakers [17,18] is the least understood of the common diseases, but does occur in fish distant from discharge sites. Additional work is warranted since they both may originate from a virus or a very specific, widespread chemical agent.

3.4. Growth, mortality and reproduction

There have been very few studies specifically designed to document the effect of coastal waste discharges on the growth, reproduction and mortality of nearshore fish or invertebrate populations or even to ascertain the rates of these processes for populations in uninfluenced areas. This is particularly unfortunate since the basic tools of fisheries research have existed for many years and have been successfully applied to understanding and managing marine resources elsewhere.

We have attempted to examine local growth rate variation in at least one common species of fish, the Dover sole. Coastal populations show faster rates of growth than those at an offshore island (Fig. 10).

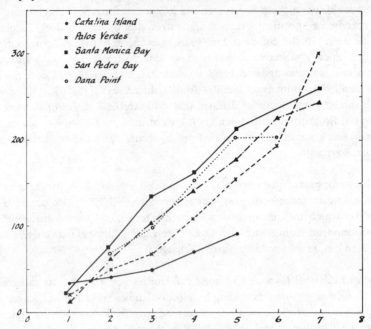

Fig. 10. Estimates of growth rates of Dover sole (*Microstomus pacificus*) from three coastal outfall sites (Santa Monica Bay, Palos Verdes, and San Pedro Bay), a coastal control site (Dana Points), and an offshore island control site (Catalina Island). The ordinate is body weight in grams; the abscissa is age in years.

And, with one noteable exception, those from major waste discharge sites appear to be growing faster than those from "control" areas. The exception to this occurs at Palos Verder (L.A. County discharge site) where growth is faster than in offshore populations, but clearly slower than at other coastal discharge areas. This is also the site of origin of the fin erosion disease and is otherwise an abnormal discharge site because of the previously high DDT input and its persistance in nearby marine sediments.

The Dover sole feeds primarily on infaunal organisms which are generally more abundant on the coastal shelf than at offshore islands (owing to higher silt loads and organic material) and are even more abundant at discharge sites. Therefore, there may be a direct connection between growth rates (and probably survival rates) of some resident fish species and enhancement of infaunal biomass. Locally, the effect is striking, but we do not yet know to what extent the entire resource is affected. It could be relatively small.

Natural mortality rates of common local marine resources have not been studied, so we do not know to what extent municipal wastes, fishing or other factors have affected these populations. We have never observed fish kills at the offshore sites and have only rarely encountered fish which appear to have been dead prior to capture. Application of more precise fisheries research procedures to selected species of fishes and invertebrates near and away from waste discharge sites could prove to be quite informative.

Perhaps the greatest controversy about the impact of municipal wastes on marine ecosystems results from concern over inhibition of reproduction. It is now quite certain in our region that DDT, previously released in large quantities through the municipal L.A. County discharge, was accumulated by marine birds and caused reproductive failure in pelican populations. But the effects of this material or of others on fish resources has not been established. We do find seasonally many species of local fish bearing eggs or young and subsequent appearance of large numbers of juvenile fish in trawl surveys near discharge sites.

Whether or not these represent recruitment of both parents and young from outlying areas is not at all certain, but could be tested by adequate mark and recovery programs.

4. DISCUSSION AND CONCLUSIONS

Obviously much more field research is required to fully assess the impact of municipal wastewaters on marine resources of the open coast. The application of established fisheries models to measure recruitment, survival, reproduction and productivity of nearshore species would be particularly useful.

Despite our still limited knowledge, it is obvious that many species of marine fish and invertebrates are abundant near and away from the major discharge sites in southern California. The sites themselves are not "biological deserts" but the organisms do show obvious changes in community structure, variety, disease frequency and growth. Beyond the discharge sites, it is difficult, if not impossible to demonstrate effects of the discharges on major fishery resources due to their wide dispersal, natural fluctuations and intense and very selective fishery. However, DDT, a persistent synthetic organic compound capable of source control has shown a wide coastal distribution and possible wide ranging sub-lethal effects.

Trace elements (metals) originating from discharged municipal wastewaters do not appear to be accumulated in the tissues of fishes and large invertebrates despite high concentration in sediments and suspended particulates [20]. Most of the effects we have studied, including disease and changes in community structure, appear to be more likely explained by the degradable characteristics of wastewater (fine particulates, high concentration of organic material, nutrients) or persistent synthetic toxins than by the trace elements themselves. It is quite possible that many metals are presently not being discharged in large quantities in biologically available forms (e.g. soluble or organically bound), a factor which is rarely considered in laboratory bioassays and toxicity tests.

The populations we have studied in some detail include only a few species of present economic or recreational significance. Most of the presently exploited species in California represent either very large stocks which reside near outfall sites for only brief periods or are as yet insufficiently sampled in existing programs. Moreover these resources exhibit fluctuations in real abundance associated with unpredictable natural changes and overfishing and fluctuation in actual landings associated with changes in fishing effort and gear. Proper monitoring and wise management of the lesser known unexploited populations provide water quality criteria which can be evaluated in the field. New management models, such as those offered by Regier [21], can provide guidelines to managing both desirable resources as well as the not-so-obvious resources which are an equally intimate part of the coastal environment.

REFERENCES

1. SCCWRP, *The Ecology of the Southern California Bight: Implications for Water Quality Management*; Technical Report 104, Southern California Coastal Water Research Project, El Segundo; California (U.S.A.), 1973.

2. SCCWRP, *Coastal Water Research Project Annual Report for the Year Ended 30 June, 1974*, Southern California Coastal Water Research Project, El Segundo, California (U.S.A.), 1974.

3. Hendricks, T.J. and Young, D.R., *Modeling the Fates of Metals in Ocean-Discharged Wastewaters*, Technical Memorandum 208, Southern California Coastal Water Research Project, El Segundo, California (U.S.A.), 1974.

4. Horn, Michael J., Fisheries; 20-1 to 20-58. In: *A Summary of Knowledge of the Southern California Coastal Zone and Offshore Areas*, Southern California Ocean Studies Consortium, Long Beach, California, 1974.

5. Murphy, G.I., Population biology of the Pacific sardine (Sardinops caerulea). Proc. Calif. Acad. Sci. 34 1-84 (1966).

6. Soutar, A. and Isaacs, J.D., Abundance of pelagic fish during the 19th and 20th centuries as recorded in anaerobic sediment off the californias. Fishery Bull. 72 257-273 (1974).

7. North, W.J. and Pearse, J.S., Sea urchin population explosion in southern California coastal waters. Science 167 209 (1970).

8. Mearns, A.J. and Stubbs, H.H., Comparison of otter trawls used in southern California coastal surveys. Technical Memorandum 213, Southern California Coastal Water Research Project, El Segundo, California (U.S.A.) 1974.

9. Mearns, A.J. and Greene, C., (Eds.), a comparative trawl survey of three areas of heavy waste discharge. Technical Memorandum 215, Southern California Coastal Water Research Project, El Segundo, California (U.S.A.) 1974.

10. Holme, N.A. and McIntyre, A.D., Methods for the Study of Marine Benthos; IBP Handbook No. 16; International Biological Programme, London; Blackwell Scientific Publications, Oxford, 1971.

11. Reish, D.J., Effect of pollution abatement in Los Angeles harbours; Mar. Poll. Bull. 2 71-74 (1971).

12. Pearce, J.B., The effects of solid waste disposal on benthic communities in the New York Bight. In: Marine Pollution and Sea Life, pp. 400-401. Fishing News, London, 1972.

13. Watling, L., Leathem, W., Kinner, P., Wethe, C. and Maurer, D., Evaluation of sludge dumping off Delaware Bay. Mar. Poll. Bull. 5 39 (1974).

14. Fager, E.W., Communities of organisms. In: The Sea, Vol. 2, pp. 115-132. Ed. M.N. Hill, Interscience, New York, 1963.

15. Valentine, D.W., Soule, M. and Samollow, P., Asymmetry analysis in fishes: a possible statistical indicator of environmental stress. Fishery Bull. 17 357-370 (1974).

16. Mearns, A.J. and Sherwood, M.J., Environmental aspects of fin erosion and tumors in southern California Dover sole. Trans. Am. Fish. Soc. 103 799-810 (1974).

17. Russel, F.E. and Kotin, P., Squamous papilloma in the White Croaker. J. Nat. Cancer Inst. 18 857-861 (1957).

18. Young, P.H., Some effects of sewer effluent on marine life. Calif. Fish Game. 50 33-41 (1977).

19. Sniezko, S.F., The effects of environmental stress on outbreaks of infectious diseases of fish. J. Fish Biol. 6, 197-208 (19).

20. de Goeij, J.J.M., Guinn, V.P., Young, D.R. and Mearns, A.J., Neutron activation analysis trace element studies of Dover sole liver and marine sediments. In: Comparative Studies of Food and Environmental Contamination, pp189-200. International Atomic Energy Agency, Vienna, 1974.

21. Regier, H.A. and Henderson, H.F. Towards a broad ecological model of fish communities and fisheries. Trans. Am. Fish. Soc. 102, 56-72 (1973).

TOXICOLOGICAL STUDIES IN THE FIELD OF OCEANOLOGY

Maurice Aubert

Centre d'Etudes et de Recherches de Biologie et d'Océanographie Médicale,
Parc de la Cote, 1 avenue Jean-Lorrain, 06300 Nice (France)

Summary — Toxicity phenomena are present in the marine medium. They may be
natural and concern oceanic species with regard to one another, as in the "red tides",
or they may endanger the consumers of marine species as is the case with "ciguaterra".
But they may also be artificial, in connection with the introduction of chemical products into sea water. The action of these toxic products may be exerted directly upon
the species, entailing a loss of nutritional capital. It may also be indirect and concern
the end consumer of marine products, due to the possible transfer and concentration
of some remanent pollutants. Such processes can now be studied on trophic marine
chains simulated in the laboratory, on which transfer factors are measured. Lastly, it
should be mentioned that exogenous chemicals are likely to entail ecological drift of
the medium by damaging the chemical telemediators that control interspecific relationships. This is also a toxicological aspect, whose action is slow but may prove dangerous
to the preservation of the natural environment on the long range.

Toxicity phenomena have been known for a long time, and the oral or written tradition of many cultures,
including the most primitive ones, tells what is good or bad for Man, who feeds on natural resources.
Animals, too, seem to have gained some experience in this regard, enabling them to avoid such plants as
could be detrimental to their health. Toxic natural products are not restricted to land, since they also
occur in the marine world. There are many examples of interspecific relationships that result in some
species destroying others through toxic processes.

Aside from these natural toxicological phenomena which, for millenia, had been controlling the biological
equilibrium of plant and animal life, toxical phenomena artificially created by Man have appeared since
the beginning of the industrial era, and have developed through the biological processes of nutrition. The
large creative possibilities of industry now release into our environment a whole range of chemical substances which are digested by and become part of the living organisms, whose vital behaviour they modify,
sometimes entailing their death. Thus, we can observe either massive destruction of living species, or
transfers of remanent chemical substances that can concentrate in the biomass and bring about toxicity
in the last stages of the biological chains.

We shall describe later in this report the various toxical phenomena that may develop in the oceanic
environment, and we shall have to consider them under their twofold aspect: those of natural origin, and
those that are artificially created by Man.

Among the natural phenomena, what immediately comes to mind is the wholesale slaughter of fish killed
by the "red tides". Such an alteration of the usual biological characteristics of sea-water may appear under
certain physico-chemical conditions: increase of the temperature, decrease of the salinity, increase of the
nitrogen content, etc. Where these conditions are met, a sudden proliferation of some species of Dinoflagellates may take place, superseding the habitual phytoplanktonic species and releasing substances that
are poisonous to the marine species which live in the surrounding waters. The species responsible for "red
tides" most frequently encountered belong to genera *Conyaulax, Cochlodinium, Gymnodinium,*

Exuviella. They are found in all warm and temperate oceans, and the areas most often affected are situated in certain regions of the Pacific and Atlantic oceans, particularly off the Florida coast. The Mediterranean is not secure from that threat, as several epidemics of red tide have been seen along the Algerian coast and, even on the French coast we witnessed such an outbreak ourselves. Microbiological inspection of the waters evidenced a proliferation of either *Gonyaulax polyhedra,* or *Cochlodinium Sp.*, or still *Chattonella subsala*, a chloromonadina. Owing to the proliferation of these microorganisms, the water takes on a characteristic discoloration of which the Bible already gave us a striking description, this being one of the Seven Plagues that smote the people of Egypt at the request of Moses. Here is this description: "And all the waters that were in the river (the Nile) were turned to blood, and the fish that was in the river died, and the river stank, and the Egyptians could not drink of the water of the river" (Exodus 7. 20–21).

The consequences of the phenomenon are not restricted to the microbiological aspects, because the products released by those species prove to be toxic to the fish and molluscs that live in the vicinity, so that the surface of the sea is strewn with the bodies of dead fish following an outbreak of those planktonic forms.

In addition to the damage suffered by the oceanic capital, toxic phenomena may affect the consumer of infected fish, and the literature supplies a number of observations on the intoxications resulting from the consumption of fish that have fed on such toxic flagellates. The clinical picture is the following: the symptoms appear within half an hour after ingestion. They begin with a sensation of burning and prickling of the lips, face and tongue, sourness of the mouth, constriction of the throat, drowsiness. Then comes a feeling of intense discomfort in which cold, thirst and sweating dominate. The digestive syndrome includes abdominal pain, alimentary and bilious vomiting, and diarrhea. Urticarial rash sometimes happens. There is no fever. The pulse is quick, the blood pressure low. Faintness and dizziness occur. The patient is anguished. Itching is severe and often localized on the palms of the hands and foot soles. It is accompanied with impaired sensitivity, prickling and muscular pain. Insomnia is total, and the sufferer experiences intense weariness, especially in the legs. In the most severe cases, one observes some degree of ataxia with paralysis of the peripheral type. The eyes are involved: diplopia, strabism, mydriasis, loss of accommodation. The achilles and patellar reflexes are weakened. After a course of 12 h, the disease is rarely fatal. The digestive disorder quickly subsides, while the cardiovascular symptoms remain for 2 or 3 days before subsiding. It takes some 10 days for the sensorimotor disorders to improve. However, fatigability of the legs and dysesthesia persist for several weeks, even months, and convalescence is slow, marked by utter asthenia. Death occurs in 1–3% of the cases, in the extremely severe forms, with a choleriform picture including suffocation, collapse of the blood pressure, paralysis of the legs, eyes and chest muscles [1]. This is why strict surveillance is exercised in many countries upon the potential red tides, especially in the United States where fish catches are rigorously checked in this respect.

But the nuisance to Man may follow a much more direct path through airborne infection, as each microdrop contains either a few of the microorganisms or some of the toxical substance they produce, so that observations can be made inland on pathological consequences of the proliferation of the Dinoflagellates at sea. The picture is of the toxo-anaphylactic type, with rhinitis, bronchitis and even pulmonary capillarity [2].

We shall not leave this natural aspect of the toxical phenomena coming from the Ocean without mentioning a disease which mainly develops in Polynesia and Melanesia, the scientific name of which is "ichtyosarcotoxism" and is called "ciguaterra" by the inhabitants of those islands [3]. Some fishing grounds will, at times, produce animals which, when eaten, bring about a picture suggestive of severe food poisoning with a neurotoxic attack. Adynamia, headache, vomiting and itching are often present, together with temperature increase, sensorimotor disorders and this collapse may end up with death in some cases. This disease is rather puzzling in that the areas in which it appears are unpredictable. For years and years local fishing can be safe in a given area and suddenly, for some unknown reason, the catches will induce "ciguaterra" and the area will be banned to fishing for several years, until it becomes healthy again. It is no wonder that many laboratories on the Pasific, including the American Laboratory in the Hawaiian islands and the French laboratory in Tahiti, are anxious to elucidate a pathogeny of this disease and to find out the natural mechanisms that bring it about. So far, the results of these studies have not brought all the explanations wanted, and it is now thought that the poisonous evolution of the fish would be due to the consumption of certain algae capable of synthesizing, at given periods of their life, substances that are not toxic to the fish themselves but give rise, in their bodies, to a complex chemical substance, "ciguatoxine", which is detrimental to mammals.

Some authors observed that the population increase − resulting in a pollution increase − seems to bring about "ciguaterra" in the surroundings. Actually, it is quite possible that the increase of the organic matter content results in a proliferation of the algae that are responsible for "ciguaterra" [4].

This latter fact brings us to the other aspect of our study, namely the toxic phenomena due to human activities. Chemical pollution of the Ocean is on the increase. More numerous and bigger industries release to the sea waste water that is often highly toxic and, in addition, Man uses more and more products such as detergents and pesticides that act as poisons to the marine flora and fauna [5].

A great part of the chemical substances is metabolized by the organisms living in the sea and, owing to concentration processes, they may reach high proportions in the tissues of some species. The concentration process is multiplied because of the passage through the successive stages of the marine trophodynamic chains; thus, with certain products, this ends up with phenomena that are highly toxic to the human populations who consume these products from the sea. This is the case, in particular, with metallic salts of industrial origin like mercury, cadmium, chromium, etc. and some pesticides [6].

All these facts are demonstrated by chemical as well as biological measurements, which evidence an aggravation of the contents of toxic substances, in the marine species and in the consuming populations. These data have been highlighted in the General Study of the Chemical Pollution on the French Coasts which we have carried out [7]. It should be stressed that, at present, the Industry is definitely not in a position to remove the highly toxic substances from waste waters.

Oil pollution deserves a special mention. The growing power requirements and the development of petrochemistry make it necessary to produce and transport increasing quantities of petroleum products. Whether it be due to offshore drillings, transportation accidents, ruptured lines, shipwrecked tankers, etc. to unpurified process water from the refineries, or to degassing at sea, a great part of the sea has become covered with an oil film whose action toward marine life is not negligible as it destroys part of the surface species. As a counterpart, hydrocarbons, when metabolized by marine bacteria, bring extra organic matter which has a beneficial effect on productivity. Due to this metabolic activity, the biodegradation capacity of which is not totally known yet, one could feel a little relieved, although there is little information available on the way some of their components are released and fixed, especially the benzopyrenes which are indisputably carcinogenic [8]. Among the products from the petroleum industry, detergents, owing to mass consumption, account for much of the pollution and their biodegradability characteristics do not seem to guarantee protection against the hazards they represent. Among the most dangerous chemical pollutants, we must mention heavy metal compounds which are likely to become metabolized and concentrate in marine organisms, making them a hazard to the consumers. Everybody remembers the Minamata affair for which methyl mercury was responsible [9].

What are the means available to us in studying chemical pollution of the Ocean? We mentioned them at the start of this report. We shall now discuss them in detail.

The methods of analytical chemistry have, for a long time, been developed and adapted to the analysis of drinking water, fresh water, waste water and sea water. Such inspection is a necessity because it gives an overall *a priori* view of the major elements contained in waste waters, quality − and quantity − wise. Yet, when it comes to investigating the toxicity of an efflent in order to decide whether it can be disposed of at sea, several gaps appear: a systematic study does not permit assessment of certain substances that are not actually part of the manufacture. This is the case for disinfectants, used in some paper-mills to prevent moulding. On the other hand, it is not possible, even by asking the industrialists, to get to know the nature of the additives they use, because these are covered by trade secrets which are not to be divulged. In addition, in composite effluents, some substances may be masked or occur in so low proportions that they cannot be determined. Even assuming that the composition of an effluent is exactly known, this is going to become markedly altered according to the nature of the receiving medium. For instance, in a marine environment, the effluents are bound to undergo a number of actions:

> variation of the pH, sea-water acting as a buffer;
> variation of the salinity;
> precipitation by salts;
> complexing action of the natural organic substances of sea-water;
> influence of the organisms living in the marine medium.

Chemical analysis may lead to assumptions as to the possible toxicity of waste waters, but the combined effect of the various elements or the simultaneous action of a mixture of effluents cannot be predicted.

Biological tests are common practice in every field, for controlling drinking water (by raising gudgeons in it), river water − the tests being made on both plant an animal species − and sea water. There are many examples of such bioassays in the literature [10, 11]. They supply valuable information concerning the direct ecological consequences of industrial disposals at sea. Yet they give little indication of the indirect consequences that such disposals may have on the consumers of marine products.

While the methods described above make it possible to anticipate the damage to the productivity of the marine medium, hence the downgrading of fishing or sea-farming activities, they take no account of all the phenomena that take place when chemically polluted effluents are disposed of at sea.

Once the chemical substances are present in the marine medium, their biological fate is submitted to a twofold mechanism:

On one hand − this is the case for hydrocarbons − they are attacked by specifically marine bacteria which feed upon them. The intial product gradually disappears as it is converted to various substances the toxicity of which may be either nil, unchanged, or increased [12]. In like manner, some detergents are biodegradable in a marine environment whereas others, which are non-biodegradable, stay there for ever. Recent studies have shown that biodegradable detergents would in some cases give rise to new compounds that are more toxic to sea life than the initial product [13].

On the other hand, there is another phenomenon which antagonizes the action of biodegrability. It is concentration which, following the mathematical laws of nutrition in marine environment, may end up with extremely high factors of up to 10,000 at the planktonic level. This phenomenon is now quite well studied by the specialists of radio-active pollution. Marine organisms, vegetable or animal, concentrate certain products in their tissues at rates that are much higher than can be found after diffusion in sea water. These concentrating effects result, with some products, into severe pathological symptoms. They are especially severe if the organism consumed is high in the biological chain; so that, in view of this dynamics of chemical products at sea, one must consider the concentration phenomena to anticipate any possible extension. To this end, we have simulated in the laboratory a system of trophodynamic chains, all the way from marine microorganisms to higher mammals [14].

This unique methodology enables us:

to quantify the extent of the damage undergone at the level of typical organisms representing some of the most characteristic stages of the marine trophodynamic chains. This way, it is possible to assess the destruction, in terms of capital, of the biological resources of the ocean;

to consider all the phenomena of concentration and biodegradation of the chemical substances submitted to those tests, and to appreciate the biological effects induced by the initial toxicity or appearing at a later stage, since these products travel through the medium and are successively absorbed by the various living organisms.

Lastly, by introducing a mammal as an end consumer, we can judge of the toxic phenomena at the end of the chain and measure the potential hazard to a man feeding on polluted marine products. The sanitary aspect opens up on the impacts of marine pollution upon Public Health.

This is how we were led to develop four types of trophodynamic chains:

a general pelagic chain, i.e. phytoplankton, zooplankton, fish, mammals;
a benthic chain with molluscs, i.e. phytoplankton, molluscs, mammals;
a benthic chain with crustaceans, i.e. bacteria, invertebrates, benthic fish, mammals.

A biological chain can be used with miscellaneous purposes.

Determining toxicity thresholds. By putting each stage of the chain in contact with the pollutant (pure chemical product or industrial effluent) at different concentrations, it is possible to establish with a fairly good accuracy the sensitivity of each species to the pollutant. Thus, a toxicity threshold can be defined at each stage.

The results obtained permit short range forecasts of the consequences of pollutant disposal at sea upon survival of the species and are a means for the determination of the potential loss of nutritional capital.

Determining possible transmission of the toxicity. After determining the direct toxicity threshold of an effluent, we select the maximum concentration that allows for total survival of all the stages of the marine chain: this way, the toxic substance can be carried over from one stage to the next and eventually become concentrated in the successive stages without the biological chain being broken by the death of one of the intermediates.

Once the starting dose is established, the experiment consists in feeding the higher stages with the lower stages that have already been contaminated with chemical polluted water. All the stages are raised, however, in a polluted medium in order to simulate the natural conditions of a polluted medium.

The results permit assessment of the risks incurred by human consumers of marine products, as well as possible hazards to Public Health.

Determining non-degradability and biodegradability of pollutants. When disposed of at sea, a pollutant may act as a substrate for marine bacteria and, therefore, undergo biodegradation. Such an effect can entail a variation of the toxicity of the pollutant toward the diverse marine stages of the biological chain.

In order to appraise this biodegradability, we put the pollutant in contact with sea water with added marine bacteria for periods ranging from 0 to 9 days. The pollutant is added at three different rates: a non-toxic concentration and two toxic concentrations close to the threshold.

The results give a measure of the variation of toxicity of a pollutant after it has been in contact with sea water.

So far we have been discussing the direct or indirect effects of the more or less toxic pollutants in the field of nutrition. There remains a last point, which is the action of pollutants on inter-species relationships.

It seems that the marine biological equilibrium is maintained through the action of chemical substances that are released into the environment by the creatures that live in it. Such substances, which we have called chemical telemediators, are synthesized by marine plants or animals, then released into the environment and act remotely upon the behaviour or the biological functions of the same or other species [15].

Experiments made in this field give us some idea of the chemical nature of a number of such "messages", which are essential in maintaining ocean life. The very small rates at which they are active as well as the complexity of their chemical structure (for instance, the protein that we have isolated is released by a Peridinian and controls the antibiotic producing capacity of some Diatoms) indicate that the influence of chemical products in contact with them may entail an lateration of the messages, hence a severe ecological drift. Some recent tests which we have carried out have shown that certain poisons – like certain pesticides and hydrocarbons – introduced into the marine medium alter the message and provoke a distortion of the biological equilibrium instead of maintaining it [16].

Some have a specific action which may affect the self-purifying biological factors of the medium toward terrestrial bacteria disposed of in the sea. Their introduction in the medium stops or alters the natural release of certain substances by given species of phytoplankton or marine microorganisms. Thus, the sanitary condition of the water is impaired, as a side effect of direct chemical pollution [17].

To conclude, we can see how important toxicological processes are in studying ocean life. Toxicology helps understand natural phenomena but should also be used to evaluate the hazards resulting from the action of Man toward the marine medium.

REFERENCES

1. Ehrhardt, J.P., Les phénomènes d'eaux rouges. *Rev. Corps Santé Armées.* **9** 333-350 (1968).

2. Woodcock, A.H., Note concerning human respiratory irritation associated with high concentration of plankton and mass mortality of marine organisms. *J. Mar. Res.* **7**, (1948).

3. Bagnis, R., Contribution à l'étude de l'ichtyotoxisme en Polynesie Francaise. *Rev. Int. 'Oceanog. Med.,* **6-7**, 89-110 (1967).

4. Ehrhardt, J.P., Le problème de l'eau, facteur de déclenchement et d'aggravation de l'ichtyosarcotoxisme tropical. *XXXIe J. Scient. Nat. S.F.M.P.S.* (1970).

5. Aubert, M. (avec la collaboration technique de Charra, R.), Etude des effets des pollutions chimiques sur le phytoplancton. (Compte-rendu du IIIe Colloque International d'Océanographie Médicale, Nice, 1967) *Rev. Int. Océanog. Méd.* **10**, 81-91, (1968).

6. Aubert, M., Pollutions chimiques et chaines trophodynamiques marines. *Rev. Int. Océanog. Méd.* **28** 9-25, (1972).

7. Aubert, M., Aubert, J., Donnier, B., Gambarotta, J.P., Barelli, M. and Daniel, S., Edute Générale des pollutions Chimiques rejetées en mer. Inventaire et Etudes de toxicité. Suppl. *Rev. Int. Océanog. Méd.* **I**, 72 p. (1969); **II**, 135 p. (1969); **III**, 225 p. (1970); **IV**, 113 p. (1971).

8. Aubert, M. and Aubert, J., Pollutions marines et aménagement des rivages. *Rev. Int. Océanog. Méd.* (Suppl), 309 p. (1973).

9. Ui, J., Minamata disease and water pollution by industrial waste. *Rev. Int. Océanog. Méd.* **13-14**, 37-44 (1969).

10. Sprague, Measurement of pollutant toxicity to fish II. Utilizing and applying bioassay results. *Wat. Res.* **4**, 3-32 (1970).

11. Portmann, J.E. Trace metals in fish and shellfish from around England and Wales. Aquatic Microbiology Group, Plymouth, (1972).

12. Aubert, M., Aubert, J., Daniel, S. and Gambarotta, J.P., Etude des effects des pollutions chimiques sur le plancton. Dégradabilité du fuel par les micro-organismes telluriques et marins. *Rev. Int. Océanog. Méd.* **13-14**, 107-123 (1969).

13. Foret-Montardo, P., Problemes biologiques posés par la dégradation des détergents issus de la pétrochimie. *J. Etud. Pollut. Mar. Aménage, Littoral*, 49-56 (1970).

14. Aubert, M., *Pollution des océans Conférence à la Faculte des Sciences de Paris dans le cadre du cycle "Les Océans"*. Gauthier-Villars, Paris, Volume V, pages 247-268, 1970.

15. Aubert, M., Télémédiateurs chimiques et équilibre biologique océanique. Théorie générale. *Rev. Int. Océanog. Méd.* **21**, 5-16 (1970).

16. Aubert, M., Gauthier, M., Donnier, B., Pesando, D., Pincemin, J.M. and Barelli, M., Effets des pollutions chimiques vis-à-vis de télémédiateurs intervenant dans l'écologie microbiologique et planctonique en milieu marin. *Rev. Int. Océanog. Méd.* **28**, 129-166 (1972).

17. Aubert, M., Gauthier, M. and Pesando, D., Effets des pollutions chimiques vis-à-vis de télémédiateurs intervenant dans l'écologie microbiologique et planctonique en milieu marin. 2e Partie. *Rev. Int. Océanog. Méd.* **37-38**, 69-88 (1975).

NUTRIENTS AND BIOSTIMULANTS IN THE MARINE ENVIRONMENT: APPLICATION TO MUNICIPAL WASTEWATER DISCHARGES INTO THE SEA

Maurice Aubert

Centre d'Etudes et de Recherches de Biologie et d'Océanographie Médicale,

Parc de la Cote, 1 Avenue Jean-Lorrain, 06300 Nice, France

Summary — The ocean constitutes one of the most important food resources for mankind and hence it is necessary to maintain its integrity. Its productivity is based on the existence of nutrients and oligoelements, which are practically recycled and utilized by the primary biomass, bacteria and plankton. Some regulatory factors permanently intervene in order to preserve the biological balance of the marine environment. The intervention of Man may modify the biological cycles: this is the reason why the methods for marine disposal of wastes from inland must be thoroughly examined.

At a time when continuous population growth in the world menaces that the earth's food resources will become insufficient, we must remember that in theory the ocean might provide far larger amounts of nourishment. Actually, according to the calculations reported by J. Stirn [1], photosynthesis in the ocean produces 120×10^9 t of organic carbon per year, whereas earth ecosystems can produce only 20×10^9 t of it.

1. NUTRIENTS

When dealing with productivity, i.e. the gradual evolution of a mass of living matter, it is necessary to consider the sources of this life, the fundamental substances at the basis of its occurrence. It must be equally noticed that, without the intervention of man, the cycle of such substances is naturally closed, with regard not only to the water environment, but also to the earth and atmospheric ones, since the biosphere constantly acts on the three.   That is why the study of the nutritional phenomenon in the Ocean involves a prior short description of the natural cycle of the substances either dissolved in water or in suspension. This cycle on one side involves organic materials and on the other mineral salts. However, for a complete description, the influence must be considered of the living organisms of our planet — either vegetable or animal, mankind included; as a matter of fact, they significantly affect the water balance and its consequences for biosphere maintenance. We shall later describe the natural recycle of organic substances and of mineral salts, which also play a significant role.

The presence of organic material in the sea is twofold: one form is part of water itself as a more or less diluted solution; at the highest concentrations, it acquires a particular form. On the other hand, the organic material appears in the form of living organisms, i.e. microbiological (bacteria, yeasts, fungi), vegetable (phytoplankton, algae), and animal (benthic or pelagic fauna). The evolution of such materials is essentially characterized by the fact that no barrier exists between the two forms, but there is only an evolution, a transformation with time from one form to the other in an ever recurring cycle. Furthermore, since organic substance constitutes the essence of ocean life, the importance is stressed of the knowledge that may be acquired concerning its origins, evolution, degradation and reutilization.

Before describing this evolution, a short — and necessarily incomplete — inventory must be made of the sources of organic substances in the sea. According to paleontological discoveries, we know that the sea contained, even in the most ancient geological eras, an important biomass, in the Cambrian period, the ocean was rich in varieties of living forms, which are supposed to be at the origin of life on the continents.

After the geologic restructuring of the globe and the settlement of the oceanic structures as we presently know them, pelagic and benthic life acquired a specific feature; its characteristic — unlike continental life — is an evolution of its own. However, the ocean has a twofold connection with the continents: on one side, run-off waters and rivers flow from continents to the sea; on the other side, water comes back to the land through evaporation and rainfalls. However, though the latter does not participate in the cycle of the organic material, the contribution from inland, enriched with the wastes of the earth cycle, which leaves the continent to join the immense marine receptacle, constitutes one of the most important sources of enrichment in such materials. For example, large rivers, such as the Nile or the Mississippi, supply the marine environment with an extra amount of nutrients, which is of some importance as proved, for example, by the decrease in productivity occurring when such an addition stops. For instance, this took place when the Nile was dammed and consequently the sedimentary contribution to the East Mediterranean was stopped. The immense contribution of rivers all over the world to the marine environment may be easily imagined. Independently of such contributions, run-off waters due to coast washing, supply an extra amount of mineral and organic materials, which is not negligible though it is difficult to evaluate [2].

If the earth origin of such materials may be easily imagined, processes exist within the ocean itself that spontaneously supply organic material starting from gaseous atmospheric carbon dissolved in water, as well as marine mineral components, thanks to a part of the oceanic biomass. Actually, at this level, solar energy acts through photosynthesis and causes a considerable release of oxygen. By this mechanism, a primary series of organic materials is formed, which essentially consists of carbohydrates, such as starch, cellulose and all the polysaccharides that constitute the essential part of the vegetable biomass. Such a biomass needs elements naturally present in sea water and in the atmosphere; starting from them, the elements are formed that, after death, return to the ocean through biochemical processes connected with microbial activity (cellulolytic bacteria). Owing to this activity, the substances produced circulate again in the sea where they will be subjected to other biosynthetic mechanisms, which, as we will see, are involved in the formation of the animal biomass. Such a destructive process consists in mineralization, occurring by subsequent catabolic processes operated by microorganisms; instead, the constructive process brought about by the vegetable biomass is an anabolic process.

Along with the cycle that merely concerns hydrocarbon chains, there exists a constructive mechanism concerning the nitrogenous chains, i.e. one of the main characteristics of living phenomena. The living cell, which mostly consists of aminoacid chains, or proteins, takes the elements necessary for its growth and reproduction from the environment through metabolic processes.

Nitrogen, a fundamental substance of the living matter, has two sources: one, as for carbon, is the atmosphere; the other is water, where its compounds are found in solution; nitrogen is then taken by nitrogen-fixing bacteria, thus producing the amino compounds necessary for their life. However, independently of this source, and thanks to the activity of other specific microorganisms, a considerable amount of organic nitrogenous materials is recirculated and, in its turn, will be retaken by cell life. This general process of the life of the ocean is so well regulated that in the natural state there is a constant equilibrium, which allows life itself to permanently regulate the productivity of the initial stock and the conditions of its maintenance.

Some figures will allow the evaluation of the stocks of organic materials or of its constituting elements in the marine environment. According to Vinogradov, the planktonic biomass is on average evaluated at 100 mg/m^3 for the layer from 0 to 4000 m. Some points are much richer: as a matter of fact, according to Peres, a density of $21 - 22$ g of phytoplankton/m^3 has been observed on the south-east coast of Kamtchatka, and $2 - 4$ g/m^3 on the east coast of Greenland. Other figures give a more dynamic picture of this vital transformation. The primary production, i.e. the amount of atmospheric carbon fixed by sea vegetables per m^3 and *pro die* ranges between 0.2 and 2.5 g [3]. On a yearly scale, in temperate or subtropical seas, the primary production seems of the order of 120 g carbon/m^2. With regard to nitrogen, according to Emery *et al.* [4], the present reserve of the oceans may be evaluated at 920×10^9 t, 9600 of which are consumed every year by phytoplankton; at the same time, rainfall provides the ocean with about 59×10^6 t of nitrogen and rivers 19×10^6 t (in the dissolved state). According to these authors, the ocean reserve of phosphorus is of 120×10^9 t, with a yearly consumption by phytoplankton of 1.3×10^9 t. As to this substance, the exogenous contribution of rivers is of 14×10^6 t.

We have oversimplified the biochemical cycles of organic substances; reality is far more complex. Actually, the organic matter may be bound to mineral substances, with formation e.g. of organometallic complexes, the degradation of which requires more specific and varied means. Similarly, the degradation of lipid substances, e.g. hydrocarbons, involves other complex biochemical processes, which are brought about by a particular microbic flora.

As already mentioned, through the action of bacteria, such a living matter — either vegetable or mineral — after its death, regains its solution state, i.e. it is brought back to its original constituents, essentially minerals, electrolytes, and some soluble organometallic compounds. Such a process of mineralization may be brought about only by bacteria. But, as may be easily imagined, considerable amounts of oxygen are required for such a bacterial metabolism.

However, the ocean does not seem to contain a yearly oxygen amount that is sufficient to provide for natural oxidation of organic materials and oxidation articially brought about by human activity. As a matter of fact, it is traditionally assumed that the oxygen content/year in the ocean surface is 8 mols/m^2, which is just enough for the metabolism of the beings living there. Hence it appears that any artificial addition of organic materials is a contribution towards the occurrence of a deficit in a system that is already almost in equilibrium. Thus the destructive process of decomposition involving the uptake of oxygen is opposed to the constructive synthesis created by vegetable life through photosynthesis. The preservation of a biological equilibrium in the sea calls for the quantitative comparison between these two actions, the constructive and destructive. The biomass increases with the sufficient increase in nutrients, mineral salts and organic materials. But, after an optimum of productivity, such a biomass dies and becomes a stock of organic material, the mineralization of which requires oxygen comparable in amount with that evolved during its synthesis. Through bacterial degradation, the stock after mineralization is recycled [5].

It soon appears that the biological equilibrium of the sea may be maintained only if important variations between additions and losses of this initial stock of organic materials on the planet do not occur [6]. In such a dynamic situation additions from land sources and from the atmosphere have to be taken into account. However, there is a natural loss of organic materials, i.e. the slow sedimentation on sea bottom of species living in the pelagic and benthic regions, after their death. The inert sediments and the mineral substances brought by run-off waters gradually bury such dead matter and in this way a non-negligible stock is lost. According to Emery *et al.* [4] this loss caused by sedimentation reaches a yearly total of 9×10^6 t of nitrogen, 13×10^6 t of phosphorus and 3.8×10^9 t of silica. However, it may well be imagined that a part of this stock actually is recycled by the action of benthos: bacteria, annelids and other invertebrates, which are utilized by vagile benthonic animals, and which after their death undergo mineralization and dispersion of their organic components.

Independently of such biological activity, a quasi constant restructuring of the bottoms occurs by dynamic sedimentary phenomena. Washing of the bottoms and conveyance of sediments cause stocks of accumulated materials to circulate again, by the action of submarine streams, some of which move at a non-negligible speed. Furthermore, frequent breakdowns of submerged reefs, through the turbidity currents they cause, bring huge amounts of sediments into circulation again. Finally, what is most significant, submarine eruptions of gas or materials throw considerable amounts of substances accumulated on the bottom toward surface ocean layers.

In spite of such mechanisms, part of the organic matter is lost due to nutritional processes.

However, it is known that sea life is in equilibrium. Additions of nitrogen, phosphates and nitrates from inland regularly occurring by rivers and run-off waters permanently compensate such a loss and maintain the life of the ocean as it is known and with a stability that, according to geology, dates back to the most ancient times of our planet. This is also the case and to a much larger extent starting from atmospheric carbon, which is an inexhaustible source of carbon.

Along with organic materials, mineral salts exert an important impact on the water environment and consequently on the life of earth. The vast water reserve of the world, the ocean, contains different mineral salts in solution, most of which consist of sodium or magnesium chloride, as well as a large number of oligoelements, which are essential to maintain the metabolism of cells and organisms. Phytoplankton evidently draws the chemical components for its solid parts from the sea, and it is equally evident that terrestrial beings need calcium for their bones and some halogens, such as fluorine and iodine, are essential for the most complex organisms.

The importance of the mineral salts dissolved in the sea is evident not only on a nutritional level, but also from the point of view of the ionic equilibrium of living tissues. Nutrients, either organic or mineral, constantly circulate in the ocean, both due to the biological activity and to the physical mechanisms of settlement of sea bottoms by the impact of streams, breakdowns of submerged reefs or volcanic eruptions in sea bottoms.

2.   BIOSTIMULANTS

The approach we have just used, which consists in establishing a balance of the productivity possibilities as a function of the nutritional sources and of the mechanistic laws of thermodynamics, is quite artificial: as a matter of fact we have not taken into account an essential factor, i.e. the specific and baffling aspect of the biological phenomenon. This involves very fine processes that either spur or hinder the metabolic activities and may make illusive the overall long-term forecasts.

Such processes ruling the bloom of life are both biostimulating and inhibiting.

A regulating action of this type is exerted both by the oligoelements that are present in the sea and are essential for cell growth, and by the substances of biological origin playing a role on the vital functions of the beings that consume them.

The presence of oligoelements, such as cobalt, copper, manganese, etc., is essential to the vital or reproductive functions of cells. Several proofs exist concerning the need of such a presence and among them the rigorous ionic equilibrium that must exist to maintain productivity in phytoplanktonic cultures. We do not insist on such classical and well known notions.

We may remind ourselves that such a biostimulating action is quite often due to the intervention of chemical substances in the mechanisms of photosynthesis, in which they act as electron carriers.

Some, like iron, simultaneously act as stimulants and as nutrients, since they participate in the general metabolism of organisms.

However, it must be pointed out that such chemical substances, in particular copper, may exert a higher inhibitory effect than in nature, as our co-worker, F. Laumond, demonstrated in her thesis [7].

To summarize, the essential micronutritional elements have different functions: some, such as zinc, may act as catlysts, some others, such as iron, copper, cobalt, may participate in oxyreductions; others may be co-factors of enzymes responsible for cell biochemistry, or contribute to the structure of respiratory pigments.

Therefore, we think it interesting to thoroughly consider the action of the biological substances that regulate productivity. Such substances have been previously described by us and called "chemical mediators": "These substances are synthesized by marine species, either animal or vegetable, and evolved in the environment; they act at a distance on the behavior or on the biological functions of the same species or of other species" [8].

Although such interspecific mechanisms are present at all levels of the oceanic domain, we will just consider a few examples of acts exclusively affecting the primary biomass, plankton and bacteria.

In this domain, the best known example concerns the biological equilibrium existing among some phytoplanktonic species. For example, the antagonism that may be found between communities of diatoms and of peridins, unicellular photosynthetic organisms that constitute the largest amount of sea phytoplankton. A yearly phytoplanktonic cycle exhibits, generally in the sea, some development maxima of diatoms, followed by a development of peridins, the ones systematically excluding the others.

More recently, at CERBOM, our co-worker J.M. Pincemin [9] demonstrated this interspecific opposition *in vitro*. He described the decrease of the diatoms *Asterionella japonica*, after being contacted with *Glenodinium monotis*; after a series of laboratory tests, he also succeeded in demonstrating that the culture environment were the diatom mentioned had lived favourably influenced the growth of Peridin. By taking into account that the reproduction of Dinoflagellates is slow in comparison with that of diatoms, the regulation effected by the substances evolved by each species and released in the environment where the other species lives, corrects the possible proliferation of diatom communities in the environment; in this way, a rational biological equilibrium between the two species is maintained. As shown by this example, this is a two-phase mechanism, by contacting two types of organisms at least connected by two chemical mediators. They have not yet been isolated and their chemical nature has not been determined.

Another example has been obtained from our work: an assistant professor of our Centre, M. me D. Pesando [10], succeeded in demonstrating that *Peridinium trochoideum* exerts an analogous action to that of *G. monotis* on the growth of *A. japonica*.

The growth of the diatom is inhibited both when a diatom culture is contacted with a culture of Peridin and when the diatom is cultivated in the Peridin culture environment, freed from cells by filtration, either when an environment is enriched with nutritional substances or not. The existence of this primary productivity mediator is evidenced in the dialysate of a cellular extract of Peridin and an active fraction is obtained after fractionation of such a dialysate on an anion exchanger resin. Such preliminary work shows that this is not a nutritional competition, but the action of a chemical mediator.

Other authors have shown analogous facts: D.M. Pratt, for example, observed an antagonism in the development of a Diatom *Skeletonema costatum* and of a Xanthophy *Olisthodiscus luteus*, although a chemical mediator could not be isolated. Similar observations have been made in fresh-water environment by Rice, and by Lefèvre, Jakot and Nisbet.

Another example is that reported by Fontaine on the antagonism that seems to exist between some zooplankton communities and others of phytoplankton that do not allow a permanent co-existence of the two communities. The authors have supposed that the phytoplankton evolves some substances, and this emission, connected with photosynthesis would reject zooplankton from the phytoplanktonic biomass, thus rhythmically creating, due to the alternation of sunlight and darkness, a mechanism of defence of the phytoplanktonic biomass and a nutritional regulation of zooplankton, which is its usual nourishment. Such an explanation is still a hypothesis, since, as in the previous examples, the substances issued by those phytoplanktonic species do not seem to have been isolated or analysed. This fact seems to be established from the biological point of view, but chemical characterization of the active aubstances has not been effected.

In a neighbouring domain, it is equally demonstrated that some phytoplanktonic species, such as the Diatoms, exert an inhibitory action on the development of bacteria.

By a more thorough examination of the interactions between telluric bacteria, Diatoms and Peridins, it appears a true turnover of their communities: the bacteria proliferating in the waters rich in organic substances evolve growth mediators, such as vitamin B_{12} that favours the reproduction of diatoms, which follows a 2-week cycle; the diatoms evolve substances such as fatty acids or polysaccharides that inhibit telluric bacteria. The process would end with a diatom invasion of the environment, if in their turn, they were not inhibited by Peridins, which follow a longer cycle (about 30 days). Vitamin B_{12} favours the growth of Peridins, which evolve mediators by stopping proliferation of diatoms, but Peridins, by their own secretion, are naturally self-inhibited, starting from a given cellular density.

We could check that our experimental findings are also true in the natural environment. Actually, we have shown that the communities of these three types of microorganisms exhibit a maximum nearly every 30 days in the following order: bacteria, diatoms, peridins, each maximum being separated from the following by a $\sim$ 10-day interval.

Thus, very fine biological mechanisms modify the forecasts that may possibly be made starting from existing knowledges on nutrients in the marine environment. But, independently of such natural biological processes, the gradual growth of human activity will possibly have an ever increasing impact on the biological equilibrium of the ocean; hence, the problems arising from recycling artificial substances produced by man in the natural environment must be evaluated.

Pollution exists which does not involve industrial activity, but is exclusively due to population growth in limited areas. This is the case of pollution caused by municipal waste waters from important coastal towns; it consists of microorganisms and particularly of organic materials. This subject will not be stressed further; it is connected with natural process and has already been thoroughly considered. However, more frequently, massive pollution phenomena are caused by the substances that chemical industries discharge into the marine environment either directly or through rivers, involving considerable modification of the biological water quality. Several substances are directly toxic and their destructive action is a function of their initial concentration, both when they are mixed with water and are detrimental to marine life, and when the normal pH of waters is largely modified by disposal of acids or bases. A loss of biomass follows, i.e. a loss of the nutritional capital available to man. This initial consequence has been long since known. Studies devoted to this phenomenon are at the basis of legislation promulgated all over the world, for the purpose of protecting aquatic flora and fauna.

A further consequence of pollution was discovered about 30 years ago: it is that resulting from fixation and accumulation of some products by aquatic species. This phenomenon, which had been initially described for radioactivity, is of importance also for some substances, such as pesticides, or some metal salts. The serious consequences that occurred after the introduction of mercury and cadmium into the sea is common knowledge. Such toxicity indirectly reveals itself through the various levels of the aquatic biological chains up to the final consumer, Man, causing pathological consequences. Hence, it constitutes a public health hazard. Such phenomena could be demonstrated by experimentally reproducing the concentration processes by using marine trophodynamic chains.

Finally, a third consequence has been evidenced quite recently: it consists in the impact that some chemical substances may have on the mediators ruling the interspecific relations. The work recently carried out at the CERBOM has shown the importance of such messages in the field of microbiological or planktonic ecology; furthermore, they have shown the possibilities of ecological drift caused by detrimental effects on such mediators, which act at a distance, are conveyed by waters and rule the biological equilibrium of the sea [11].

3. APPLICATION TO WASTE WATER OUTFALLS

From biological and biochemical data, a water-tight sewer may be designed so as to minimize the polluting effects of waste waters from coastal towns [12]. Outfalls are responsible for macroscopic and microscopic pollutions and hence technology must strive for their control; furthermore, it is worth determining which are the conditions that result in the smallest change in the original state of the marine environment. Visible pollution is not seriously detrimental, but it damages the aesthetics of coasts and therefore is of touristic importance. It is fairly easy to control by simple and well known methods, such as screening, systematic purification, the prohibition of sewage discharges into rivers or along sea coasts, and possibly primary settlement of sewage before discharge. Suspended faecal material in effluents also causes offensive conditions but by comminution a fairly easy and cheap solution is provided for this problem.

On the other hand, microscopic pollution is far more dangerous and insidious. Several authors, and in particular Buttiaux (1968), have described the manifold dangers involved in the diffusion of microorganisms into the sea; the bacterial diffusion laws described by us interfere in the dispersion of such more or less pathogenic elements. Two types of purification are possible: natural purification, i.e. self-purification of the sea, and artificial purification, obtained through common processes (activated muds, bacterial beds, oxidation tunnels, etc.). However, such technical sterilization processes do not provide perfect results and when they operate under satisfactory conditions, which does not always happen, sterilization falls short of 100% success and often only reaches 50% efficiency.

The installation of sewage purification works is rather expensive and they often constitute a financial burden for countries with small populations. This is why the natural self-purification of the sea is often exploited, which may have serious consequences since the marine environment has a limited capacity to deal with certain pollutants. Natural self-purification is often considered enough and, if favoured by currents carrying waste waters away from the coast, it may in some cases provide permissible hygienic conditions.

In any case, with regard to both a fairly well purified effluent and comminuted sewage with a considerable bacterial load, the problem of sea disposal is the subject of laws, which are becoming recognized.

In order to prevent the return of pollutants to the bathing and touristic areas, the point of the effluent outlet will be calculated as a function of the hydrological, current velocity and bathymetric situation of the discharge area by a mathematical formula, like those established by Pearson or Aubert and Désirotte.

If discharges of chemical substances constitute an often underestimated danger, bacterial discharges are much less hazardous and, in the long run, complete sterilization takes place by self-purification in the sea. However, beyond the point of view of the sanitary expert, it is interesting to know the action of such discharges at fairly long distances from the effluent outlets.

Recent investigations carried out in the United States and in France have shown that organic materials and growth factors present in sewage favourably affect the productivity of the primary biomass, which further influences secondary and tertiary ocean productivity; hence it can be foreseen that such discharges — made under the required conditions — will increase owing to sea productivity.

Hence it seems that the disposal of waste waters into the sea — with the exception of all industrial effluents — does not adversely affect marine life on the whole, and does not destroy the equilibrium factors of the oceanic biomass.

However, in spite of such beneficial effects, a few less favourable consequences should not be forgotten. The effects of benthos causing the regression of some species and the proliferation of others have been well described by American and French schools; they have succeeded in showing that the consequent ecological modifications may be used as reliable tests of water pollution.

Some authors and in particular P. Koch insist on biological transformations occurring at the effluent discharge zone, where within a quite limited area there is a true "azoic region where vegetation is destroyed and where fish reproduction disappears".

However, sewage may be loaded not only with microorganisms, but also with chemical substances of industrial origin; their detrimental effect varies depending on the products which may damage the marine environment. The changes that occur are the subject of further studies. Such phenomena differ from those just described for bacterial pollution. The diffusion of these toxicants occurs by processes not described in this paper.

The technical considerations on the choice of the point of discharge of bacteriologically polluted waters show that, if only the self-purification power of the sea is available, effluents may be disposed without giving rise to pollution of neighbouring coasts. The conditions under which such achievements are possible are far from being general, both when currents flow towards the coast too frequently or too rapidly, and when the coastal structure partitions off the hydrological masses, when depth is too little, and finally when the quality of waste waters is particularly rich in organic materials or in chemical domestic products. This is the reason why natural self-purification must frequently be assisted by artificial purification processes. In such instances effective help is given by sewage purification stations. They are of considerable value in removing organic materials from the sewage, but less active with regard to the bacterial content. In any case, although a purification works may be installed, effluent discharges into the sea must occur and the length of the outfall which conveys the effluent must be designed according to the methods of calculation such as those quoted above.

To conclude, the manifold biological, physical or chemical aspects involved in marine waste disposal plants require great care concerning the consequences of the discharges on the sanitary situation of the sea as well as on the possible long-term modifications of the biological equilibrium of the sea.

REFERENCES

1.   Stirn, J., Ecological consequences of marine pollution. *Rev. Int. Océanog. Méd.* **24**, 13-46 (1971).

2.   Raymont, J.E., *Plankton and Productivity in the Oceans.* Pergamon Press, Oxford, 1963.

3.   Postel, E., "Les resources biologiques": généralités sur l'écosystème océanique, son potentiel de production. *Vie Méd.* **50**, 25-30 (1969).

4.   Emery, K.O., Orr, W. and Rittenberg, S.C., *Nutrient Budgets in the Ocean.* University Southern California Press. Los Angeles, pp. 299-309, 1955.

5. Rittenberg, S.C., *Marine Bacteriology and the Problem of Mineralization.* Symposium on marine microbiology. Oppenheimer, 1963.

6. Aubert, M., Aubert, J. and Gauthier, G., Le milieu marin et les matieres organiques. *Rev. Int. Océanog. Méd.* **28**, 181-193 (1972).

7. Laumond, F., Etude de la répartition d'un élément trace, le cuivre, en milieu océanique. Son influence sur la biomasse marine. Thèse de Doctorat d'université Nice. 137 pp., 1974.

8. Aubert, M., Télémédiateurs chimiques et equilibre biologique océanique. Theorie Générale, I. *Rev. Int. Océanog. Méd.* **21** 5-16 (1971).

9. Pincemin, J.M., Télémédiateurs chimiques et équilibre biologique oceanique. Etude *in vitro* de relations entre populations phytoplanctoniques, II. *Rev. Int. Océanog. Méd.* **22-23**, 165-195 (1971).

10. Pesando, D. and Aubert, M., Effets des pollutions chimiques vis-à-vis de télémédiateurs intervenant dans l'écologie microbiologique et planctonique en milieu marin, III. *Rev. Int. Océanog. Méd.* **39** (1975).

11. Aubert, M., Recyclage des dechets et sauvegarde des cycles biologiques dans le domaine de l'eau. Colloque Mondial *Biologie et devenie de l'Homme.* 18 pp. La Sorbonne, Paris, 1974.

12. Aubert, M. and Aubert, J., *Pollutions Marines et Amenagement des Rivages.* Edition Revue Internationale d'Océanographie Médicale. 1973.

Progress in Water Technology, Vol. 4, pp. 51—58.　　　　　Pergamon Press, 1978. Printed in Great Britain.

EFFECTS ON PHYTOBENTHOS OF MARINE DOMESTIC WASTEWATER DISPOSAL

G. Giaccone

Department of Botany, University of Trieste, Trieste, Italy

Summary — The study of the effects on phytobenthos of domestic wastewater disposal into the marine environment involves a detailed knowledge of the composition and typology of phytobenthonic communities under conditions of ecological equilibrium. The understanding of the mechanisms of action of such effects requires surveys *in situ* of the structure and evolution models of both stages and phases of indigenous phytocenoses. Toxicity tests on benthonic algal cultures in sea water added to sludge supply useful information not only with regard to the toxicity of the discharge in different parts of the diffusion area, but also suggest the characteristics required for the purification plants and outfall diffusers. Research on phytobenthonic communities, both on studying the feasibility of a purification plant and during the control of its efficiency, are to be preferred to those on plankton and on nekton. As a matter of fact, phytobenthonic populations — owing to their close connection with the substrate and with the sedentary habit of their organisms — not only express the host biotope, but also constitute essential elements of it.

1.　INTRODUCTION

In order to understand the mechanism of action of the ecological factors connected with domestic wastewater disposal, it is necessary to know the biocenoses under natural conditions as well as their evolution as a function of the dynamism of biotic and abiotic factors both endogenous and exogenous. Biocenoses in marine environment may be planktonic, nektonic and benthonic. The set of ecological factors that, under natural or artificial conditions, constitutes the marine environment, conditions all such living communities. Benthonic populations more clearly and lastingly show this quality and extent of the complex interactions between biotic and abiotic factors and, owing to their close connection with the substrate, not only express the host biotope, but are also essential elements of it. Phytobenthos is the most complete expression of the environmental factors in the different units of the benthonic domain and particularly of the phytal one, in the microclimates and in the facies of marine biotopes. Hence, when stress occurs, they yield the most reliable data on the type and extent of the damage and on the consequences in the short- and long-term. In benthonic biocenoses of the coastal system, which practically comprises the whole continental shelf, vegetal components condition the quality, vitality and consistency of a few essential levels of the trophic chain, on which the whole living community of the involved biotope is based. In view of such preliminary remarks, the study of the effect of marine disposal of polluted waters — both on checking the feasibility of a disposal design and in the control phase — must consider the composition and evolution of the phytobenthonic populations that are present in the outfall diffusion area. Such surveys *in situ* must be both preceded and followed by laboratory toxicity tests in order to determine the effects of sewages at different concentrations on some significant types of natural populations. This study is based on tests done in various areas of the Mediterranean [5, 16], and particularly in the Gulf of Trieste [4] as well as in our laboratories [10]; accordingly, this paper is divided into three sections. In the first section an outline is given of the composition and typology of the various phytobenthonic communities of the Mediterranean under natural equilibrium conditions with biotic and abiotic factors. In the second the effects are considered of polluted discharges on the structure of indigenous communities and with the evolution mechanisms occurring after such alterations. In the third section a description is given of the usefulness of toxicity tests on cultures of benthonic algae in the presence of sewage added to sea water at different concentrations.

2. COMPOSITION AND TYPOLOGY OF PHYTOCENOSES OF THE MEDITERRANEAN UNDER CONDITIONS OF ECOLOGICAL EQUILIBRIUM

2.1. Composition

From a comparative study of the data reported in literature it appears that, in a geographic area subjected to homogeneous climatic factors and in a 200—300 km coastal region, ca 50% of all phytobenthonic species of the Mediterranean may be found through accurate investigations. Nearly 1000 species of benthonic vegetable organisms live in this sea, i.e. about 13% of the species existing all over the world [7]. Most species consist of algae, of which five are Angiosperms. Among algal species ca. 200 exist in the Mediterranean only, forming the endemic element, while the others are common to the temperate and subtropical Atlantic regions. Among angiosperms, *Posidonia oceanica* only is endemic to the Mediterranean and draws its origin from the Indo-Pacific domain. The endemism-richest biogeographic areas are the central basins of the western Mediterranean, the Aegean Sea and the Central Adriatic. The regions where Atlantic species predominate are the Alboran Sea, the Strait of Sicily, the Strait of Messina and the Upper Adriatic. Indo-Pacific elements predominate in the southern regions of the east basin and after the opening of the Isthmus of Suez and the construction of the Assuan dam, they increased both in number and in expansion capacity. Some species, and in particular a number of *Cystoseira* live in so well defined areas that they may be considered as true biogeographic indicators. Others show a well defined ecological valence and their presence indicates environmental conditions determining the completion of their development cycles. The presence of species and varieties having a precise biogeographic and ecological meaning in the various phytobenthonic communities of the Mediterranean is the main reason for their diversification in several geographic and ecological variants.

2.1.1. *Biological indicators.* Efforts have been made by a number of researchers to find benthonic species whose presence might indicate given types of pollution or more generally particular environmental factors. The concept of indicator species has been recently replaced by the more reliable concept of community; as a matter of fact, the composition and evolution of the latter more completely and truly show the changes in both quality and quantity of a biocenosis. The choice of benthonic communities as biological indicators is based on their constant presence in the involved biotope, on the sedentary habit of their species and on the possibility of showing, in the interactions between biotic and abiotic factors, different evolution tendencies as a response to both occasional stresses and even to slight and continuous modifications in the ecological parameters.

2.1.2. *Specific diversity.* For more than 50 years, efforts have been made to look for specific diversity indices capable of evidencing the living conditions of a biocenosis. As a matter of fact, the specific diversity of a living community decreases when this undergoes ecological stress. However, also natural events — and not only pollution — may be responsible for such an effect (salinity and temperature increase or decrease in bays and estuaries, violent sea-storms on the coasts subjected to predominant winds, distance of the general circle streams from the coast, location and duration of local countercurrents owing to meteorological phenomena, etc.). Pollution, at least at given concentrations, sometimes increases the specific diversity to the detriment of the species bound to the indigenous communities and in favour of euryeceous species. These latter set up on the empty spaces left by other species that, more than a diversity fall, at least initially undergo a fall in the abundance-predominance values. Marine botanists working in the Mediterranean adopt, as an index connected with the effects of latitude, the ratio between the numbers of red and brown algae (R/P = Rhodophyta/Phaeophyta). Such an index may also be applied for evidencing thermohaline pollutants. The number of green algae has been sometimes used to evidence organic and industrial pollutants [16]. However, according to the investigations done in the whole Mediterranean, Clorophyta are not significant in number. Furthermore, in nitrophilic communities the number of green algae is not too high and only some galenophilic species in convenient environments reach such abundance-predominance values as to influence the specific diversity of the biocenoses.

2.2. Typology

2.2.1. *General remarks.* The phytal system consists of some fundamental units called vegetation planes [15]. Characterizing elements of a plane are the associations of vegetable species that are either statistically faithful or preferential to the ecological conditions singling out the vertical space covered by the plane. Among all species, some may be considered as "monitoring" not only because, by their presence, they single out the plane or the bionomic units below it (subplanes and horizons), but also because they condition thallus growth, with a vegetative periodism and with the emission of chemical mediators, as well as the presence or the absence of other species generally with a wider ecological valence than that of the monitoring species. In the Mediterranean, monitoring mesolittoral species mainly are the blue algae, some encrusting species and some calcareous algae; brown algae or Phaeophyta and particularly *Cystoseira, Sargassum* and *Laminaria* are the monitoring infra- and circalittoral populations on solid substrata, while Angiosperms and free Melobesia are on the mobile ones. The percent frequency of green algae or Chlorophyta is almost constant from the meso- to the circalittoral and this discredits the belief of the predominance of such species in the upper planes and particularly in polluted environments. Red algae or Rhodophyta show a bipolar distribution with two frequency maxima of the preferential species, one situated in the mesolittoral and one in the circalittoral. A number of red algae adjust themselves quite easily and may be found almost everywhere. They are normally found in polluted environments and their presence markedly increases the specific diversity of such environments especially by colonizing the irregularities of the substrate where polluting effects are generally lower.

2.2.2. *Coastal vegetation.* Coastal vegetation comprises phytobenthonic communities either constantly or periodically immersed of both upper- and mesolittoral, and the communities present in lagoons and coastal ponds with an average depth not above 2 m. The knowledge of the typology of such populations is fundamental for the study of the ecological effects of discharges on the shoreline or into coastal lagoons. The physical nature and the morphology of the substrate, the highest and lowest values of temperature and of salinity, the intensity of light, the effects of exposure and of the tidal range are the main abiotic factors determining both composition and distribution of the vegetable organisms along the coast. Vegetation in lagoons and coastal valleys may be divided [7] into three groups: (1) haptophyte vegetation attached to natural or artificial solid substrata; it comprises impoverished aspects or nitrophilic and galenophilic facies of associations of Acrochaetietalia in the mesolittoral and of Cystoseiretalia in the infralittoral; (2) rhizophyte vegetation on mobile substrata with sand and mud; the associations are of three orders Zosteretalia, Charetalia and Ruppietalia; (3) secondary pleustophyte vegetation with aeropleustophytes floating on the surface of water and with benthopleustophytes lying or wallowing in the muddy or sandy bottoms. The study of marine vegetation on solid substrata in upper- and mesolittoral has been tackled by using as monitoring species either almost exclusively *Cyanophytes* [2, 11] or all organisms present in the two planes [1, 5, 12]. In the classification done by these authors as regards the upperlittoral only one association based on lichens is reported, while as regards the mesolittoral, eight associations are described, which however, after more recent studies [3] seem to be only a few statistically homogeneous and bionomically significant populations.

2.2.3. *Submersed vegetation.* Submersed vegetation populate both infra- and circalittoral. The typology of such phytobenthonic communities on rocky as well as on muddy and sandy substrata must be known in order to study the ecological effects of discharges through offshore outfalls, and of polluted coastal effluents, whose output may involve a sufficiently large area as to cover the biotopes of the planes under examination. The abiotic factors determining both distribution and association of phytobenthos in the upper infralittoral and in the upper layer of the lower one are the amount of light reaching the various horizons, the type and intensity of hydrodynamism occurring in them. Abiotic factors in lower infralittoral and circalittoral are rather levelled, while the biotic ones, resulting from the interactions among the various components of biocenoses regulate the association processes. In the Mediterranean, vegetation of rocks is given by a succession of populations in which the various species of *Cystoseira* and *Sargassum* genera characterize the different bionomic units as well as facies, seasonal aspects and geographic and ecological variances of the various phytobenthonic units [8]. *Laminaria* populations are widespread on the circalittoral solid substrate. Infra- and circalittoral phytobenthos consists of seven associations, of which one is formed by nitrophilic and galenophilic species. By computer analysis of the surveys done in differentiated biotopes, these species were found to be independent of the factors that, as previously indicated, govern the processes of vertical and horizontal distribution in the different bionomic units of the phytal system. Angiosperms and partcularly *Posidonia oceanica*, predominate on mobile substrata. The equilibrium of a *Posidonia* prairie depends on complex biotic and abiotic factors, which will be discussed in the following section. Under particular hydrodynamic conditions, populations of free *Melobesiae* and of encrusting calcareous algae grow on infra- and circalittoral mobile substrata. Such associations are quite sensitive even to slight variations of the ecological parameters and biotic ratios. Biocenoses where building organisms predominate, show signs of suffering and of regression phenomena even when the remainder do not show any reaction to incipient pollution. Such a behaviour derives from the fact that equilibrium in these biocenoses is determined mainly by biotic factors, which are immediately affected by the change of specific composition and by the alteration of the covering capacity of the single species.

3. STRUCTURE AND EVOLUTION OF PHYTOBENTHONIC COMMUNITIES IN THE PRESENCE OF SLUDGE

3.1. General remarks

As the determination of the composition and typology of a natural community under equilibrium requires the study of the various components on both statistical (frequency and covering) and biological bases (vitality and development cycles), likewise, the evidencing of the effects of pollutants requires, after comparative check with the composition and typology of populations present in neighbouring areas under equilibrium, the study of the structural aspects and evolution tendencies of the population itself. Such a study must be carried out by assuming the bionomic units as a reference and must consider a set of samples on homogeneous vegetation in order to evidence its possible nitrophilic or sciaphilic facies in relation to the indigenous associations or to check its lack of homogeneity and of repeating units testifying a deep crisis not only in the responses of single species to the parameters of the abiotic factors, but also the lack of equilibrium in biotic ratios. A significant indicator of the presence of municipal sewage initially is the presence of species with a large ecological valence, which tend to mask indigenous associations. Among these species, several are considered sciaphilic: actually under conditions of ecological equilibrium they live in biotopes in the shadow. This has a twofold meaning: on one hand these species — owing to their low competitive capacity — take shelter in the biotopes where the other species do not proliferate; on the other hand like all sciaphilic species, they feed on heterotrophic nourishment and hence may live with a reduced light. As a matter of fact urban pollution always leads to turbidity, which causes a reduction in penetration of sunlight, a decrease in salinity and temperature, and water enrichment with nutrients, i.e. it gives rise to the conditions that naturally occur in deep environments, though with different gradualness and intensity. Upwelling of sciaphilic species always indicates an equilibrium that tends to break. The appearance of these species is transitory and limited to the cold and less bright seasons if the polluting effluent output is not large or, within certain limits, inconstant. On the contrary, if emission is both continuous and large, there takes place a gradient of situations that improve more and more as dilution becomes acceptable, provided that accumulation phenomena or toxic effects do not take place. In the next section laboratory data on the effects of various urban sewages on benthonic algae are given. But already in nature, population structure and evolution in the diffusion area of an urgan effluent clearly show the effects of the different concentrations and the preferential directions of the pollutant-dispersing currents.

3.2. Solid substate populations

3.2.1. *Coastal vegetation.* Under equilibrium conditions, coastal vegetation tends to arrange itself in belts singling out ecological horizons that may be perfectly characterized by abiotic parameters. If belts are lacking and the phytobenthonic succession is simplified either in two bands or in one band with two lateral fringes, one factor predominates, which is superimposed to the climatic and edaphic ones of equilibrium. This may also be a natural event (a strong sea-storm or a prolonged summer calm, etc.), but if the phenomenon becomes constant there must be a source of pollution, which, if toxic, destroys the biotope and if eutrophicating, causes an outbreak of euryvalent species and the presence of several transgressive elements coming from lower levels. In such biotopes the most resistant species are the perforating blue algae, which inside the rock succeed in creating microenvironments that are less subject to pollution effects. Upperlittoral lichens and misolittoral calcareous algae are the most vulnerable elements and are the slowest in recolonizing lost biotopes. Several red algae with a soft thallus withstand urban pollution and are recruited *ex novo*, whereas among green algae, the predominance is abserved of *Ulvales* and *Cladophorales*. Encrusting brown algae show a considerable resistance and, under certain conditions, may compromise mussel populations by covering their valves, which thus become weaker and may be easily attacked by perforating forms. In the infralittoral, *Cystoseira stricta* are initially replaced by juvenile or reduced forms of *Cystoseira fimbriata* and then by a belt with species of the *Ulva, Pterocladia* and *Corallina* genera. After completion of evolution, the mesolittoral physionomy consists of a prairie of Enteromorphae and the infralittoral of an *Ulva* belt that also invades the bionomic units below colonized by ever submersed vegetable populations.

3.2.2. *Submersed vegetation.* Fully developed infra- and circalittoral vegetation generally is stratified. The upper layer consists of brown algae (*Cystoseira, Sargassum, Laminaria*) with several epiphytes, the sublayer consists of a few hundreds species, among which Rhodophyta predominate. In the various development phases, such populations show a linear evolution and, except few monitoring species, the remainder consists of forms with an ecological valence that is generally large and, in any case, still little known. As a general rule, urban pollution directly concerns the upper infralittoral and causes the disappearance of the large *Cystoseira* that form monitoring species of indigenous associations. These species are first vicariated by others of the same genus (*Cystoseira barbata, Cystoseira fimbriata*) endowed with a larger ecological valence, then there occurs a population of Dictyotales and sciaphilic red algae; furthermore, if ruinous situations do not occur, there may occur the formation of lasting stages of populations that are continually liable to colonization attempts by the most unvaried species that, according to the seasons, contribute to increase the heterogeneity of such phytobenthonic associations. Pollution effects on circalittoral phytobenthos are known only by the effects caused by the resulting materials or in general by sedimentation that destroys the rocky substrata by the effect of coverning. As mentioned above, in the case of massive pollution — such as, for example, in the Gulf of Naples — an upwelling of circalittoral species to the infralittoral and a spreading out of species of half-dark biotopes into the circalittoral are observed. On the other hand, with regard to animal components, a high development of filtrators is observed. A quite frequent degradation — considered both natural and aperiodical — in solid substrata populations is due to the outbreak of pasturing organisms and especially of sea-urchins. A synergic action of intense pasturing, strong sedimentation and consequent decrease in brightness may cause a quite serious impoverishment in phytobenthonic populations. In general, such a degradation tends to regress with a lower impact of pasturing animals, as happened in the Gulf of Trieste; however, in order to enhance re-establishment in coastal environment, discharges must be removed from the shoreline and sedimentation must be relieved through primary treatment of sewages.

3.3. Populations on mobile substrata

Lagoons and coastal ponds are the biotopes in which mobile substrata mesolittoral is wider and more richly peopled with phytobenthonic forms. Instead, adlittoral psammophilic populations are more convenient for studying the effects of pollution on marine beaches. In lagoons the study of adlittoral and mesolittoral phytobenthos is more significant than that of the infralittoral one. The effects of urban pollution or comparable effects (e.g. fertilizers for agriculture) cause a reduction in the haptophytic and rhizophytic forms and a spreading out of aero-, meso- and benthopleustophytic species. An abnormal development of Chaetomorpha, Gracilaria, Vaucheria and Ulvales foreshadows secondary eutrophication phenomena and oxygen depletion situations, especially in summer nights with consequent death of animals that are useful to valley culturists. Typical marine environments shelter wide Angiosperms prairies in the meso- and infralittoral. Like all aquatic weeds, even marine angiosperms absorb nutrients both through leaf- and rootage. Hence their development cycle is not only affected by the quality of water but also by the composition of sediments. *Posidonia oceanica* is the most vulnerable species as well as the most extensively widespread in the Mediterranean; as a matter of fact its prairies cover for 70% the whole vegetable cloak of the Mediterranean infralittoral bottoms. In the economy of the submersed life of the Mediterranean, this phytobenthonic community is of fundamental importance being the place of reproduction and of nourishment of a large variety of species, which are the basis of profit for coastal fishers. Onshore discharges are extremely noxious to *Posidonia* prairies, owing to the excessive emission of nutrients in milited areas and to the consequent appearance of turbidity. Canal digging, anchorage apparatus, ploughing and drag-net operations constitute further sources of damage; all mechanisms start erosion processes that may lead to fragmentation and then to the disappearance of the biocenosis unit, which may be guaranteed only by a well developed prairie. A pollution-liable area shows a lack of balance in the cyclic evolution of this community. Erosion areas (intermattes), whose evolution under natural conditions is similar to that of the "clearings" of woods, in an altered environment tend to broaden and to become stable. *Posidonia oceanica* colonizes neighbouring substrata quite slowly by the tracer rhizomes system, though intermattes repopulating occurs more rapidly by seeds. Sediment alteration may prevent seed germination and in the long run may also prevent propagation of tracer rhizomes. Pollution exerts a more immediate action on the composition of epibionta species on leaves and rhizomes of this species. Sediment alteration leads to a lasting replacement of *Posidonia* by *Cymodocea nodosa* and under more serious conditions by *Caulerpa prolifera*. This latter belongs to the large-ecological valence sciaphilic algae and its behaviour is analogous to that of Dictyopteris membranacea on solid substrata. The only difference is that, since *Caulerpa* is a tropical genus, its presence is not widespread in the Northern Mediterranean. Angiosperm with the largest ecological valence and hence with the highest resistance to pollution is *Zosterella noltii*, which, however, never forms wide prairies, but small mesolittoral and infralittoral colonies. Mobile substrata of the eastern Mediterranean are also colonized by *Halophila stipulacea*, an Angiosperm coming from the Red Sea and showing a good resistance to polluting factors. The effects of pollution on free Melobesiae populations are unknown, even because the presence of bottom currents in the biotopes chosen by these species, creates unfavourable conditions to the accumulation of polluting substances.

4.   TOXICITY TESTS ON CULTURES OF BENTHONIC ALGAE IN THE PRESENCE OF SLUDGE

Such tests constitute a part of a more extensive investigation concerning the ecological valence of the phytobenthonic species toward some abiotic factors of both climatic and edaphic nature. The application of such basic research to toxicity tests has led to positive results. By these tests it is possible to check the species tolerance toward simple or complex substances entering in the sewages as constituents. The tests are complex since they necessarily consist of several steps: (1) species acclimatization in order to obtain generations adjusted to the artificial breeding conditions; (2) control tests on the effects at different concentrations using both sewage *in toto* and single components; (3) choice of a standard series of concentrations and of the sample features that must be checked in order to relate their variations to the different substances or to the different concentrations. After completing, together with my co-workers, such preliminary research [10], I have used ten species of benthonic algae: four green algae (*Bryopsis bypnoides, Enteromorpha intestinalis, Ulothrix flacca, Ulva rigida*), three brown algae (*Ectocarpus siliculosus* v. *arctus, Giffordia sandriana* and *Scytosiphon lomentaria*) and three red algae (*Acrochaetium secundatum, Bangia fusco-purpurea* and *Chondria coerulescens*). Recently the number of species and research presently in progress has been increased. Such species have been chosen in order to have available a complete range of the various types of organization of the thallus, of plastidial apparatus, of growth, of reproduction elements, of development cycles and of growth forms. The standard dilutions adopted as a consequence of the results of Phase 3 of the preliminary study are: 1/10, 1/5, 1/2.5, 1/1. Controls were carried out both on samples cultivated in filtered seawater and in seawater diluted with distilled water at the same ratio. Details on methodology are reported in the quoted paper. After the results obtained, I have concluded that in a laboratory possessing a species stock in a continuous culture, toxicity tests may be carried out within 3—5 weeks at most. Information are numberless and suggest various operative solutions for designing purification plants and outfall systems. To quote a few examples only, from such tests it is possible to obtain some indication for the calculation of sewage dilution in the outfalls equipped with diffusers, for the determination of acceptable values of NH_3, NO_2, NO_3, PO_4, MBAS and of some heavy metals, such as Cu, Zn, Pb [13]. The preparation required for toxicity tests with phytobenthonic elements certainly is more specific than with phytoplanktonic ones; however, if phytobenthonic communities are taken as a reference for the study of the effects of such discharges on natural environments, they must be used also for toxicity tests. To conclude, the study of the problems connected with pollution caused by sewage discharges cannot exclusively consist of a microbiological investigation or of a complete chemical analysis. The effects concern biocenoses and hence the qualified expert for their study is the biologist. On the other hand, the ever more numerous examples of significant results of multidisciplinar investigations both in the design phase and in the control phase on the effects of purification plants of polluted discharges, prove the indispensable role of marine biologists for the solution of problems resulting from pollution.

REFERENCES

1. Boudouresque, Ch.F., Recherches de bionomie analytique, structurale et expérimentale sur les peuplements benthiques schiaphiles des Méditerranée occidentale (fraction algale). Thèse, CNRS, Paris, N° A.0.4693, 1-624 (1970).

2. Ercegovic, A., Etudes écologiques et sociologiques de la cote yougoslave de l'Adriatique: *Bull. int. Acad. Youg. Sci. Arts classe Sci. math. et nat.* 26, 33-56 (1932).

3. Fabiani, A., Gabucci, G., Ghirardelli, E., Giaccone, G., Mosetti, F., Olivotti, R., Orel, G. and Volpe, S., Esperienze su uno scarico a mare della città di Trieste. In: *Inquinamento marino e scarichi a mare*: 2nd International Meeting Sanremo, Politecnico di Milano, Istituto di Ingegneria Sanitaria, Preliminary edition, 36-37 (1973).

4. Feoli, E. and Giaccone, G., Un'indagine multidimensionale sulla sistematica dei popolamenti fitobentonici del Mediterraneo. *Mem. Biol. Mar. Ocean.* 45-67 (1975).

5. Ghirardelli, E., Orel, G. and Giaccone, G., L'inquinamento del Golfo di Trieste. *Atti Mus. Civ. St. Nat. Trieste,* 28, 431-450 (1973).

6. Giaccone, G., Elementi di Botanica Marina. I: Bionomia bentonica e vegetazione sommersa del Mediterraneo. *Pubbl. Ist. Bot. Univ. Trieste, Ser. didat.,* 1-41 (1973).

7. Giaccone, G. Lineamenti della vegetazione langunare dell'Alto Adriatico ed evoluzione in consequenze dell'inquinamento. *Boll. Mus. Civ. Venezia.* 26 87-90 (1974).

8. Giaccone, G., Tiplogia delle comunità fitobentoniche del Mediterraneo. *Mem. Biol. Mar. Ocean.* 25-44 (1975).

9. Giaccone, G. and Bruni, A., Le Cistoseire e la vegetazione sommersa del Mediterraneo. *Atti Ist. Ven. Sc. Lett. ed Arti* 131, 59-103 (1973).

10. Giaccone, G. and Rizzi-Longo, L., Structure et évolution de la végération marine dans les environments polluées; *Rev. Int. Oceanog. Méd.* 34, 67-72 (1974).

11. Giaccone, G., Rizzi-Longo, L. and Princi, M., Effets des eaux polluées sur cultures d'algues marines benthiques: Méthodes et résultats préliminaires. *Rapp. Comm. int. Mer Médit.*, 24ᵉ *Congres — Assemblée plénière de Monaco,* 1974.

12. Le Capion-Alsumard, Th., Contribution à l'étude des Cyanophycées lithophytes des étages supralittoral et medio-littoral (régio de Marseille). *Tethys.* **1**, 119-172 (1969).

13. Molinier, R. (thèse soutenue en 1958), Etude des biocénoses marines du Cap Corse. *Vegetatio.* **9**, 121-312 (1960).

14. Morris, O.P. and Russell, G., Inter-specific differences in responses to copper by natural populations of *Ectocarpus Br. phycol. J.* **9**, 269-272 (1974).

15. Peres, J. and Picard, J., Nauveau Manuel de bionomie benthique de la Mer Méditerranée. *Rec. Trav. St. Endoume.* **31**, 5-137 (1964).

16. Pignatti, S. and Cristini, P. de, Associazioni di alghe marine come indicatori di inquinamento delle acque del Vallone di Muggia presso Trieste. *Arch. Oceanog. Limnol.* **15** suppl. 185-191 (1968).

17. Rizzi-Longo, L. and Giaccone, G., Le Ulvales e la vegetazione nitrofila dei Mediterraneo. *Quad. Lab. Tecnol. Prsca Ancona,* **5**, 1 suppl. 1-62 (1974).

PERSISTENT ORGANICS

John E. Portmann

Marine Environment Protection Division, Ministry of Agriculture, Fisheries and Food,

Fisheries Laboratory, Burnham-on-Crouch, Essex, England

Summary — Municipal and industrial waste waters contain a wide variety of organic chemicals many of which are relatively persistent in the marine environment. Not all of these are currently recognized; this paper attempts to summarize the present state of knowledge on persistent substances which are recognized as such and to indicate what problems can result. Also included is a reference to the special situations which can arise when normally non-persistent materials are discharged under particularly unfavourable conditions. Among the substances of which special mention is made are PCBs, organochlorine pesticides, halogenated substances, organo-metals, phenols, detergents and persistent oils. In general none of these persistent substances create problems in the marine enviornment provided disposal is properly conducted. A brief section is included on the methods by which detrimental effects of persistent organic substances in the marine environment can be alleviated or prevented. Present controls at both national and international levels are briefly reviewed and some predictions are made as to the likely future needs for control and how the needs for these can be recognized.

1. INTRODUCTION

For millions of years the marine environment has been receiving inputs from the land by a variety of processes, including the natural ones of weathering and run off. Over this period a dynamic equilibrium has been established for most of the elements, and probably also for the natural organic materials which enter the sea either directly from living marine resources or indirectly from land. However, in the last 100 years or so, since the dawning of the industrial age, an increasingly bewildering array of organic compounds has been synthesized, and has passed into everyday production and use. Many of these organic substances have pathways in the air and land environments which eventually lead to their entry into the marine environment. Some are relatively resistant to the normal degradation processes, whether physical, chemical or biological, which can take place in the marine environment. A few such as PCBs and DDT are well known and I will discuss these and a few other examples in more detail later. It is very easy to fall into the triple traps of believing that all synthetic substances are persistent, that no natural substances are persistent, and that all persistent substances are harmful. A moment's reflection will tell you that none of these assumptions is correct. We must also recognize that persistence is in any case a matter of degree.

The theme of this Congress is the disposal of municipal and industrial waste water to the marine environment, and this confines the coverage of this paper to the effects of persistent organics which might be present in such waste waters. As mentioned earlier, an essential first step in dealing with this topic is to establish precisely what we mean by the term "persistent organics". I think it will also be helpful if I explain where I stand with respect to whether or not marine disposal of waste waters, especially those containing persistent organics, can or should be, practised.

As regards waste disposal at sea my personal position is that I believe quite firmly that, in the same way that the land environment can tolerate limited quantities of waste, the marine environment also has a capacity to accept some waste. I also believe that it is logical to utilize this capacity of the marine environment to accept wastes. This capacity will vary locally according to the fishery resources and other interests of the sea area and will also depend on local conditions of depth and water movement etc. The capacity to receive wastes may be quite large for non-persistent materials which do not survive long in the sea, but it extends even to the most persistent and toxic substance, although clearly in local terms the permissible quantity of such a substance would be small, and, in practical terms, may be completely valueless in the context of waste disposal.

The term "persistent" presumably relates to substances which are not readily degradable. A definition of the meaning of these two ways of expressing the same thing has exercised the parties to the Oslo Convention on the dumping of wastes at sea for about 3 years. It would therefore be presumptious of me to claim a solution, but, in the context of this paper, I have adopted the view that persistent organics are those which are only slowly degraded in the course of the effluent treatment or in the marine environment, or which, under particular circumstances, may persist in the marine environment.

2. ROUTES OF ENTRY

It almost goes without saying that the capacity of the sea to absorb wastes must be used with care, especially as our predictive capacity has, on a number of occasions, been demonstrated to be woefully inadequate. In this connection it is relevant to examine how, and where, waste waters enter the marine environment.

By far the biggest volume of waste waters enters the sea by rivers and streams, and in many instances in Europe a very high proportion of what is euphemistically called freshwater flow is in fact water which has been used at least once, i.e. is waste water. Fortunately, however, the rivers have a self-cleaning capacity and in the better situations this means that the waters are largely clean and free of pollution by the time they enter the sea. It must also be recognized that whatever the pollutant load entering the estuary, not all of it will reach the open sea. However, the role of estuaries as fish nursery grounds, etc., renders them very important to the well-being of the marine environment as a whole, and it is often in such areas that the first clear signs of pollution are observed.

There is also the direct route of entry via pipelines to the sea. Until fairly recently this meant discharges either across a beach or to just below low water mark. The unsatisfactory aspects of such practices were probably first drawn to our attention by the public complaints of aesthetic nuisance. However, it is now recognized that, where sewage is concerned, waste is better disposed of at a distance far enough from the beach for the discharge not to be obviously apparent and in such a way as to provide good dilution and dispersion possibilities. Indeed, in recent years we have seen a steady increase in both numbers and length of marine sewage outfalls.

Certain special categories of waste waters may also be dumped at sea from specially constructed or modified vessels. Generally, though in our experience by no means always, these wastes are too strong for disposal on land or to freshwater rivers, but by taking advantage of better dilution, available in the wake of a moving vessel compared to that of a pipeline, disposal of the waste into the marine environment can be safely executed. An additional factor is the increase in land transport costs which has led to sea disposal being economically attractive compared to land transport and disposal in landfill sites.

Economic factors are also responsible for another feature of waste waters which is likely to affect the methods by which their safe and effective disposal at sea can be brought about, namely the increasing strength of industrial wastes. Although with increasing standards of living domestic usage of water is increasing and the concentration of wastes in municipal waste waters is decreasing (though less than proportionately), the concentration of potential pollutants in industrial waste waters is increasing. This is largely being caused by economies in use of water, brought about by the very rapidly increasing costs of both raw water and effluent treatment. We are now experiencing, at least in the United Kingdom, large scale recovery of water from effluents. This is perhaps only the first stage in a developing process and, if so, it could be a temporary phase followed by decreasing concentrations as pollutants are recovered for recycling or re-use.

3. DOES PERSISTENCE MATTER?

The disposal of waste waters by long sea outfalls or from ships at sea has received much attention in recent years. This practice takes advantage of the fact that dilution can be a solution to pollution. The solution is not however infallible and we have a number of examples where the calculations have either assumed that the characteristics of the waste are unimportant or have failed to take full account of unexpected hydrographic features in the area of disposal. In the latter case reconcentration or deposition of the waste can occur and this can lead to undesirable accumulation of organic or mineral matter.

Unusual characteristics of the waste include persistence. Clearly problems can arise if in the marine environment the input of the pollutant is such as to exceed the rate of removal via dilution, dispersion and degradation, or by adsorption and immobilization on the sediments.

If a substance is not readily degraded in the sea, possibilities of problems arising are increased, since with time, a toxic concentration may build up in the area of discharge or disposal. If the persistent substance is bio-accumulated then, although it does not necessarily follow that bio-accumulated substances are toxic, the probability of ecological damage/harmful effects is increased. It is for these reasons that persistent organic substances and the toxic elements such as lead, cadmium, mercury, selenium, etc. are the cause of particular concern to scientists working in the field of marine pollution and its control.

I do not propose to deal separately with industrial and municipal effluents since, although industrial effluents are sometimes discharged through separate pipelines, municipal waste waters do not usually consist only of materials of domestic origin. Certainly in the United Kingdom it is not uncommon for up to 50% of the flow in a municipal sewerage system to be of industrial origin. This appears to apply generally to northern Europe at least, and the recent ICES survey of pollution in the North Sea [1] revealed that the proportion of industrial effluent could be very high — 83% for the Bremen area of Germany.

Although the majority of organic substances present in either industrial or domestic effluents will be readily degradable, a few are persistent. Unfortunately the number of substances currently recognized is quite small. One of the problems arises because the nature of industrial effluents is complex and precise details, at least in the past, were largely unknown. All that can be said is that the content of an industrial effluent will be very different, in both composition and content, from the mix of materials which are used in the process, or the product, which the factory manufactures. For these reasons it is uncommon for industrial discharge consent conditions to stipulate the quality of an effluent in more than fairly general terms such as BOD, COD, extractable matter, total solids, loss on ignition etc. However, a few substances, most of which are considered persistent, are recognized as being liable to create problems, e.g. PCBs, chlorinated solvents and phenols, and in the United Kingdom limits are often set for these substances. As mentioned earlier, under particular circumstances even substances normally considered non-persistent can resist degradation and a few comments on these are also appropriate.

4. SUBSTANCES EXHIBITING PERSISTENCE IN THE MARINE ENVIRONMENT AND THEIR ATTENDANT PROBLEMS

4.1. Degradable materials

Although soluble organic material is unlikely to accumulate, if the area in which disposal takes place is one in which the currents are either slow or of a circulatory nature, it is quite likely that solid organic material will accumulate in or around the discharge area. Under such circumstances even normally degradable materials can be persistent, since once the process of accumulation on the bottom is started there is a tendency for the sludge bank to persist and enlarge. The most commonly quoted example of this is the so called "sludge monster" in New York Bight, which, it is claimed, is increasing in size as a result of sewage sludge dumping. It should, however, be stated that the evidence for the increase in size is very flimsy, although it is admitted that a substantial area of the seabed has been affected for many years [2].

The problem is particularly likely to occur where waste waters containing a high proportion of solid material have discharged into partially enclosed bays or fjord type systems. Degradation of most substances is temperature dependent and is much reduced at low temperatures, thus accumulation is more likely to occur in high latitudes than in tropical regions. Of the various industrial wastes known to lead to such problems, pulp and paper effluents are probably the best known [3]. Apart from being toxic in their own right the fibrous material which they contain tends to settle on the bottom. The fibres are only slowly degradable, and in sludge banks this rate is reduced as anoxic conditions are approached. Ultimately anoxic conditions set in and degradation of the organic matter proceeds via anaerobic bacteria and highly toxic hydrogen sulphide is evolved. Such areas are devoid of marine life on the bottom, although some life usually remains in the water, which will only rarely be completely de-oxygenated.

4.2. Organochlorine pesticides

Organochlorine pesticides are present in most sewage effluents and especially in sewage sludges where they tend to be accumulated by adsorption onto particulate material. A range of reported values in sewage sludge and components of the marine environment are given in Table 1. Among the first to recognize the presence of organochlorine pesticides in sewage were Holden and Marsden [4] who reported

Table 1. Levels of DDT and dieldrin in sewage and the marine environment

	DDT	Dieldrin
Sewage	36 − 130 mg/l	100 − 300 mg/l
Sewage sludge	10 − 500 mg/kg*	1 − 2500 mg/kg*
Plankton	0.2 − 100 μg/kg*	1 − 23 μg/kg*
Molluscs	0.008 − 0.10 mg/kg*	0.001 − 0.02 mg/kg*
Fish (Cod)	0.003 − 0.05 mg/kg*	0.001 − 0.02 mg/kg*
Seals	0.01 − 20 mg/kg*	0.06 − 2.8 mg/kg*
Marine birds	0.6 − 3.1 mg/kg*	?

* on a wet weight basis.

in 1966 that Scottish sewage effluents, of largely domestic origin, contained at least 100×10^{-6} mg/l dieldrin and rather less of DDT. In a few instances, where the sewerage system accepted industrial effluents from wool processing industry, the dieldrin value could be as high as $10\,000 \times 10^{-6}$ mg/l and 500×10^{-6} mg/l of DDT. Lowden [5] reported similar concentrations in twenty sewage effluents from England and Wales, with an average of 55×10^{-6} mg/l dieldrin and 95×10^{-6} mg/l of DDT.

Although these early values might have been slightly high, due to interference by PCBs, values recently reported for a number of marine sewage outfalls in California are of a similar order, ranging from 20 to 98×10^{-6} mg/l [6]. As the *per capita* water consumption in North America is somewhat higher than Europe, this represents a larger quantity of organochlorine pesticides.

The source of organochlorine pesticides in sewage of a purely domestic nature is something of a mystery. Certainly where industrial discharges are involved the levels can be quite high. Croll [7] conducted a survey of organochlorine pesticide residues in English freshwater rivers and found by far the highest concentrations in rivers passing through wool processing areas in northern England and, as mentioned earlier, Holden and Marsden [4] reported a similar situation in Scotland. By far the best example of the importance of industrial input is afforded by the White Point (California) discharge, which was found to contain between 59 and 97×10^{-3} mg/l DDT — approximately one thousand times that normally found in sewage effluents; this high input was directly attributed to a DDT manufacturing plant [6].

In the absence of any industrial input the levels found probably arise from the presence of organochlorine pesticides in food and water, in both of which they are ubiquitous. In the past home usage of DDT against flies and of dieldrin in mothproofing of woollen garments must have led to some input to sewers, but several countries have now prohibited the use of these compounds in the home and levels can be expected to fall.

The solid component of sewage-sludge, where separation and treatment is conducted, appears to concentrate organochlorine pesticides and Schmidt *et al.* [6] reported that the sludge discharged from the Hyperion treatment plant contained at least four times the concentration of DDT present in the treated effluent. Similar values for the United Kingdom do not appear to be directly available, but comparison of the effluent concentration reported by Lowden [5] and Holden and Marsden [4] with our own values for sewage sludge suggests a concentration factor in the sludge of between 40 and 400 fold.

Because DDT and other pesticides are readily adsorbed by particulate matter, if the effluent is not adequately dispersed on discharge, they might be expected to accumulate in the bottom sediment. This has been found to occur in the Glasgow sewage sludge dumping area. Similar accumulations have also been found in Keratsini (Athens) municipal outfall in Greece where concentrations of up to 1.5 mg/kg DDT was found in sediments immediately adjacent to the outfall, decreasing with distance to less than a detectable level of about 0.01 mg/kg DDT [8]. Appreciable accumulation of organochlorine pesticide residues has also been reported off the Californian outfalls [9]. Work conducted by my own laboratory has however failed to reveal a similar accumulation in either the Thames Estuary or Liverpool Bay; the lack of accumulation in these two areas can undoubtedly be attributed to the good dispersion characteristics: the Thames Estuary area, for example, is an area of high tidal velocity. However, in an area like the Mediterranean, which has limited tides and water movements, there is a possibility that accumulation of DDT or other organochlorine pesticides could occur.

The scale of damage is however, likely to be relatively small unless accumulation takes place on a scale well in excess of that documented elsewhere. The species most likely to be affected by such accumulation are those animals which live on the sea bed, but this will depend on their ability to absorb organochlorine pesticide residues from the sediment particles. There is not a great deal of evidence to show whether or not this is a realistic mode of uptake but if it does occur there is a risk of accumulation in the food chain when these benthic organisms are consumed by bottom-feeding fish.

The only evidence I have been able to find in the literature where fish have been affected by organochlorine pesticides in the marine environment, relates to inlets or bays and has been associated with agricultural run off, from intensively cultivated areas. Even then recovery was fairly rapid when organochlorine pesticides usage was stopped [10]. Avian fish predators, some of which accumulate quite high levels of organochlorine pesticide residues in their tissue, appear to be somewhat more at risk. Their breeding success can suffer, and in periods of starvation when their fat reserves are being mobilized, previously safely stored organochlorine residues may be liberated in large amounts to the blood supply, causing death. Although in most cases where definite effects have been recorded, e.g. in brown pelicans off California [11], the main source of organochlorine pesticide resudes is attributed to agricultural use rather than municipal waste waters.

In short, provided accumulation of organochlorine pesticide residue bearing material is not allowed to occur or is limited in scale, the dangers posed by the presence of organochlorine pesticides in municipal waste waters is probably minimal.

A greater danger can, however, be posed by industrial waste waters since, although the concentrations of organochlorine pesticides allowed by the discharge licences ensure that there are no detrimental effects under normal conditions, the possibility of accidental releases cannot be ruled out. There are a number of instances where this has occurred, ranging in size from quite small incidents involving fish kills in rivers, to the large scale accident in the Netherlands. In that case [12] telodrin was accidentally discharged from a manufacturing plant over a considerable period of time. The levels were not high enough to cause a fish kill and the incident was not revealed until several species of marine birds, including eider ducks and terns, suffered heavy mortalities. Such incidents are fortunately rare but, because of the quantities which can be released in relatively short periods of time, special measures must be taken to avoid their occurrence.

4.3. Polychlorinated biphenyls

Polychlorinated biphenyls (PCBs) are usually associated with organochlorine pesticide residues, and because they tend to interfere with organochlorine pesticide residue analyses the two groups of materials are usually determined together. Although PCBs have some pesticidal activity they have never been used for that purpose on any scale. They have, however, been used in a very wide range in industrial applications since they were introduced in the 1930s. Like the organochlorine pesticides they are strongly lipophilic and behave in a similar way in the marine environment. There have recently been suggestions that PCBs can be produced by a degradation of DDT [13], but there is no evidence that this could occur on any scale and the molecular rearrangements necessary look rather unlikely.

The presence of PCBs in municipal waste waters was first reported by Holden [14] who found an average concentration of 1 mg/kg in sewage sludge dumped in the Clyde estuary from Glasgow. He also reported that my laboratory had found between 0.1 and 5 mg/kg of PCBs in sewage sludge from Manchester and 0.2 mg/kg in sewage sludge from London. More recent analyses on these sludges put the normal levels

Table 2. Levels of PCB in sewage and in the marine environment

	PCB
Sewage sludge	0.04 – 5.0 mg/kg*
Plankton	1 – 1000 µg/kg*
Molluscs	10 – 100 µg/kg*
Fish (Cod)	10 – 400 µg/kg*
Seals	1 – 3000 mg/kg*
Marine birds	1 – 10 mg/kg* body
Marine birds	1 – 1000 mg/kg* liver

* on a wet weight basis.

at around 2.0 and 0.4 mg/kg respectively. However, these seem to be abnormally high for the United Kingdom and analyses of eleven different sludges ranged in content from less than 0.04 to 2.4 mg/kg. The levels in sewage sludge discharged by the Hyperion outfall in California are of a similar order, 0.8 to 0.9 mg/kg [6], but in sludges which receive only primary treatment the levels were reported to be approximately ten times those found in the Hyperion effluent which received secondary treatment. A summary of these and other typical values for PCBs and sewage sludges and the marine environment are summarized in Table 2.

PCBs are complex mixtures of compounds made by several different levels of chlorination of a biphenyl molecule. A typical commercial formulation may contain up to 100 different compounds or isomers; a typical gas chromatographic record will show at least fourteen well defined peaks. PCBs are bioaccumulative in the marine food chain and several of the compounds appeared to be very persistent. There is, however, a steadily increasing amount of evidence to suggest that the lower chlorinated compounds can be degraded by loss of chlorine atoms, either in effluent treatment plants or in the environment. This is supported by the fact that most United Kingdom environmental samples are analysed on the basis that the PCB is Aroclor 1254 or Clophen A50 (i.e. corresponds to a formulation with 50% chlorination), whereas the greatest use was always of lower chlorinated formulations.

PCBs were first produced in the 1930s and since then have found a wide range of industrial uses. Many of these were in completely open systems and in applications which led to a very wide distribution of products containing PCBs. Consequently, the source of PCBs in municipal waste waters is extremely diverse and a fair proportion probably enters through the human diet. In the early 1970s the main manufacturer of PCBs voluntarily imposed restrictions on their use and other manufacturers have since followed this lead. The only uses for which PCBs can now be employed are in closed systems where the risk of release to the environment is greatly reduced. At least one of the manufacturers has also instituted a recovery system by which unused new formulations are taken back, and old formulations are recovered e.g. from transformers. It is difficult to see how release to municipal sewerage systems can now occur on any scale. However, although there have been some notable successes in the reduction of sewage effluent levels, e.g. in Glasgow, [15] the concentrations do not appear to have declined appreciably in most sewage effluents or sludges in the United Kingdom or in the United States.

Where reductions have occurred it appears to have been as a direct result of control measures taken by regulatory agencies rather than restrictions in use. There can however be little doubt that the restrictions must have reduced inputs to the marine environment by direct input of industrial waste waters or via the atmosphere. It has recently been claimed that the effect of the reduced inputs can already be detected in North Atlantic waters and plankton [16]. However, the extent of the claimed reduction (tenfold) is so large that in relation to the expected half life of PCBs the claims appear very unrealistic and an examination of the supporting evidence reveals that the data which are compared refer to samples taken in different places and analysed by different methods. Nevertheless reductions can be expected but the $T_{\frac{1}{2}}$ will probably be of the order of 10 years.

As with organochlorine pesticide residues, PCBs are accumulated on particulate matter, and in areas of discharge of waste waters or sludges, they are likely to be found in the sediment. Concentrations of up to 2.9 mg/kg dry weight have been found in the surface layers of sediments in the Glasgow sewage sludge dumping are [15] and similar concentrations have been found in the sediments in the major California outfall areas [9]. Off Athens sewer outfall the sediments have been found to contain up to 0.7 mg/kg dry weight of PCB.

The ecological significance of such levels in sediments is uncertain and will depend to some extent on the nature of the sediment and strength of adsorption. Nimmo et al. [17] have however reported that PCB was absorbed from sediments by fiddler crabs and pink shrimps, although they were not able to establish whether this was indirectly from interstitial water containing leached PCB, or directly from the sediment particles.

At the concentrations presently found in seawater and most marine species, there is little likelihood of harmful effects from PCBs. Phytoplankton are affected by PCB concentrations in the water of the order of 0.205 to 0.1 mg/l, but the highest concentrations reported in open ocean waters are considerably below this [18] and although PCBs are accumulated by fish and seals there is little evidence that the levels normally encountered are detrimental. However, in certain restricted areas, e.g. off California, levels in marine animals and especially fish-eating birds are high enough to be implicated, with DDT residues, in population declines.

4.4. Other halogenated organics

There have been a number of reports of the presence of halogenated hydrocarbons from the marine environment both in sea water and in marine organisms [19–21]. A summary of the typical concentrations is given in Table 3. Most of the reports refer to chlorinated and fluorochlorinated aliphatic compounds

Table 3. Typical concentrations of chloroform, carbon tetrachloride, trichloroethylene

Surface water	10 − 1000 mg/kg*
Sewage sludge	100 − 1000 mg/kg*
Sea water	100 − 1000 mg/kg*
Molluscs	1 − 10 μg/kg*
Fish	1 − 10 μg/kg*
Marine mammals	1 − 100 μg/kg*
Marine birds	1 − 100 μg/kg*

* on a wet weight basis.

which are produced in very large amounts. The earliest estimates put global production at more than 3×10^{-6} t [22] but more recently global production capacity has been estimated at 3.5×10^{-7} t/year [19]. Much of this is of substances used in plastics manufacture, but capacity for production of solvents such as trichloroethylene and perchloroethylene exceeds 2×10^{-6} t and, as these are used largely as degreasing agents and in cleaning fluids, it is not unreasonable to assume that losses to the environment are equal to production.

Although, because of their volatility much of the transfer of these substances to the marine environment is likely to be via the atmosphere, significant quantities are known to be present in both industrial and municipal waste waters. In the sludge dumped at sea by one major municipal authority in England, the concentration of tri- and perchloroethylene averages 4 mg/kg dry weight. Montgomery and Conlon [23], in the course of an investigation into the possible causes of sewage works failures, found carbon tetrachloride, trichloroethylene, chloroform etc. in two crude sludges, a digested sludge and a digester overspill. There was little difference between the concentrations found in crude and digested sludge but these were about double that found in the overspill effluent, e.g. 2 mg/l and 0.7 mg/l respectively for chloroform. In certain areas where industrial inputs are particularly high the concentrations in municipal waste waters are likely to be much higher than those quoted above.

Other chlorinated compounds have also been reported in sewage effluents and sludges, these include hexochlorobenzene and trichlorobenzenes, and in some cases quite high concentrations have been reported, e.g. several hundreds of ppm of trichlorobenzenes; again the source is almost certainly industrial.

There is some controversy over the ecological significance of the simpler halogenated compounds. Acute toxicity data for a range of simple chlorinated solvents has been presented by McConnell et al. [19] for unicellular algae, barnacle larvae, and dabs (a small flatfish). With the exception of hexachlorobutadiene, the LC_{50} or EC_{50} values are all in the range 10—350 mg/l. The lethal dose concentrations of hexachlorobutadiene were 0.5 mg/l to dabs and 0.87 mg/l to barnacle larvae.

Most of the halogenated hydrocarbons appeared to follow the pattern shown by DDT and PCBs and are accumulated by marine organisms. In general however, the degree of concentration relative to seawater is much less, usually only about 10 to 100 fold and there is little evidence of biomagnification with progression of the food chain [19]. The uptake process appears to be a rapid but entirely passive process, and it is probably associated with the relative solubilities of the compound in water and in the animal lipids. Loss proceeds almost as rapidly as uptake, if the animal is removed from polluted seawater and placed in a clean environment.

Although chlorinated acetic acids have been shown to be degradable by microorganisms in seawater [20], there is little evidence that biodegradation is an important factor in the removal of halogenated hydrocarbons from the marine environment. McConnell et al. [19] go so far as to say that microbiological degradation has little direct part to play in their degradation in the sea. A few compounds are very rapidly broken down by physical-chemical action of the pH of sea water, e.g. pentachloroethane 1—2 days and 1,1,1-trichloroethane 9 months [24, 19] but generally these reactions are slow, e.g. McConnell et al. [19] quote a $T_{\frac{1}{2}}$ of about 6 years for perchloroethylene in sea water.

As a general rule it is safe to say that most of this group of substances fall into the definition of "persistent" adopted for this paper, but few are likely to be harmful in the marine environment at their present concentrations. In this general context it is perhaps worth noting that, on the basis of the concentrations of chloroform and carbon tetrachloride found in sea water, it seems unlikely that the entire marine burden arises from man's activities and it has been suggested [21] that these two compounds may be formed naturally by reaction of chlorine and methane in the atmosphere.

Halogenated hydrocarbons in waste waters pose a particular problem to the scientists involved in pollution control, in that the Oslo and London Dumping Conventions, and the Paris Convention for discharge from land-based sources, place them on the so-called blacklist unless they are present in trace quantities, or are non-toxic, or readily degradable, i.e. non-persistent. The conduct of toxicity tests is reasonable simple with present techniques, but as yet we have no really satisfactory system for the determination of degradability under marine conditions. There is little information in the literature to assist the decision maker, and what data there are refer to tests under laboratory conditions where most of the conditions are adjusted to optimum. Consequently degradation rates tend to be optimistic and half life prediction is extremely tricky.

However, a few general guidelines can be established [25] and it is possible to state that even under conditions favourable to microbial attack, i.e. a large surface area exposed to suitable bacteria and with an adequate nutrient supply, certain compounds will be resistant to microbial degradation. These include chlorinated aliphatic hydrocarbons with four carbon atoms or less, and any aliphatic or aromatic hydrocarbon containing three or more chlorine substitutions. With aliphatic compounds the inhibiting effect of chlorine substitution decreases with increase in length of the unsubstituted chain between the site of substitution and the terminal carbon atom and, if six or more carbon atoms intervene, biodegradation is relatively easy. Although mono- and dichloro-aromatic compounds can be degraded more easily than trichloro aromatics the rate depends upon the relative positions of the chlorine substitute and the nature and presence of other substitutes or side chains.

4.5. Organo-metals

All municipal and industrial waste waters contain some metals. Generally these are ionic or insoluble inorganic forms and can be excluded from the coverage of this paper. However, a few metals are discharged in the effluents from industrial processes in the form of organic compounds or are converted into organic forms in the marine environment. Since one of the most notorious cases of marine pollution involved such a situation this paper would hardly be complete without some mention of the dangers.

The most toxic form of mercury is methyl mercury and this is formed by microbial action in the sediments. The methyl mercury so formed is readily taken up by marine organisms in which it has a very long biological half life. Although there is little evidence of biomagnification, there is strong evidence of correlation between age of the organism and concentration in that organism [28]. Very high concentrations of methyl mercury can be built up in heavily contaminated areas, and in Minamata Bay concentrations of up to 40 ppm were found in fish and shellfish. Apparently even at these high concentrations there was little evidence of harm to marine life, but methyl mercury is extremely toxic to mammals and several fishermen and their families were killed or seriously maimed as a result of eating the contaminated fish.

In that instance the mercury was discharged from a factory where is was being used as a catalyst. The other major use of mercury in industry is in the chloro-alkali industry where metallic mercury is used as a cathode in an electrolytic cell. Ionic and elemental mercury escapes both to the atmosphere and in the effluent and once in the sediments can be converted to methyl mercury. The process of methylation is a dynamic one, but the rates of methylation and demethylation are such that methyl mercury is a persistent product of ionic mercury discharges.

Wong et al. [27] have recently published evidence of methylation of lead in sediments from a freshwater lake and it is believed that methyl lead probably behaves in a similar way to methyl mercury. It is volatile and insoluble in water, probably accumulates in aquatic life and is more toxic than inorganic lead. The main source of inorganic lead in the environment at the present time, is believed to be motor fuel anti-knock compounds, which decompose in the internal combustion engine and lead is evolved as an aerosol. Although lead tetremethyl and tetraethyl are likely to be released in small quantities in the waste waters from plants manufacturing motor fuel anti-knock compounds, the concentrations involved are regulated by licences and the number of plants producing such compounds is so small that the input is only of importance locally.

4.6. Detergents and detergent additives

When synthetic detergents were first introduced there were many problems as a result of their resistance to degradation. A great deal of attention was paid to these substances and much information was accumulated on their persistence and toxicity. Since the mid-1960s there has been a gradual swing away from the branched chain type of compound which was biologically "hard". Most countries have now adopted the use of the "softer" linear alkyl benzene sulphonate type of compounds. These are much more readily degradable in effluent treatment plants and presumably in the marine environment. However, for certain industrial processes, especially in the textile industry a change to soft detergents has been a slow process and it is possible that significant concentrations of "hard" detergents enter the marine environment whenever particular industries discharge to municipal sewerage systems.

Generally, the content of an ionic detergent entering a sewage works ranges from 8 to 25 mg/l and removal during treatment is usually of the order of 90%. In the United Kingdom undigested sludge normally contains less than 1000 mg/l and digested sludge about 150 mg/l. Only a very small proportion of non-ionic hard detergent is removed in a sewage works and a typical effluent from textile industries contains about 2 mg/l [28]. If the concentration factors in sewage sludge are the same for non-ionics as for an ionic detergent, then the concentration in the sludge component of undigested and digested effluents would be about 100 and 15 mg/l respectively.

In the main, detergent compounds will be associated with the solid matter and not with the liquid fraction, i.e. conditions which will be most favourable for bacterial attack. It is therefore unlikely that significant accumulation, of even the "hard" detergents, will be observed, and, as the lowest concentrations reported to be toxic to fish are at least 1 mg/l it seems unlikely that the concentrations of detergents found in either industrial or municipal waste waters will pose a problem except in the immediate vicinity of the discharge. Synthetic detergents as sold to the public and as used in the industry contain a number of other compounds and one, an optical brightening agent 2,5-di-(benz oxole-2-yl) thiophene has been reported to occur in concentrations of up to 40 mg/kg dry weight in sewage sludge from certain Swedish towns [29]. In laboratory studies fish were found to accumulate this compound by a factor of 1000 relative to the water concentrations and in a freshwater river high concentrations found in fish were attributed to discharges from textile processing factories. This suggested some degree of persistence in the aquatic environment and Jensen and Petterson [29] estimated the $T_{\frac{1}{2}}$ in water to be between 30 and 60 days. Sturm et al. [30] have recently confirmed the findings of Jensen and Petterson but failed to find any accumulation with three anionic optical brighteners. All four compounds have a comparatively low acute toxicity, the lowest value being found for the accumulating substance studied by Jensen and Petterson [29] — an illustration of the fact that bioaccumulation does not necessarily equal harm.

Synthetic detergents usually contain substantial quantities of "builders" added to the improved performers, in most cases these are polyphosphates and the large quantities involved have led to problems of eutrophication in freshwater lakes. In an attempt to alleviate these problems alternatives have been sought and one which found some use was nitril-tri-acetic acid (NTA). The use of this compound was never allowed in the United Kingdom but it was used in some countries and may still be. It is readily degradable [28, 31] at temperatures above $20°C$ but much less so at low temperatures, e.g. less than 3% degraded at $5°C$. Thus in lower temperature environments some accumulation in the environment may occur. Since it is of relatively low toxicity (96 h LC_{50}) values for diatoms and bluegills range from 185 to 487 mg/l [32] and toxicity decreases with water hardness, there is unlikely to be any direct effect on marine life. However, since NTA is a chelating agent it is possible that the solubility of compounds of toxic metals might be increased, rendering them more available to marine life and thus increasing their toxicity. Work by Sprague [33] suggests that the latter is unlikely, because NTA is an extremely strong chelating agent and the metal although solubilized is still not available to the animal.

4.7. Phenolics

Phenol does not occur in significant amounts in municipal waste waters unless these include a significant proportion of industrial wastes. However, a wide range of industrial wastes contain phenol, ranging from those of oil refineries to those of drug manufacturers. Phenol is degradable in effluent treatment plants and there is no evidence that phenol persists in the marine environment.

Controls are usually imposed on the discharge of phenol at about 0.5 mg/l and, since the toxic levels to marine species are variously reported to be between 1 and 100 mg/l, there appears to be little danger of acute toxicity. However, phenols are accumulated by marine fish and shellfish and tainting of marketable species is a very real possibility. The danger is greatly increased if phenol is exposed to active chlorine, as could happen in the chlorination of sewage effluents, since the organoleptic threshold for chlorinated phenols is lower than that of phenol, and due to the presence of the chlorine in the molecule degradation is likely to be inhibited.

4.8. Persistent oils

A comprehensive discussion of the effect of oils on the marine environment is beyond the scope of this paper. Vegetable and animal oils are usually considered to be degradable and it is only petroleum oils which are usually considered to be of concern. Petroleum oils are composed of a very wide range of compounds and there is considerable controversy within organizations such as IMCO as to which compounds should be regarded as persistent. For the time being however, the range of materials included in the definition is extensive and includes a number of low boiling refinery fractions.

Until comparatively recently it was generally believed that the major source of oil pollution in the sea was from tanker discharges in the course of tank washing at sea. It is now recognized that a substantial amount of oil enters the sea via the atmosphere from incompletely burnt fuels, from land-based sources such as factories, and in the municipal waste waters. Of the industrial waste waters which are most likely to contain oils, the most obvious are those from oil refineries. However, at least in the United Kingdom, these are now subject to fairly tight control and other sources, e.g. engineering and engine repair facilities, probably contribute at least as much.

Municipal waste waters can also contain a surprisingly high concentration of oil. Analyses of oils in municipal waste waters is conducted rather crudely by simple solvent extraction and weighing the residue after drying. This underestimates volatile components but includes detergent and vegetable and animal oils. More sophisticated techniques usually reveal that much of the so-called oil is not of mineral origin and does not fall into the definition of persistent as usually adopted. Even so, the majority of sewage sludges contain well in excess of 100 mg/l of persistent oil and we have encountered concentrations as high as 3000 mg/l. According to the IMCO Convention [34], the maximum concentration allowed for an oily discharge from a vessel at sea is 100 mg/l or 60 l/mile and it can be seen that many sewage effluents now contribute more oil to the marine environment than a ship is allowed to do. The sources of oil in municipal waste waters are diverse, but industry is no doubt responsible for some of the oil; home servicing of motor cars is also believed to be a major source.

At the concentrations normally encountered in municipal or industrial waste waters, the worst that is likely to be seen is a faintly irridescent film but no fouling of beaches or sea birds is likely. There are, however, suggestions that some components of oil are extremely toxic to marine life and that certain carcinogenic compounds of oil are persistent in the marine environment. The true significance of oil in the sea is being examined by a working group of GESAMP (Group of Experts on Scientific Aspects of Marine Pollution) and it is hoped that this will resolve many of the more controversial claims or counter-claims.

4.9. Persistent plastics

There have been a number of reports recently of the presence of persistent plastics in the marine environment [35—38]. In most cases at least a proportion of the particles were of obviously non waste-water origin and related to synthetic ropes, netting, plastic cups, and polystyrene and other plastic foams. My colleagues at the Fisheries Laboratory, Lowestoft and Morris and Hamilton [35] however, reported finding large numbers of small plastic spherules in the waters and sediments of the Bristol Channel. These small beads were either neutrally buoyant or just negatively buoyant and were more or less uniform in size. The beads were analysed and were found to consist of polystyrene and styrene butodiene copolymer. No specific source was identified, but the particles were clearly of industrial origin and were found in larger numbers in the inner part of the Bristol Channel. Colton and Knapp [36] also identified a plastic production or processing plants as being responsible for the input of some of the particles.

The ecological significance of these plastic particles is not absolutely clear but there is no evidence of harm. Hays and Cormons [37] found that small plastic particles were to be found in tern pellets and Carpenter et al. [38] found plastic particles in the gut of fish and fish larvae. Although in none of these cases was there any evidence of harm, both sets of authors expressed some disquiet at finding these clearly alien objects in the marine life even if they were harmless. Morris and Hamilton [35] found that the beads have been used in the construction of tubes by a small polychaete, indicating that they may even be useful as well as harmless.

5.   ALLEVIATION OF PROBLEMS ASSOCIATED WITH PERSISTENCE

The possibility that persistent organic compounds will be harmful is clearly greatly increased under conditions in which the concentration of that substance is allowed to build up. Since an inherent feature of such a compound is its reluctance to degrade and disappear the most obvious solution is to try and disperse it as widely as possible. Thus no industrial or municipal waste water containing persistent organic

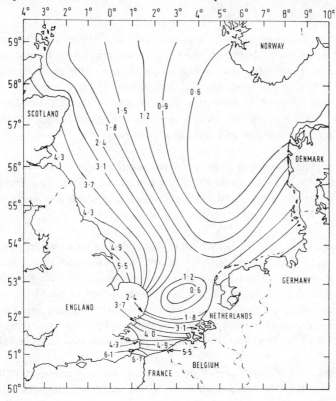

Fig. 1. North Sea — co-range chart.

compounds, whether dissolved or solid, should be discharged to the marine environment without precautions being taken to ensure good dilution and dispersion.

In the United Kingdom we are fortunate in that we have large tidal ranges (Fig. 1) and strong tidal currents. There is also good residual water movement in most of our coastal areas, which helps to ensure that no accumulation of persistent substances occurs. In the past we have encountered problems as a result of discharges for organic materials to estuaries which have as a result become deoxygenated. However, major efforts have been made to overcome these by treating effluents before discharge to the estuary, and substantial recovery has been possible, e.g. in the Thames Estuary.

It is not always possible to rely on natural water movements and under such circumstances special measures must be taken in order to try to obtain good initial dilution on discharge. One method mentioned earlier is to dispose of the waste water from a moving vessel. Where this is not feasible, e.g. because of volume, special diffuser sections can be used to spread the discharge over as wide an area as possible. That is the subject of a further session at this Congress and it is not appropriate to give the matter further consideration at this stage.

As mentioned earlier, dilution is not necessarily enough to prevent problems with persistent substances. DDT and PCBs are extremely widely dispersed but because they are accumulated by marine organisms and some form of biomagnification occurs in the food chain, toxic levels have been reached in the higher predators. These toxic levels are not encountered world wide but it is clear that, at least in some areas, the quantities which have been discharged have been too great.

6. CONTROLS — PRESENT AND FUTURE

In order to reduce the quantities of persistent organics discharged to the marine environment controls are necessary. Equally important are alternative methods of disposal which do not impose unnecessarily severe economic penalties. In the case of an industrial waste water, it is generally relatively easy to trace the major source of say PCB or oil input, and with a minimum of effort the input can be diverted from the main waste water flow. Disposal of the waste water containing the PCB or oil may still present difficulties since, e.g. if the waste arises from a rinsing process, it may not be easy to achieve a concentrated effluent or it may not be easy to dispose of a concentrated effluent.

The problem posed by persistent organics in municipal waste waters is even more complex since it is often very difficult to trace the major sources of the compound in question. Nevertheless, this can be done and significant improvements can be achieved, e.g. a ten-fold reduction in the PCB content of Glasgow sewage was achieved by tracing one major contributor [15].

In some cases reduction of inputs can be achieved simply by better housekeeping, i.e. prevention of wastage and input altogether. In others, however, recovery of the material may have to be attempted, often from fairly dilute solutions. The technology for this is not always readily available and the process may be costly. Once the organic substance has been concentrated it may be possible to recycle the material for further use, or failing that, to incinerate it. All the persistent organics considered in this paper are amenable to incineration, although the chlorinated substances require high temperatures and, at least if conducted on land, water scrubbers to remove hydrochloric acid from the exhaust gases.

At the present time control of the discharge of persistent organic materials in either industrial or municipal waste waters lies largely with national authorities, and is of differing stringency. Recently three Conventions have been negotiated which seek to harmonize controls of input of pollutants to the marine environment on an international basis. Two referred to the dumping of wastes at sea, the so-called Oslo and London Conventions [39, 40] and one to land-based sources, the so-called Paris Convention [41].

Of these only the Oslo Convention is in force and, although this applies to the Atlantic Coast of Europe, it does not apply to the Mediterranean and outside the north Atlantic. The Paris Convention covers a similar area. A composite Convention for the Mediterranean area is currently being negotiated.

All the Conventions list organohalogen compounds and persistent plastics, and the London and Paris Conventions also include persistent oils. In the dumping Conventions these substances are described as substances which cannot be dumped at sea at all, unless they are present in trace concentrations or can be shown to be non-toxic. The Paris Convention requires the elimination of pollution by such substances.

Total prohibition of disposal such as that required by the dumping conventions is unrealistic, since it is necessary to exclude trace concentrations, and may be the only feasible method of safe disposal. It is also unnecessary, since the disposal of relatively large quantities of material could, with appropriate dispersal, be conducted in the oceans without leading to concentrations in the water which would be either directly harmful or even indirectly harmful via accumulation. What is required is the definition of standards of acceptability so that the safe quantities can be determined and if necessary allocated on a quota and area basis.

Such quotas would recognize that different areas of the sea have different capacities to accept wastes. They would take account of existing pollution loads and the uses of the marine environment in and around the discharge areas. It is therefore apparent that different standards would be applied to waste waters with the same characteristics but discharged in different areas. This is still within the policy of harmonization of controls and elimination of pollution. It is clearly ridiculous to require the same standard for a discharge of waste water on the exposed northern coast of Scotland as that required for a coastal area of the Mediterranean, which relies heavily on tourism or which is already heavily industrialized.

Future control measures should stem more from a scientific basis than from public or emotively induced expediency, but they should be based on the promise that the detrimental effects of pollution on any interest in the marine environment, must be prevented or eliminated. Ideally they should be backed by properly established standards and well designed monitoring systems. Finally they should be flexible enough to allow rapid updating in the light of changing knowledge regarding known or newly recognized pollutants.

7. CONCLUSIONS

Municipal industrial waste waters contain a wide variety of organic substances. Some of these are persistent in the marine environment, either as a result of their special characteristics, or the characteristics of the disposal area. Persistence does not necessarily mean a substance will be harmful in the marine environment but experience with, e.g. DDT, PCBs and oil, shows that harmful effects can occur if the level of the substance exceeds a specific danger level. The definition of that danger level will differ according to the ecological structure of the area and man's interest in that area.

Experience has shown that persistent organic compounds can cause harm in the marine environment but this has always been limited in scale and has generally been associated with what in retrospect were unsatisfactory conditions of discharge or quantity. At present it is still difficult to establish safe standards and recourse is usually made to the total prohibition of discharge. This is unnecessarily restrictive and strenuous efforts are required to allow appropriate standards to be developed such that some disposal can be allowed without causing pollution. Our capacity to recognize pollution is improving steadily but the number of compounds we classify as persistent organic pollutants is small. It is almost inconceivable that the present list is complete and it is highly likely that as analytical expertise, techniques and man power all improve we will detect other compounds which are potentially damaging. At the present time however, in spite of the very wide distribution of some compounds, there is, with the exception of a few local trouble spots, little evidence of damage to marine organisms, and no risk to man through consumption of marine foods.

REFERENCES

1. ICES. Report of Working Group for the international study of the pollution of the North Sea and its effects on living resources and their exploitation. Co-operative Research Report No. 39, 1974, pp. 191.

2. Dewling, R.T., Spear, R.D., Anderson, P.W. and Braun, R.J., EPA's position on ocean disposal in the New York Bight. *Pretreatment and Ultimate Disposal of Wastewater Solids.* Ed. Freiberger, A. Publication E.P.A. — 902/9 — 74—002, pp. 283-330, 1974.

3.   Waldichuk, M., Effects of pulp and paper mill wastes on the marine environment. *Biological Problems in Water Pollution.* Technical Report No. W60—3 of Robert A. Taft. Engineering Centre, 1959.

4.   Holden, A.V. and Marsden, K., The examination of surface waters and sewage effluents for organochlorine pesticides. *J. Proc. Inst. Sewage Purif.* **3**, 3-7 (1966).

5.   Lowden, G.F., Saunders, C.L. and Edwards, R.W., Organo-chlorine insecticides in water: Part II. *J. Soc. Wat. Treat. Exam.* **18**, 275-343 (1969).

6.   Schmidt, T.T., Risebrough, R.W. and Gress, F., Input of polychlorinated biphenyls into California coastal waters from ur ban sewage outfalls. *Bull. Environ. Contam. Tox.* 6, 235-243 (1971).

7.   Croll, B.T., Organochlorine insecticides in water: Part I. *J. Soc. Wat. Treat. Exam.* **18**, 255-274 (1969).

8.   Dexter, R.N. and Pavlov, S.P., Chlorinated hydrocarbons in sediments from southern Greece. *Mar. Poll. Bull.* **4**, 188 (1973).

9.   S.C.C.W.R.P., The ecology of the Sourthern California Bight: implications for water quality management. Southern California Coastal Water Research Project. California TR104, pp. 180, 1973.

10. Butler, P.A., Childress, R. and Wilson, E.J., The association of DDT residues with losses in marine productivity. *Marine Pollution and Sea Life.* Ed. Ruivo, M. pp. 262-266, Fishing News, England, 1972.

11. Anderson, R.W., Hickey, J.J., Risenbrough, R.W., Hughes, D.F. and Christensen, R.E., Significance of chlorinated hydrocarbon residues to breeding pelicans and cormorants. *Can. Field Naturalist* **83**, 91-112 (1969).

12. Swennen, C., Chlorinated hydrocarbons attacked the Eider population in the Netherlands. *Side Effects of Persistent Pesticides and other Chemicals on Birds and Mammals in the Netherlands.* Ed. Koeman, J.H., pp. 556-560, TNO, Netherlands, 1972.

13. Waugh, T.H., DDT an unrecognized source of PCB. *Science* **180**, 578-579 (1973).

14. Holden, A.V., Source of polychlorinated biphenyl contamination in the marine environment. *Nature* **228**, 105 (1970).

15. Halcrow, W., Mackay, D.W. and Bogan, J., PCB levels in Clyde marine sediments and fauna. *Mar. Poll. Bull.* **5**, 134-136 (1974).

16. Harvey, G.R., Steinhauer, W.G. and Miklas, H.P., Decline of PCB concentrations in North Atlantic surface water. *Nature* **252**, 387-388 (1974).

17. Nimmo, D.R., Wilson, P.D., Blackman, R.R. and Wilson, A.J., Polychlorinated biphenyl absorbed from sediments by fiddler crabs and pink shrimps. *Nature* **231**, 50-52 (1971).

18. Keil, J.E., Priester, L.E. and Sandifer, S.H., Polychlorinated biphenyl (Aroclor 1242): effects of uptake on growth, nucleic acids and clorophyll of a marine diatom. *Bull. Environ. Contam. Tox.* **6**, 156-159 (1971).

19. McConnell, G., Ferguson, D.M. and Pearson, C.R., Chlorinated hydrocarbons and the environment. *Endeavour* **XXXIV**, 13-18 (1975).

20. Pearson, C.R. and McConnell, G., Chlorinated C_1 and C_2 hydrocarbons in the marine environment. *Proc. R. Soc.*

21. Lovelock, J.E., Maggs, R.J. and Wade, R.J., Halogenated hydrocarbons in and over the Atlantic. *Nature* **241**, 194-196 (1973).

22. Goldberg, E.D., Atmospheric transport. In: *Impingement of Man on the Oceans*. Ed. Hood, D.W. Wiley-Interscience, New York, p. 75, 1971.

23. Montgomery, H.A.C. and Conlon, M., The detection of chlorinated solvents in sewage sludge. *J. Inst. Wat. Poll. Cont.* **66**, 190-192 (1967).

24. Portmann, J.E., Persistent organic residues.

25. Norton, M.G., Personal Communication.

26. Scott, P.D. and Armstron, F.A.J., Mercury concentration in relation to size in several species of freshwater fishes from Manitoba and north-western Ontario. *J. Fish. Res. B. Can.* **29**, 1685-1690, 1972.

27. Wong, P.T.J., Chau, Y.K. and Luxon, P.G., Methylation of lead in the environment. *Nature* **252**, 263-264 (1975).

28. Dept of the Environment. *Thirteenth Progress Resport of the Standing Technical Committee on Synthetic Detergents*. H.M.S.O., London, 1972.

29. Jensen, S. and Petersen, O., 2, 5 di- (Benzoxazole-2-yl), an optical brightener contaminating sludge and fish. *Environ. Poll.* **2**, 145 (1972).

30. Sturm, R.N., Williams, K.E. and Macek, K.J., Fluorescent whitening agents; acute fish toxicity and accumulation studies. *Wat. Res.* **9**, 211-219 (1975).

31. Rudd, J.W.M. and Hamilton, R.D., Biodegradation of trisodium nitrilotriacetate in a model aerated sewage lagoon. *J. Fish. Res. B. Can.* **29**, 1203 (1972).

32. Sturm, R.N. and Payne, A.G., Environmental testing of trisodium nitrilotriacetate: Bioassays for aquatic safety and algal stimulation. *Bioassay Techniques and Environmental Chemistry*. Ed. Glass, G.E. p. 403. Ann Arbor Science. Publishers, Wiley, New York, 1973.

33. Sprague, J.B., Promising anti-pollutant: chelating agent NTA protects fish from copper and zinc. *Nature* **220**, 1345 (1968).

34. Intergovernmental Maritime Consultative Organisation, *International Convention for the Prevention of Pollution from Ships 1973*. IMCO, London, 1973.

35. Morris, A.W. and Hamilton, E.I., Polystyrene spherules in the Bristol Channel. *Mar. Poll. Bull.* **5**, 126-127 (1974).

36. Colton, J.B., Knapp, F.D. and Burns, B.R., Plastic particles in surface waters of the north western Atlantic. *Science* **185**, 491-497 (1974).

37. Hays, H. and Cormons, G., Plastic particles found in tern pellets, on coastal beaches and at factory sites. *Mar. Poll. Bull.* **5**, 44-46 (1974).

38. Carpenter, E.J., Anderson, S.J., Harvey, G.R., Miklas, H.P. and Peck, B.B., Polystyrene spherules in coastal waters. *Science* **178**, 749-750 (1972).

39. Her Majesty's Stationery Office. *Convention for the Prevention of Marine Pollution by Dumping from Ships and Aircraft.* Science and Technology Series. Miscellaneous No. 21, Cmd 4984 (1972).

40. Her Majesty's Stationery Office. *Final Act of the Inter-governmental Conference on the Convention on the Dumping of Wastes at Sea.* Science and Technology Series. Miscellaneous No. 54 Cmnd 5169 (1972).

41. Her Majesty's Stationey Office. *The Convention for the Prevention of Marine Pollution from Land-based Sources.* Pollution Series, Miscellaneous No. 1 Cmnd 5803 (1975).

RECEIVING WATER STUDIES FOR PRELIMINARY DESIGN

Lawrence A. Klapow and Robert H. Lewis

California State Water Resources Control Board,

1416 Ninth Street, Sacramento, California 95814, U.S.A.

Summary — The objectives of predesign surveys are predictive, in the sense of identifying the most favorable site(s) for wastewater disposal, and descriptive, in terms of providing an assessment of environmental conditions prior to discharge. The first objective requires the identification of measurable waste properties which may affect the marine environment and an analysis of receiving water dispersion characteristics which diminish the effective concentrations or magnitudes of identified waste properties. The second objective requires the acquisition of physical, chemical, and biological data in a manner which will allow for the early detection and correction of unacceptable water quality impacts.

1. INTRODUCTION

The State Water Resources Control Board and nine Regional Water Quality Control Boards are responsible for regulating wastewaters produced through the activities of California's twenty million residents. One billion gallons per day (approximately 50% of the municipal wastewater produced in the State) is disposed of in open coastal waters by discharges ranging in size from a mere 25,000 gallons per day to substantial discharges on the order of 400 million gallons per day. The steep slope of the adjacent continental shelf, with deep waters relatively close to shore, in combination with a highly developed deep water outfall technology, have produced conditions for marine disposal which are generally more favorable than those encountered in other coastal states or nations. Natural and technological advantages notwithstanding, the value of the coastal resource and an environmentally concerned public have supported an active regulatory program. The issuance of discharge permits containing both effluent and receiving water standards and monitoring requirements, which are legally binding on the discharger, is an important part of this program. In addition, the State and Regional Boards provide guidance and direction to local agencies which seek to modify treatment or disposal practices. Guidance takes the form of prescriptions on the information needed to evaluate disposal alternatives rationally, and if marine discharge is found to be the best alternative, assistance is provided in designing pre-discharge monitoring programs.

2. COMPREHENSIVE PLANNING

Until recently, the general practice has been to provide separate treatment facilities for municipal wastes from each coastal city or metropolitan area, each with its own outfall extending, more or less, immediately offshore. This situation is changing in California as the result of the State's Comprehensive Planning Program which attempts to analyze treatment and disposal options from a regional perspective. Comprehensive planning often involves consideration of alternatives for consolidating treatment facilities in a cost-effective manner, which may require the transport of raw or treated wastewaters over substantial distances. Another important task of the planning program is the assessment of the comparative advantages of marine discharge and reclamation. The growing tendency to consolidate waste treatment facilities (requiring the transport of wastewaters over large distances) means that information on the marine environment must be obtained on a larger geographic scale than in the past. Furthermore, since ocean disposal is no longer considered an absolute necessity but rather an option, more refined environmental information is required to compare the suitability of marine discharge with reclamation alternatives.

3. MARINE SURVEYS

The need to look at a larger coastal zone in greater detail can often be dealt with effectively by dividing predesign surveys into what we term "reconnaissance" and "intensive" phases. The reconnaissance survey is a screening procedure whereby the most environmentally favorable discharge areas are identified within the coastal waters under consideration. The intensive survey serves to identify the most favorable discharge site and also provide baseline data to be used in assessing post-discharge effects on water quality.

4. RECONNAISSANCE SURVEYS

A reconnaissance survey should cover the portion of the coastal environment appropriate to the most expansive consolidation alternative which is considered practicable. Reconnaissance surveys should include a number of elements:

(a) a description of broad-scale patterns of hydrologic circulation;

(b) a description of the geographic distribution of biological communities, including an assessment of their current and potential values as harvestable resources and as objects of humanistic or scientific interest;

(c) a description of the pattern and intensity of human activity in the coastal environment including such uses as bathing, boating, fishing, diving, casual and scientific observation and study;

(d) a description of the more obvious terrestrial or submarine features which might bear on the cost or structural integrity of outfall facilities.

A reconnaissance survey may require extensive field work such as that which was recently completed in Monterey Bay [1], a 100 mile wide, open embayment with diverse oceanographic characteristics and patterns of human use. However, this may not always be the case. A review of the existing technical literature and consultation with academic institutions, government agencies, and knowledgeable private citizens can sometimes provide most, if not all, of the information required for a reconnaissance survey. The objective of the reconnaissance survey should be to reduce the universe of potential study areas to one or a few candidate areas which can then become the objects of intensive predesign surveys. The choice of candidate disposal areas is a matter of both science and sociology. We cannot suggest any definite rules for assessing the sociological element other than our belief that the consensus of knowledgeable disinterested persons will, more often than not, indicate the appropriate choice of potential discharge sites.

5. INTENSIVE SURVEYS – OUTFALL SITING

The location of appropriate disposal sites and the characteristics of a given wastewater (as measured by the concentrations of its constituents and flow) are, of course, related. The concept, while hardly profound, does indicate that we must consider one or the other as a starting point. From the designer's standpoint, potential marine disposal sites are generally less constrained than the characteristics of a waste discharge which are determined to a degree by population size, intensity of industrialization, and regulatory policies which specify treatment levels or otherwise control the quality of wastewater discharges. For these reasons we will consider waste characteristics to be determined and from that point attempt to outline how the most acceptable disposal sites may be determined.

5.1. Regulatory standards governing waste discharge

Our approach should not be interpreted to mean that any wastewater, of whatever quality, would automatically be judged acceptable for disposal in marine waters. A necessary precondition should be the existence of effective regulations reflecting good waste control practices. Such regulations are being implemented in California through the application of what has been termed "best practicable control technology" and discharge standards contained in the Water Quality Control Plan for Ocean Waters of California (a policy that severely limits the discharge of toxic materials to the marine environment).

5.2. Conceptual scheme for outfall siting — prerequisites for predictive modeling

The prerequisites for developing quantitative predictions regarding the suitability of potential disposal sites are as follows:

(a) the identification of the criteria which are to serve as the basis for site selection which may include public health, esthetics, or concern for the biological integrity of the marine waters;

(b) the identification of *measurable* wastewater constituents or properties which can be taken as a measure of the potential impact of the wastewater on the identified criteria;

(c) the establishment of receiving water standards for the selected wastewater constituents;

(d) delineation of the area within which the receiving water standards must be met;

(e) development of a mathematical model (or, in some cases, a physical model) which can be used to predict the distribution of the appropriate waste constituents or properties within the receiving waters. The model should contain terms that reflect the principal dispersing processes, e.g. initial dilution, subsequent dilution, and when appropriate, time-dependent decay processes.

5.3. Assessment of the magnitude of receiving water mixing processes

The objective of the predesign intensive survey (at least with regard to its predictive aspect) becomes one of establishing the characteristics of the receiving water which determine the magnitude of dispersing processes. Mathematical formulations which attempt to describe initial dilution, subsequent dispersion, and time-dependent decay processes are covered in a fairly extensive technical literature [2, 3]. We will not attempt to develop such formulae here, but rather discuss the receiving water parameters which are required as inputs for assessing the magnitude of mixing and dispersing processes.

5.3.1. *Initial dilution.* Initial dilution is a mixing process energized by the momentum with which the wastewater is discharged, and buoyancy forces resulting from the difference in density between the wastewater and the receiving water. Typically, a submerged freshwater waste discharge will rise vertically toward the surface until its momentum and buoyancy forces are exhausted. Initial dilution is completed when the wastewater plume stabilizes vertically in the water column and begins to spread in a horizontal direction. During periods of strong thermo-haline density stratification of the water column (as is generally encountered year-round in tropical waters and during the summer months at temperate latitudes) the wastewater plume will remain submerged below the surface. When water column density stratification is weak or nonexistent (e.g. the general condition of polar seas year-round and in temperate regions during winter months), initially diluting waste plumes will surface before they begin to spread horizontally. Both depth and water column density structure are receiving water characteristics which must be assessed in order to estimate initial dilution. Hence, a predesign monitoring program should include a bathymetric survey (if one has not already been conducted) and should provide for an array of stations within the study area at which vertical temperature and salinity gradients are determined on a seasonal basis.

5.3.2. *Subsequent dilution.* Subsequent dilution is a mixing process driven by ambient energy, generated through wind, wave, tidal, and density forces which give rise to turbulent motion in the receiving water. Because the energy sources which drive subsequent dilution are more variable than those acting to accomplish initial dilution, the rate of subsequent dilution is generally less predictable. Vertical and horizontal eddy diffusion coefficients are a measure of the rate of subsequent dilution. Values for these coefficients can be estimated by measuring the rate of dispersion of dyes released into the recieving water. For small discharges, it has been general practice to select what are thought to be representative values of eddy diffusion from the literature.

5.3.3. *Time-dependent decay processes.* Information on elapsed travel times from prospective discharge sites to protected areas should be obtained as inputs to model components which consider time-dependent decay of waste constituents. This information can be obtained through either fixed-point current measurements (Eulerian Techniques) or drift type current measurements (Lagrangian Techniques).

Fixed-point current measurements. Fixed-point current measurements can be obtained by deploying current meters at stations throughout the study area, either from shipboard for short-term measurements, or tethered to fixed monitoring buoys for continuous measurements. Current meter measurements, whether short-term or continuous, may not always provide useful data for calculating elapsed travel times from a prospective discharge site, unless there is a reasonable degree of spatial coherence in water currents. In order to calculate meaningful elapsed travel times, it must be assumed that the current vectors measured at the fixed stations represent the motion of a larger water mass on a dimensional scale exceeding the distance between fixed stations. The assumption is invalidated to the degree that eddy currents are present on a scale smaller than that of the distance between stations. When the assumption of spatially coherent currents is suspect, it is preferable to use drift rather than fixed-point techniques.

Drift current measurements. Drift or Lagrangian techniques measure trajectories of passively drifting devices, usually either drift cards or drogues. Drift cards may be released at prospective disposal sites and recovered in periodic surveys of the shoreline. The difference in time between the release of a drift card and its recovery can then be used to estimate the minimum velocity or maximum elapsed travel time in transit. Drift card measurements are inaccurate to the degree that there may be a significant interval between the time a card is washed up on the shore and the time it is recovered. For this reason, drift card studies are used more frequently to establish the general direction of prevailing currents rather than exact trajectories and travel times. The latter sort of information is best acquired through active (shipboard) tracking of drogues. The parachute drogue is the typical device used in predesign surveys. Such drogues consist of a surface float (equipped with flags, lights, or radar reflectors, to facilitate visual or electronic tracking) attached to a weighted line to which a parachute is attached. Because the parachute offers more resistance than other structural components of the drogue, the motion of the drogue tends to reflect the movement of water at the depth at which the parachute is set. As previously indicated, horizontal spreading of the diluting waste plume will occur at a depth which is related to the vertical density stratification of the water column. In practice, it is often useful to incorporate more intensive drogue studies during winter months in the Northern hemisphere when waste plumes are likely to reach the surface where they may be rapidly driven toward shore by strong winds.

5.4. Outfall siting criteria

5.4.1. *Public health criteria.* To date, most municipal outfall siting studies have been based largely on public health considerations. For the most part, two factors are responsible for this situation. First, a measurable property of wastewaters, in this case the concentration of coliform organisms, has been identified as a useful index of potential impact on public health. And second, receiving water standards for coliform organisms have been established which provide an adequate level of public health protection for swimming and shellfish harvesting.

5.4.2. *Esthetic criteria.* When considering design criterion based on esthetics, we can at least identify the relevant and measurable wastewater properties of concern, mainly grease, oil, and floating matter. There appears to be a basis for establishing numerical receiving water standards as it is known that surface films of varying thickness present different visual impacts. However, the fact that surface films and floatables are subject to rapid displacement by surface winds may indicate that initial dilution is the only mixing process which can be relied upon to reduce effective concentrations in the discharge area.

5.4.3. *Biological criteria.* The prerequisites for predictive modeling are least complete when the design criterion is one of protecting marine organisms. In fact, design criteria reflecting this concern are most often qualitative in nature in the sense that areas of high resource value, such as fishing grounds, productive reefs, kelp beds, or marshlands, are excluded from consideration as prospective discharge sites. The first barrier to developing an adequate predictive theory is the absence of a single measurable property (or even a manageable array of properties) for a complex wastewater which is generally recognized and accepted as a measure of potential biological impact. Suspended solids, pH, BOD, temperature, nutrients, and trace toxicants are examples of wastewater constituents or properties which are known or suspected to have an effect on marine life at least in areas of restricted dilution. The problem is not a lack of relevant candidate properties, but rather one of over-supply.

Wastewater toxicity. While there is no general consensus, it is nevertheless correct to say that the biological parameter which has been most often utilized to regulate wastewater quality is the median tolerance limit (TL_m or LC_{50}*), as determined in an acute bioassay. Such assays are usually restricted to a single species exposed to various dilutions of wastewaters for periods ranging from a few hours to several days. It is well to recognize that the TL_m as determined historically, suffers from a number of conceptual limitations which most researchers acknowledge. The acute nature of the test, performed under laboratory conditions, does not provide for an assessment of the concentrations of waste constituents which are chronically tolerable to the test species under field conditions. Furthermore, the single species nature of the test does not permit an assessment of either the acute or chronically tolerable limits for other species which must occur in the receiving waters.

Interpretation of toxicity test. A narrow and perhaps the most correct interpretation of the TL_m test is that it represents no more than a wastewater parameter which can be measured readily, and which reflects in a general way an ability to disrupt life-sustaining processes. The real significance of the test does not lie in the measurement itself (that is, the particular species chosen or the duration of the test), but rather in the correlations that may be established between a wastewater's TL_m and the condition of the community of receiving water organisms. In order to develop such correlations, it would be preferable to

---

* TL_m and LC_{50} are equivalent.

standardize the test by species and duration. In the absence of standardization, each study must be considered unique; a condition which can be expected to impede progress toward the development of useful generalizations. Admittedly, the standardization of the test would be at variance to existing practices; it has long been assumed that the use of a test species which occurs in the receiving water adds a significant degree of relevance to the test. This assumption should be examined critically in view of what appears to be the inherent conceptual limitation of the test.

Modeling toxicity distribution. To attempt a correlation of the acute toxicity of a waste (as measured by its TL_m) and the response of receiving water organisms, the wastewater's acute toxicity must be expressed in units which can be incorporated readily in receiving water dispersion models. The TL_m does not lend itself to this sort of manipulation. Dr. Erman Pearson suggested the use of a derived quantity which he referred to as "wastewater toxicity concentration" (defined as $100/TL_m$ and expressed in "toxic units") which effectively overcame this difficulty [4]. One advantage of the "wastewater toxicity concentration" over the TL_m expression is that the former is a direct rather than inverse measure of waste strength. More important is the fact that "wastewater toxic concentration" can be treated in the same way as any material substance in a waste. It can be multiplied by flow to yield a "toxicity mass emission rate", which, in turn, can be used in receiving water dispersion models to predict "receiving water toxicity concentrations".

Development of toxicity criteria. The wastewater toxicity concentration expression satisfies the first prerequisite for predictive modeling by providing a readily measurable wastewater parameter from which receiving water concentrations can be predicted through the application of dispersion models. To complete the requirements for predictive modeling, a value for receiving water toxicity concentration must be determined which assures the proper species balance and function of the marine community. An interesting series of studies utilizing this approach were conducted in San Francisco Bay [4–6]. A mathematical dispersion model was used in those studies to predict receiving water toxicity concentrations throughout the Bay. Benthic species diversity was used as an indicator of biological health. Multiple linear regression analysis indicated a significant inverse relationship between elevated receiving water toxicity concentration and benthic species diversity. From these regression equations it was estimated that a receiving water toxicity concentration of 0.04 toxic units had a perceptible threshold effect on benthic species diversity and presumable community health. The significance of the San Francisco Bay studies, in our view, lies in the extension of the scope of the investigation beyond the experimental determination of wastewater toxicity (a point where most previous studies stopped) to include an analysis of the long-term response of naturally-occurring communities of species to estimated toxicity loads.

6.   INTENSIVE SURVEYS – ACQUISITION OF PRE-DISCHARGE BASELINE DATA

Once a candidate outfall site(s) has been identified, the next task is to provide an assessment of seasonal baseline receiving water conditions. Ideally, baseline survey programs should be planned with knowledge of post-discharge compliance monitoring requirements to ensure data continuity and comparability. The impact of a discharge may be assessed through a comparison of pre-discharge with post-discharge conditions, or through an analysis of the spatial distribution of receiving water properties once the discharge has commenced. Significant natural fluctuation in receiving water properties may occur over a longer time period than that which is usually available for pre-discharge surveys. Hence, the data base accumulated prior to discharge may be inadequate to characterize representative receiving water conditions, a limitation which may frustrate attempts to analyze the effects of the discharge through pre- and post-discharge comparisons. Given this limitation on temporal comparisons, we believe that it is generally preferable to design pre-discharge baseline studies in a spatially intensive rather than temporally intensive manner.

6.1. Scale of pre-discharge baseline studies

It is difficult to describe precisely the level of sampling effort which would provide an adequate description of the spatial distribution of a receiving water property. One defensible criterion is that level of sampling effort necessary to contour the parameter on a chart depicting the areal (horizontal) extent of the discharge zone and adjacent control areas. We would suggest that three to four stations located along each principal environmental gradient (i.e. onshore, offshore, upcoast, and downcoast from the discharge) with sufficient ancillary stations to provide continuity in a horizontal plain would appear to be an adequate station layout. We are not implying that rectilinear location of sampling stations will always be the most effective station pattern; when persistent currents are evident, it is advantageous to locate sampling stations along the expected trajectory of the wastewater field. When sampling biological communities or sediment properties, a distinct analytical advantage is usually obtained by locating stations along lines of constant depth (as a means of separating the often dramatic depth responses of these properties from effects which may be associated with the wastewater discharge). Obviously, the spatial scale of the station pattern should reflect the quantity, persistence and toxicity of the material which is to be discharged. It has been our experience that the most readily discernible effects which can be associated with municipal waste discharges, with flows from 100 to 400 million gallons per day, are measured on a scale ranging from a few hundreds of meters to a few tens of kilometers with the notable exception of persistent organic compounds for which even larger scales are occasionally needed.

6.2. Parameters to be considered

Aside from those receiving water properties that are governed by regulatory standards, the decision as to what is an appropriate baseline parameter becomes one of professional judgment. Rather than specify the parameters which must be considered, we will only describe those which may be considered, along with an assessment of their general usefulness as indicators of wastewater impact.

6.2.1. *Physical chemical parameters.* Obviously, the properties of a particular wastewater will help identify the relevant physical-chemical properties that should be considered in the study area. It is also helpful to have some knowledge of the likely paths which a waste constituent will follow as it disperses in the marine environment. For example, because pesticides and halogenated organics are usually associated with wastewater particulates and floatables, baseline concentrations in the sediments and surface slicks should be determined whenever such materials are thought to present a potential hazard. Toxic metals are likewise often associated with wastewater particulates, again indicating the need for sediment analysis. For materials, properties or effects that are associated with the liquid phase of effluents (i.e. temperature, pH, nutrients, dissolved oxygen), water column monitoring is indicated. Without considering the many different types of waste which may be discharged, our examples can only be illustrative.

6.2.2. *Biological parameters.* Biological assessments of the receiving waters are usually organized on the basis of what can be loosely called habitat groups. These groups include phytoplankton, zooplankton, fish, and benthic organisms.

Phytoplankton. The principal concern with phytoplankton is the possibility of over-production leading to nuisance conditions or toxicity to other marine life. Deleterious phytoplankton blooms, commonly referred to as red tides, are occasionally observed in open coastal waters. However, both laboratory and field data suggest that such occurrences are not initiated by the discharge of sewage nutrients [7]. Laboratory experiments have shown that the response of phytoplankton to added nutrient levels can be expected to take place over several days [8]; hence, the effects of accelerated growth may be realized many tens of kilometers from the actual point of discharge. Because the response of phytoplankton need not be in close proximity to the discharge, we usually do not include phytoplankton sampling in pre-discharge baseline surveys for facilities located on the open coast except in areas characterized by restricted circulation.

Zooplankton. Attempts to examine the response of zooplankton to waste discharges are often not particularly revealing. There are several reasons for this situation. Species identification is extremely time consuming and the taxonomy of marine zooplankton is very incomplete. In addition, species relationships are in most cases extremely variable. For example, the zooplankton of the California current is a mixture of endemic species as well as species originating in at least three oceanic provinces [9]. Furthermore, the well-documented patchiness in the distribution of zooplankton results in extremely imprecise estimates of abundance. Finally, the drifting motion of zooplankton results in transient exposure to waste constituents, a condition which gives little reason to expect the existence of stable species patterns reflecting an integrated response to a wastewater discharge. Normally, we do not require zooplankton sampling for facilities located on the open coast with the exception of power plants which utilize large volumes of once-through cooling water where some effects have been noted.

Fish. Fish have been considered in most baseline studies because of their obvious resource value. The mobility of fish and the tendency of many species to aggregate, or to migrate, usually result in what are often imprecise estimates of local species and their abundance. Many years of sampling effort are often required to obtain sufficient data to determine the response of fish to wastewater discharges in terms of species composition and distribution [10]. However, most baseline studies should at least contain a level of sampling effort which will identify the common species in the area of discharge. To go beyond the development of a qualitative species list to obtain meaningful abundance and distribution data requires the regular use of a survey ship (representing a receiving water monitoring capability which, in California, is usually confined to discharges in excess of 100 million gallons per day). Because fish are often a directly harvested resource, their physical appearance and physiological condition become matters of some concern. It has been general practice to examine specimens collected in baseline studies for disease syndromes or abnormalities which may be related to wastewater discharge (i.e. eroded fins, discoloration, skeletal deformities, and tumors). It is also advisable to analyze fish tissues for the accumulation of toxic materials which may be present in wastewater discharges including toxic metals, radioactivity, pesticides, and persistent organic toxicants.

Soft-bottom benthos. Studies of the distribution of soft-bottom benthic organisms have yielded some of the most interpretable data [11]. The taxonomy of these organisms is fairly complete and species groups are comparatively stable. Sampling devices are more manageable and are generally more effective than those used to collect mobile organisms living in the water column. Most important, the limited mobility of most soft-bottom benthic species results in patterns of distribution which are likely to represent an adaptive equilibrium reflecting exposure to wastewater constituents.

Rocky-bottom benthos. Submerged rocky-bottom benthic communities share many of the previously mentioned characteristics of soft-bottom communities; however, they are much more difficult to sample. For the most part, the only effective means of sampling rocky-bottom communities has been through diver observation. This means that almost all assessments of species occurrence and abundance must be made in the field within a severely limited time frame and at depths which are generally less than 35 m. To compound the problem, most rocky-bottom communities are extremely rich in species. Community descriptions must, of necessity, be rather incomplete. Given these limitations, sampling effort in most cases has been directed to the assessment of a few commercially important species, or species which are considered "habitat formers" by virtue of their numerical abundance or size. Rocky-bottom habitats are usually patchy; hence an analysis of spatial gradients in species occurrence may not always be possible. Under such conditions an intensive effort should be directed toward characterizing rocky-bottom communities prior to discharge to facilitate pre- and post-discharge comparisons. Where nature has not conveniently located rocky-bottom habitat in a pattern which would lend itself to the analysis of spatial trends, it may be possible to supplement the natural occurring habitats with either settling plates or artificial reefs.

Artificial substrates. The use of artificial substrates, which serve as the site of attachment and growth of benthic organisms, is a promising tool for assessing the impact of wastewater discharges. The State Water Resources Control Board recently funded a study to demonstrate this technique [12]. In this study, settling plates attached to the anchor lines of monitoring buoys were set at various levels of the water column at an array of stations located around a major municipal outfall. The plates were maintained in the field for periods of up to 3 months. Upon recovery, the organisms collected on the plates were assayed for a number of parameters, including biomass, species abundance, and the occurrence of toxic materials in their tissues. Spatial gradients were observed for several assay parameters making it possible to identify the geographic limits of the effect of the wastewater discharge with a degree of precision not normally associated with the analysis of naturally occurring communities.

7. CONCLUSION

To conclude, we recognize that the continuing trends toward higher levels of treatment and more stringent limitations on the discharge of toxic materials may diminish the need for extensive predesign surveys and environmental analyses. It is our hope that the predesign programs which we are presently suggesting will test the validity of this presumption and narrow the range of environmental parameters which will be considered in future studies.

REFERENCES

1. Oceanographic Services, *Oceanographic Survey*. Pp. 236, OS1, Santa Barbara, California, 1974.

2. Brooks, N.H., *Dispersion in Hydrologic and Coastal Environments*; EPA-660/3-73-010, 136 pages, U.S. Environmental Protection Agency, Washington, 1973.

3. Okubo, A., Horizontal and vertical mixing in the sea. In *Impingement of Man on the Oceans*; pp. 89-168. Ed. D.W. Hood, Wiley-Interscience, New York, 1971.

4. State Water Resources Control Board, *A Study of Toxicity and Biostimulation in San Francisco Bay-Delta Waters*, Summary Report, Volume I; pp. 86, SWRCB, Sacramento, 1972.

5. Kaiser Engineers, *San Francisco Bay-Delta Water Quality Control Program*, Final Report; Chapter XII (23 pages), Biological and Ecological Studies; Prepared for the State Water Resources Control Board of California, Sacramento, 1969.

6. Department of Water Resources. *Dispersion Capability of San Francisco Bay-Delta Waters*, Final Report; pp. 107, State Water Resources Control Board, Pub. No. 45, Sacramento, 1972.

7. Southern California Coastal Water Research Project, *The Ecology of the Southern California Bight: Implications for Water Quality Management*; pp. 531, SCCWRP, El Segundo, 1973.

8. Brown, R.L., Varney, G. and Chadwick, H.K., *A Study of Toxicity and Biostimulation in San Francisco Bay-Delta Waters*, V ol. VIII, Algal Assay, 1972.

9. Fager, E.W. and McGowan, Zooplankton species groups in the north pacific. *Science* **140**, 453-460 (1963).

10. Carlisle, J.G., Jr., Results of a Six-Year Trawl Study in an area of heavy waste discharge, Santa Monica Bay, California; *Calif. Fish Game* **55**, 26-46 (1963).

11. Reish, D., The use of benthic animals in monitoring the marine environment; *J. Environ. Plan. Poll. Cont.* **1**, 32-38 (1973).

12. Biome Company, *The Demonstration and Standardization of a Method for Monitoring Ecological Effects of Marine Waste Discharges*; pp. 88, State Water Resources Control Board, Pub. No. 54, Sacramento, 1974.

STUDY METHODOLOGY AND DATA EVALUATION

Philip N. Storrs

Engineering-Science, Inc., Berkeley, California, U.S.A.

Summary — Oceanographic surveys conducted in connection with marine waste disposal programs differ in significant aspects from the usual oceanographic research investigations. This is due to the different objectives of the two types of investigations as well as to the natural differences between open ocean waters and coastal waters. Ocean surveys for waste disposal are intended to develop information needed for outfall siting and design and to provide data on receiving water conditions. Ocean surveys are expensive, and intensive planning and data management are necessary to ensure that the data required are obtained in a cost-effective manner.

1. OCEAN SURVEY OBJECTIVES

There are a number of reasons for conducting nearshore oceanographic surveys, but most of these can be encompassed in the following objectives:

to determine differences in receiving water characteristics in different locations to aid in outfall siting;

to obtain data needed for outfall design;

to obtain background or "base-line" data for later use in evaluating outfall performance;

to monitor outfall performance;

to obtain data needed for studies and research.

With the exception of research studies, most actual ocean surveys for wastewater disposal must satisfy all of the objectives listed above to one degree or another. Many of the data collected to meet one objective are useful and necessary for other objectives.

The various types of surveys and their objectives have been discussed in more detail in a previous conference in this series [1].

2. SURVEY DESIGN PRINCIPLES

2.1. Cost and the need for planning

Ocean data collection programs are expensive in terms of the cost per unit of information obtained. A survey vessel and a skilled crew are required. In some cases sophisticated and expensive equipment is necessary or desirable. Shore facilities may be required for accurate location of the vessel. Bad weather may result in aborted cruises or loss of equipment.

The variability of nearshore ocean conditions — hourly, daily, seasonally, as well as spatially — require large amounts of physical, chemical and biological data if a reasonable understanding of the nearshore processes is to be obtained.

These two factors of cost and large data requirements make it imperative that every effort be given to ensuring that all necessary data are collected and that resources are not expended in collecting unnecessary data.

The entire survey and design program should be planned in detail before starting data collection.

2.2. Data collection methods

Over the years oceanographers have developed a number of sophisticated methods and devices for studying the ocean. Many of these are not suitable or desirable for nearshore ocean surveys intended to obtain data for waste disposal design. This is understandable when it is realized that most oceanographic research and investigation is for the purpose of studying large-scale, open-ocean phenomena.

In the coastal waters the depths are generally shallow. Currents are extremely variable and complex due to the interaction of winds and tides with the shore. The fauna and flora are usually more rich and diverse than in the open sea. Salinity and other constituent concentration gradients are more pronounced.

Equipment and methods used in the open ocean are often designed to detect very small differences in concentrations or other characteristics. In the coastal waters this accuracy is usually unnecessary because of the large variations in characteristics normal in these waters. For example, if a water sample is to be obtained at a depth of 1000 m and the *in situ* temperature of the sample is desired, an elaborate thermometer with provision to "lock" the temperature reading at the time the sample is collected is required. This is because the temperature of the sample will change significantly during the time required to bring the sample to the surface. This is not the case, however, if the sample is to be collected from a depth of 100 m. In this situation a simple, clear plastic Kemmerer sampler with a mercury thermometer mounted inside will be more than satisfactory.

Data collection methods and equipment should be selected after considering the required and useful accuracy needed. This assessment should be made on the basis of the expected ranges in the values to be measured and the likelihood of accurate evaluation of the natural variability in those values.

2.3. Data evaluation

Because of the costs of data collection and the inability to remeasure conditions, every effort should be made (1) to evaluate the validity of data as it is being collected and (2) to integrate newly collected data into the body of previously collected data as soon as possible.

The first of the above points has a bearing on the choice of equipment and measuring techniques. For example, an equipment package with sensors for current speed and direction, salinity, temperature and pH is normally equipped with a deck read-out device. In the usual mode of operation the meter readings are read and transcribed onto a data sheet. This system, however, can be equipped with an analog-digital converter and tape punch to store the data in a form convenient for machine processing.

The automatic tape punching would appear to be a real advantage as compared to manual data processing and subsequent card punching. However, there is a very real danger that the use of the automatic recording system will result in less attention being paid to the data as it is being recorded. As a result, anomalous results, equipment malfunction, signal cable failure, etc. may not be detected in time to make repairs or replacements to obtain valid data.

This is not to say that automatic data recording should not be used; it can represent a real convenience. The point is that its use does not reduce the necessity for professional evaluation of the data as it is being collected. To the extent possible, data collection techniques should be chosen so that each unit of information can be assessed as it is being collected.

There is always the tendency in the earlier stages of a survey program to concentrate on the logistics and problems of field procedures and data collection. This often results in little attention being given to data analysis. As a consequence, the data may not be reviewed or analyzed for some time.

It is extremely important to collate and analyze data as soon as it is collected. If this is done, initial assumptions on which the survey design was based can be verified or modified if necessary; areas requiring additional data can be identified; data anomalies can be identified and plans formulated to resolve them.

Another advantage of immediate data analysis is in identifying areas of redundant data. For example, if it is found that a given parameter is virtually unchanged from time to time or from one location to another (and there is no reason to expect changes), it may be possible to reduce the sampling or measurement of that parameter.

3.   THE DESIGN OF OCEAN SURVEYS

In preparing for an ocean survey all available data should be collected and reviewed. In most areas the data will be scanty, and in some areas none will be available. However, in areas near cities there are usually data, often unpublished, collected by a wide variety of workers. Many of these data will be incomplete and will have been collected for purposes other than the design of a marine waste disposal system. They are almost always insufficient to use directly, and usually the available reports will be inconsistent and sometimes even directly contradictory. Nevertheless these data are useful if their limitations are recognized.

3.1. Selection of sampling locations

The first problem in preparing a survey plan is to decide tentatively on the number of locations at which samples will be collected or parameters measured and on the frequency of sampling and measurement. One might measure all parameters continuously at a large number of stations, but this approach is usually neither possible nor desirable, primarily because of cost. Consequently, compromises must be made between the amount of data desired and the resources available for their acquisition.

The first step is to lay out, on the basis of what data are available, a tentative array of sampling locations. These will be based on a variety of factors such as underwater topography (from charts), probable location of the outfall on the basis of landward factors, estimated outfall length from a desk-top analysis using estimated (or guessed-at) numbers, known or presumed regulatory requirements, estimated area of discernible wastewater effects, etc. There may be more than one area that should be sampled.

Stations are usually located at the estimated terminus of the outfall, near the shoreline at locations of public health or recreational significance, and over the area in such a pattern as to provide, hopefully, a reasonable, coherent picture of the movement, behavior and characteristics of waters and sediments in the potentially affected areas. A typical pattern for an outfall for a medium-sized city might be comprised of ten to twenty sampling locations. Usually the pattern will not extend much beyond a water depth of 100 m because of the practical limitations and costs of outfall construction in deeper waters. The sampling locations are usually selected considering the area of most probable impact of the discharge as determined by available information on currents. Biological habitats may also be a factor to be considered.

Reference or background stations should also be selected. These are stations where the habitat conditions approximate those in the vicinity of the outfall diffuser with regard to depth, distance from shore, wave action, substrate type and other physical conditions. They should be far enough removed from the site of the discharge so that they will remain essentially unaffected after waste discharge begins.

3.2. Survey frequency

The selection of survey frequency is usually governed by the available budget, but surveys are usually conducted from four to eight times in a year. Normally, three surveys would be the absolute minimum if any estimate of seasonal changes is to be made. In view of the changes from one year to the next, more than eight or ten surveys in a year generally will not add significantly useful data.

The frequency of sampling will not be the same necessarily for all parameters measured. Some parameters — sediment particle size distribution, for example — normally change very slowly so that two or three measurements in a year will usually provide adequate characterization. This, however, must be decided considering each parameter separately as well as the conditions in the receiving waters.

3.3. Selection of parameters to be measured

Table 1 lists most of the parameters measured in ocean surveys designed for marine outfall systems. This list is not complete for all situations, but it does cover most.

From the standpoint of outfall location and design, the physical water and sediment characteristics are of primary importance. The density structure of the water (as determined by temperature, salinity and turbulence), hydrodynamic characteristics and the geologic and stability characteristics of the sediments are needed for the physical design. Outfall location and length will be affected strongly by these factors as well as by bacterial disappearance rate and the location and extent of important biological habitats.

3.3.1. *Temperature and salinity*. The temperature and salinity characteristics are the primary determinants of the density structure of the water. The density structure, in turn, is needed to estimate plume rise height and the presence or absence of a pycnocline. Temperature is affected by currents, insolation, air temperatures, runoff from the land and upwelling. Temperatures often vary strongly with depth as well as seasonally. Salinity depends on land runoff, upwelling, surface evaporation, etc. Consequently, temperature and salinity should be measured at least four to six times per year, and measurements should be made at frequent intervals (1–3) vertically through the water column.

Temperature and salinity can be measured using discrete samples collected with standard sampling gear such as Kemmerer samplers, Nansen bottles, etc. With this equipment temperature is measured with a thermometer, and salinity is determined by electrical conductivity or by titration standardized against Standard Seawater (Eau de Mer Normale).

Temperature and salinity (as well as pH, dissolved oxygen and current velocity) are often measured by automatic analyzing sensors that can be either lowered on a cable to various depths or moored in a fixed position. The data from these sensors are usually transmitted to a readout device, or they can be automatically recorded. Equipment is often used for direct translation to data processing cards or tapes for subsequent computer computations.

Table 1. Types of required data.

WATER CHARACTERISTICS

Physical and Chemical

 Temperature
 Salinity
 Transparency, transmissivity
 Nutrients (nitrogen and phosphorus)
 Dissolved oxygen
 pH
 Floatables
 Others

Bacteriological and Biological

 Coliform bacteria (concentration, disappearance)
 Phytoplankton
 Zooplankton
 Fish

Hydrodynamic

 Currents (geostrophic, wind, tide)
 Wave structure
 Special conditions (such as upwelling)

SEDIMENT CHARACTERISTICS

Physical and Chemical

 Geology and structural characteristics, stability
 Sediment type (particle size distribution)
 Sediment BOD
 Metals
 Persistent biocides

Biological

 Benthic fauna
 Rooted plants

Other Data

 Meteorological (winds, air temperatures)
 Hydrography
 Shore topography, shore stability

3.3.2. Transparency and transmissivity. The clarity of the receiving water is of concern for several reasons. From a biological point of view it controls the amount of light available for rooted marine algae and benthic animals or fish. Aesthetically, clear water is considered more desirable than turbid water. Information on the light-transmitting characteristics of the water may be necessary to make decisions concerning the effects of suspended particulate materials in the discharged wastewaters.

Water clarity can be measured in several ways, each giving a different parameter. Transparency is measured with a Secchi disc, which is a white metal disc, 20 cm in diameter, that is lowered into the water until it can no longer be seen distinctly. The depth at which is disappears is called the Secchi-disc depth, and it clearly increases with increasing transparency. Light transmittance is measured by comparing the illumination at a given depth to that at the surface as measured by a photometer. Light transmissivity is measured by determining the amount of light absorbed in a given (usually 1 m) length of water.

3.3.3. Nutrients. Although recent research in many areas has shown that the macronutrients (nitrogen, phosphorus and silica) in wastewaters have little potential for stimulating excessive algal growth in marine waters, these parameters must usually be measured to some extent in order to satisfy questioners. On a long-term basis such information is potentially valuable to develop a better understanding of marine algal kinetics.

The best available techniques for routine analysis are those described by Strickland and Parsons [2].

3.3.4. Dissolved oxygen. Dissolved oxygen is seldom, if ever, a critical problem in marine outfall design. Knowledge of the dissolved oxygen content of the bottom waters and of the waters that may be present during upwelling periods can be important, however, in assessing the significance of apparent dissolved oxygen depletion in the rising wastewater plume.

The Winkler method is used routinely in analyzing discrete samples for dissolved oxygen. Automatic methods use a dissolved oxygen probe.

3.3.5. pH. Under normal circumstances, and particularly with municipal wastewaters, pH is not a consideration in marine outfall design. The usual initial dilution, the small differences in pH between seawater and most wastewaters, and the large buffer capacity of the seawater are such that little change in pH is observable in the vicinity of an operating outfall. The only reason for measuring it routinely is the low cost of the measurement; in most cases little or nothing is done with the data.

3.3.6. Floatable materials. Floatables are difficult to measure and quantify, and there is no really satisfactory method for their sampling [3]. Nevertheless it is necessary to have an idea of the background or natural surface concentrations.

Samples of particulate floatables are collected with a trawl net sampler which skims a thin surface layer of the water. At the end of the sampling run the material collected on the trawl screen is removed and analyzed [4]. One method of sampling and analyzing oil and grease in surface films is to use a 0.1 sq m fiberglass cloth capillary screen. Samples are collected by spreading the cloth on the water surface, removing it carefully, and extracting the collected material with hexane [4].

Measurements of floatables are relatively expensive because special trawls must be run during which it is difficult to make any other measurements. A reasonable plan is to establish perhaps two trawl transects in the vicinity of the proposed discharge and two in a reference area. Measurements might be made three to four times in a year.

3.3.7. *Coliform bacteria.* In the open waters coliform bacteria are measured primarily to provide a record of background levels. Samples are usually collected from the surface waters. The survey vessel should be equipped so that the serial dilutions can be made (or if the membrane filter method is used, the samples can be prepared) on shipboard.

Of more importance, usually, is sampling in the surf zone and in and around shellfish beds. Here, standard public health sampling methods can be used.

Estimated bacterial disappearance or die-away rates are often critical in establishing the length of the outfall. These rates are not at present predictable although it appears that factors such as turbulence, sunlight and temperature have a bearing on the rates. The diappearance should always be measured *in situ*, if possible; results obtained from laboratory experiments have no relationship to those obtained in the field. The general method of measuring bacterial disappearance rate has been described previously [5].

3.3.8. *Phytoplankton and zooplankton.* Because of the concern over the possible inhibitory or biostimulatory effects of wastewaters on plankton, it is usually desirable to establish the normal or background plankton concentrations in the receiving waters. Plankton are sampled with towed nets. The volume of water sampled is determined either by calculation using the capture area of the net and the distance it is towed or by means of a current meter mounted at the mouth of the net. In nearshore waters the phytoplankton concentrations may be relatively high, in which case grab samples may be used.

If plankton nets are used, sampling transects should be established in the area of waste discharge and at one or two locations which will be presumably unaffected by the discharge. The sampling frequency should be such that major seasonal changes are identified. Since the data will have little direct use in the outfall design, the sampling effort should be consistent with the concern for possible stimulation of excessive plankton growth and the need to establish a record of "natural" conditions.

If plankton samples are to be examined microscopically, they should be preserved in 5% formalin solution; a dye, such as Bengal Red, may be used to aid in subsequent visual identification.

Plankton samples can be examined microscopically in a counting cell for identification although this is a time-consuming process. Phytoplankton concentrations are often estimated by measuring the concentration of chlorophyll in the sample. However, even if this method is used routinely, some proportion of the samples should be processed for counting and identification to provide a better basis for evaluating the routine chlorophyll data.

Zooplankton samples should be analyzed to determine the biomass or biovolume and the numbers and identification of the principal genera or species.

3.3.9. *Fish.* Knowledge of the fishery resources is important in making initial decisions concerning the location and method of marine waste disposal. It is also desirable to obtain quantitative data to provide background information relating to possible effects of waste disposal on the fishery.

Information can be obtained from records of commercial fish catches, and from interviews with commercial and sport fishermen. It must be realized, however, that these data are often only qualitative and may be biased and incomplete.

Data on the abundance and distribution of fish is obtained usually by trawling with various types of towed nets or by use of fixed or free gill nets. Fish species vary greatly in their size, feeding habits and normal location (nearshore, offshore, bottom, mid-water, surface), and significant differences in catch of many species are found between day- and night-trawls.

3.3.10. *Benthic biota.* The identification and characterization of the benthic fauna living in and on the sediments is probably the most important part of the biological sampling effort. The benthic fauna represent a central and important link in the total marine food web, and they are an important food resource for man. Being relatively nonmobile (as compared to fish), they are more likely to be affected by wastewater discharges.

Benthic animals are sampled using various types of dredges from a boat and by direct observation and sampling by divers. The samplers that have been found most effective include the box corer and the Ponar, Peterson, Shipek and Smith-McIntyre dredges [6].

Samples are processed by washing through a screen (commonly a No. 30 mesh, U.S. Standard Sieve) and by preserving the animals in 10% formalin or 75% alcohol. The materials may be stained with Rose Bengal to facilitate sorting and identification. Important biota (such as coelenterates, polychaetes, macro-crustaceans, molluscs, ectoprocts and echinoderms) should be identified to species or at least to genus. All specimens should be counted to provide information on relative abundance. Attention should be given to the statistical design of the sampling plan so that the data will have the desired level of validity [7].

Transects for diver observation are usually lines running from and normal to the beach out to a depth of 25—30 m. Usually three to five transects are established along the shore in the vicinity of the outfall at perhaps 0.5—1 km intervals. One or two of the transects might be located outside the probable area of influence of the discharge.

Samples are collected by hand by the divers, who also record descriptions of the area surveyed. There is increasing use of color photography — still and motion — to provide a record.

3.3.11. *Hydrodynamic Factors. — Currents.* Continuous current measurements throughout the receiving water mass will provide the best description of the current structure and circulation pattern, but sufficient resources are generally not available for such an extensive sampling program. Usually a reasonable assessment of the current patterns can be obtained by taking measurements of current speed and direction near the bottom and top of the water column, and at mid-depth. If the number of current metering locations is further limited because of time or other constraints, single-depth current measurements at a depth representative of the currents responsible for horizontal movement away from the diffuser following the initial dilution process will be most useful.

Meters can be classed either as moored or nonmoored, and can be direct reading or recording. Moored recording meters provide an almost continuous record of current velocity but are restricted because each individual meter is fixed in both the vertical and horizontal plane. Meters operated from a vessel, on the other hand, provide measurements from a variety of depths and locations but present only an instantaneous sampling. Coastal currents are affected by lunar tides, major open oceanic currents, wind stress and tributary freshwater discharges, which are all variable with respect to time. Measurements of current speed and direction should be made at given depths and locations for at least 1—3 h, and measurements over a complete tidal cycle should be made at selected stations and depths.

In addition to the standard propeller or rotor current meters, currents are also measured with drogues and with drift cards. Current drogues are devices suspended below the water surface and supported by a float. The drogue has large vanes or a parachute so that the water current will carry the drogue with it. A small flag or marker rises above the water so that the drogue can be followed and its position can be plotted at intervals. Thus, the drogue moves with the water, and the current velocity can be estimated by how far and in what direction if moves in a period of time.

Drift cards are used to estimate the current velocity in the onshore direction. Drift cards, or drifters as they sometimes are called, are small devices usually made of plastic and weighted so that they just barely float (surface drifters) or just barely sink (bottom drifters). Large numbers of drift cards are released at one time from a given point — usually over the project outfall — and then adjacent beaches are patrolled to determine how many cards reach the beach in what period of time.

All three of these types of current measurements are useful in outfall selection and design.

Wave characteristics. Observations of major wave characteristics are important in planning the construction of an outfall. The character of the sea surface caused by action of the local wind can be described in terms of the height, period, length and direction of the waves. Wind waves normally travel in a direction within about 20° of the local wind. The dimensions of the waves are determined by the strength and direction of the wind and the fetch or distance of the sea surface over which the wind has been blowing. The length (distance from crest to crest) of wind waves is usually 12 to 35 times the wave height.

Swell is a wave system that has moved out of the generating area into a region of weaker or opposing winds or a calm. Swell usually has a greater wave length than that of wind waves. The height to length ratio for swell normally ranges from 1/35 to 1/300, and swell normally disturbs the water to a greater depth than is the case with wind waves.

Meteorological data. Expense and available time will often limit the amount of current data that can be obtained. In most cases, data concerning the wind direction and speed throughout the year is available or can be obtained more easily at or near the proposed outfall location. Sometimes a rough correlation can be made between currents and wind direction, but often the picture is complicated by the periodic tidal currents, the hydrography and the shoreline topography. These factors make analysis and prediction of current speed and direction very difficult.

3.3.12. *Sediment characteristics.* The major reasons for being concerned with the sediment characteristics are: (1) data is needed for the physical foundation design of the outfall, and (2) the benthic fauna (invertebrates and demersal fish) depend in large measure on the physical and chemical sediment characteristics.

Design considerations require information on the nature of the bottom (sand, silt, rock, etc.) and its stability. Information on these characteristics is obtained by dredging, core sampling, sieve or settling velocity analysis for particle size distribution, probing (with and without jetting), echo sounding, etc. Measurement of bottom stability is more difficult in that it requires fairly accurate fathometer measurements taken at various times to estimate changes in bottom configuration.

Chemical analyses commonly made of the sediments include total sulfide, total nitrogen, total carbon, organic carbon, sediment BOD and hexane extractable material. Of these, the sediment BOD test is probably most useful as providing a measure of the organic material in the sediments. This characteristic is often correlated with the physical characteristics and with the other chemical characteristics. Measurements should also be made for chlorinated hydrocarbons and metals. The amount of effort expended on these analyses must be determined after finding out whether or not they are present and in what concentrations.

4. DATA EVALUATION

4.1. General considerations

As discussed above, all of the data collected should be evaluated promptly to detect anomalies and/or to identify redundant data. All of the raw or unprocessed data should be stored on punched cards in a format such that it can be retrieved and presented in a number of different forms. For example, it may be desirable to sort and present data in terms of date or season, by location, by depth, by time of collection (day or night), by tidal stage, etc.

Full use should be made of theoretical considerations in analyzing the data. For example, water temperature is influenced by insolation, which (except for variation due to cloud cover) varies sinusoidally. Thus, it is useful to attempt to relate water temperature to a sinusoidal function of time with a one-year period. If the fit is good, the relationship can be used to "extend" the data. If there are significant variations from an ideal curve, the variations may provide insight into the various mechanisms of concern.

4.2. Data analyses

The data acquired during a survey program should be analyzed and evaluated to provide the following types of information or estimates:

probabilities of pollutant transport in various directions;

time of travel of wastewater constituents to various shoreline points;

expected shoreline concentrations of various wastewater constituents;

plume rise height;

initial dilution of the wastewater;

horizontal dispersion of the wastewater plume;

disappearance or degradation rates for nonconservative constituents;

relationship between winds, tides and current velocities;

vertical and horizontal gradients of parameters in waters and sediments;

faunal and/or floral communities, community structure, diversity indices, etc.;

expected effects of wastewater constituents in the receiving water on the aquatic biota.

The above list of analyses (which is certainly not complete) is ordered generally in increasing order of difficulty or uncertainty. There are many available methods for making these estimates, but there is significant disagreement among professional workers as to which methods are best for which purposes. Nevertheless, a review of the many marine outfall survey reports should provide a broad spectrum of ideas that may be applied in a given situation.

REFERENCES

1. Feuerstein, D.L., Predesign Surveys and Monitoring of waste Disposal Systems. *Proceedings, Second International Congress on Marine Pollution and Marine Waste Disposal*, Politecnico di Milano Istituto di Ingegnaria Sanitaria, December 1973.

2. Strickland, J.D.H. and Parsons, T.R., *A Practical Handbook of Seawater Analysis*, Fisheries Research Board of Canada, Ottawa, 1968.

3.   Engineering-Science, *Determination and Removal of Floatable Material from Wastewater*, Oakland, 1965.

4.   *The Significance and Control of Wastewater Floatables in Coastal Waters*, Interim Report, Berkeley: Sanitary Engineering Research Laboratory, University of California, 1971.

5.   Storrs, P.N., Requirements for Design (of Marine Outfalls), *Proceedings, International Seminar on Marine Waste Disposal*, Politechnico di Milano Istituto di Ingegnaria Sanitaria, July 1972.

6.   State of California, *Guidelines for the Preparation of Technical Reports on Waste Discharges to the Ocean and for Monitoring the Effects of Waste Discharges on the Ocean*, State Water Resources Control Board, Sacramento, 1972.

7.   Storrs, P.N., Pearson, E.A. and Selleck, R.E., *Final Report, A Comprehensive Study of San Francisco Bay, Volume II, Biological Sampling and Analytical Methods*, SERL Report No. 65-8, Berkeley: Sanitary Engineering Research Laboratory, University of California, July 1966.

INSTRUMENTS FOR MEASURING POLLUTION IN THE SEA

Willard N. Bascom

Southern California Coastal Water Research Project,
El Segundo, California, U.S.A.

Summary — Many kinds of instruments and sampling devices are needed to determine whether areas of the sea are polluted and to quantify the various pollutants. This paper describes 15 kinds of devices that are believed to be the most effective of those available. They make it possible to measure nearly all the characteristics of the water, the bottom, and the sea life. However, one must appreciate that these instruments will only obtain reliable data if they are operated carefully by experienced technicians. Their use should be part of a well thought out plan of investigation and analysis.

1. BACKGROUND

In order to determine whether the sea is polluted it is necessary to make thoughtful measurements of its physical, chemical and biological status. Instruments are required that can quantify the motion, the temperature, the clarity and certain chemical properties of the sea. Then samples of the water, the bottom, and the biota must be carefully collected and taken to laboratories ashore for analysis and identification. Eventually the data on areas unchanged by man are compared with those where pollution is suspected.

Pollution means that there is an excess of one or more materials that tend to degrade the quality of life in the sea or decrease its usefulness to man. Thus, the scientific investigator must know the original quality of the sea, the amount of excess material, and the threshold concentration at which that material may be damaging to sea life or to man.

This paper describes the author's selection of the most useful and cost-effective instruments for measuring, sampling, and making observations at sea. Laboratory instruments as well as those used mainly for ship operations or other purposes (such as navigational or meteorological instruments) are outside the scope of this paper. Instruments for making the following measurements are discussed: physical characteristics of the sea, water chemistry, soft bottom sampling, fish populations, direct inspection of the underwater world.

The accompanying photographs and drawings show some of the more recently developed equipment or the best of a large family of instruments. Other items described are known to most investigators or can be seen in instrument catalogs.

2. DESCRIPTION OF INSTRUMENTS OR SAMPLERS

The instruments described here often fall into three general classes, as follows: those for making measurements in the water (current meters, bathythermograph, turbidity meter, floatable sampler, sediment collector, multiple measurements, microbe samplers, and mussel buoy); those for sampling or studying sea life (television and 35 mm cameras, trawls, baited cine camera); and those for examining the bottom sediments (grab samplers, interstitial water sampler, corer, and box corer).

In each case the objective of the measurement, the use of the instrument, the design logic and characteristics, and some comments on alternatives or problems is presented.

2.1. Current meters

The objective of this instrument is to measure the direction and velocity of water movement in depths to 200 m to an accuracy of 3 cm/s.

Current meters are usually placed on stands on the bottom or attached to taut-wire moorings at mid-depth. They must be operated for several weeks if they are to return useful data; therefore they are usually self-powered and internal recording. Since these instruments are subject to theft or accidental loss they must be installed so as to make it difficult for them to be fouled by passing boats or removed by unauthorized persons.

Because of the high variability of currents in nearshore waters there is little value in making very precise measurements. Moreover, since the chance of loosing the instrument is substantial, it is most efficient to buy the least expensive meter that will meet the requirements and to install as many as possible.

The Coastal Water Research Project uses the General Oceanics Model 3010 film recording data logger. The instrument is a vaned cylinder that floats upward from a universal joint. The drag of water flowing past causes the cylinder to assume an angle that is proportional to the velocity of the current. Inside the cylinder a time-lapse 8 mm movie camera photographs a calendar-watch and a sphere marked with latitude and longitude lines. The sphere is a compass with a weight at the south pole and it floats in a thin film of liquid inside a clear plastic sphere. When the film is processed and projected the longitude gives compass direction and the latitude gives current velocity (in accordance with calibration curves).

Other kinds of current measuring instruments make use of propellers and savonius rotors to obtain velocity and record the measurements on tape or teledeltos paper. Alternatively, electrical signals from the sensor can be transmitted to recorders on shore by a cable or by radio. Their results are as good as that recommended but the costs are considerably greater.

Large slow-moving masses of water are best tracked by means of drogues. These parachute-like devices tow a weight that is suspended at the appropriate depth from a surface float marked with lights and flags. Observers on the shore track water movement by measuring angles to the floats at hourly intervals.

2.2. Bathythermograph

The objective of this instrument is to determine the temperature at all levels throughout the water column to about 0.1°C.

The sensor is lowered from a ship to the bottom and immediately retrieved. Response time for both pressure and depth is essentially instantaneous.

There are two main versions of the bathythermograph: (1) the electronic variety is usable to 100 m; it records on deck on an $x-y$ recorder and it lowers either a disposable or retrievable sensor. (2) The standard (mechanical) device is self contained and can be run to 100 m; it uses a stylus to scratch a record on a smoked glass slide that automatically plots temperature against depth. The former gives a record that is easily-readable on deck; the latter must be examined in a "slide reader".

Generally the information required is the surface temperature, the bottom temperature, and the depth (or existence of) the thermocline. It is good practice to make a bathythermograph record every time biological samples are taken or trawls made.

2.3. Turbidity meter

The objective of this instrument is to determine the amount of light transmitted through seawater, which is a measure of the amount of suspended solids and opaque solutions present.

The instrument is lowered from a stationary ship by means of an electrical conductor cable so that light transmittance can be read out and recorded on deck. It is useful for obtaining the size, shape and density of a wastewater plume or turbidity due to other causes.

The turbidity meter measures the amount of light transmitted through the water over a folded (reflected) 1-m-long pathway. The light is self generated and, since the ambient light level does not effect the reading, the instrument can be used day or night. The Martek *in situ* transmissiometer weighs about 12 kg in air and comes with 200 m of cable.

2.4. Floatable sampler

The objective of this device is to systematically collect the particulate and greasy material up to 10 mm diameter floating in the upper 10 cm of sea surface.

Fig. 1.

This small catamaran (Fig. 1) is anchored in the area to be sampled with a line scope of 5 to 1. Thus it always points into the wind. At hourly intervals a timer starts the rotor which turns for 5 min. This propels the craft into the wind during which time its rotor takes a sample of an area of surface about 1 m wide and 100 m long. A helical screen inside the rotor collects the solids and a polyurethane core absorbs the oils and greases. After a day's sampling the screen is removed and the floatables washed free; oil and grease on the core is dissolved off in hexane.

This type of sampler was built to replace and make automatic a previous system that consisted of towing a plankton net supported by surface floats from a beam extending outward from the bow of a small boat. By sampling at arbitrary times in one location the catamaran device obtains a more representative sample of the floatables on the sea surface.

The obvious problem with this device is that it is attractive and is therefore subject to theft. This means it is only installed where it can be watched. It is not intended to be used for measuring oil from large oil spills.

2.5. Sediment collector

The objective of this device is to capture suspended sediments and settlables in the water for comparison with bottom composition and to determine the origin of ocean sludge.

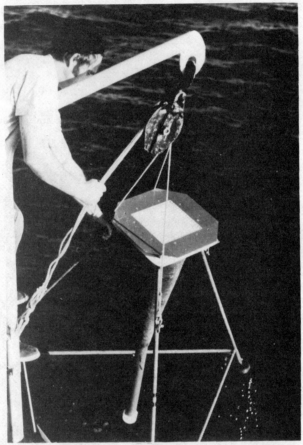

Fig. 2.

A series of these collectors are placed on the sea floor at locations that are appropriate to measuring fallout of particulate material from an outfall. After about 2 weeks on the bottom the samplers are recovered, the material removed and the collector is reset in a new location. Grab samples are taken of the bottom sediment at each collector station.

The collectors described here (Fig. 2) were modified by our Project from a deep water version developed by A. Soutar. They consist of a steep cone opening upward, held vertically by a tripod that rests on the bottom. The upper surface is a flat plate, intended to reduce turbulence, that has a crate-like opening into the cone about 30 cm square. Below the cone is a removable length of pipe or a container into which the sample falls. In 2 weeks this device often collects 5—25 g of material (dry weight) containing 12—50% volatile solids.

Much of the material collected is plankton and other natural ocean materials. Apparent sedimentation rates obtained from the collector can be as much as 50 times those determined by measuring the sedimentation rate in the bottom materials. Apparently when the natural particulates are combined with discharged materials both settle more rapidly.

2.6. Multiple measurements

The objectives of making several measurements at the same time are to (1) make certain that all data are obtained at exactly the same place and (2) increase efficiency by saving ship and technician time.

Several companies manufacture instruments in which various combinations of sensors are lowered on a cable from shipboard. TDC sensor-recorder systems for measuring temperature, depth, and conductivity (or salinity) *in situ* and displaying the data on deck are the simplest version. The Martek Mark III Water Quality Monitoring System simultaneously measures those three parameters plus dissolved oxygen and pH. Inter Ocean Systems Inc. makes *In Situ* Monitor Systems that measure many more characteristics. For example, their Model 513CSTD measures conductivity, salinity, temperature, depth, dissolved oxygen, pH, turbidity, total organic carbon, sound velocity, and various specific ions.

These kinds of devices are subject to all the usual problems of electronics at sea but if they are cared for and calibrated frequently they will give good results.

2.7. Microbe sampler

The purpose of this sampler is to obtain samples of sea water from any desired depth for microbe analysis with absolute assurance that there is no contamination from any other source or depth.

Several methods of collection have been devised. These include sterile glass jars full of air that are abruptly opened at the desired depth or ordinary sea water collecting bottles (Nansen or Niskin) filled with distilled water that open to exchange water and then close again. Our Project's microbiologists have found that the most satisfactory device is the Niskin sterile bag bacteriological sampler made by General Oceanics.

In operation sterile polyethylene bags with a capacity of 5 l are secured to a pair of flat vanes that are in contact with each other. When the sampler has been lowered to the proper depth a weighted messenger is slid down the wire. On contact it triggers a spring that opens the two vanes sucking water into the plastic bag and then sealing its opening. Once on deck the bag is removed and the water treated or stored. These convenient containers are light, unbreakable, easy to pack and are only used once.

2.8. Mussel buoy

The objective of this device is to directly determine the amounts of pollutants taken up by filter-feeding animals at various levels in the water column.

Mussels are known to be excellent concentrators of pollutant material. This device takes advantage of that fact by placing mussels where a sample of concentrated pollutants is required. A taut wire buoy is planted at the location where measurements are to be made (Fig. 3). A substantial number of native mussels obtained from unpolluted areas are put in loosely-woven sacks and suspended from the buoy at appropriate depth intervals. Each week after installation the sacks are temporarily retrieved and ten mussels recovered from each sack. Those from each level are dissected (gut and flesh); each part is homogenized and analyzed for chlorinated hydrocarbons, heavy metals, or microbes.

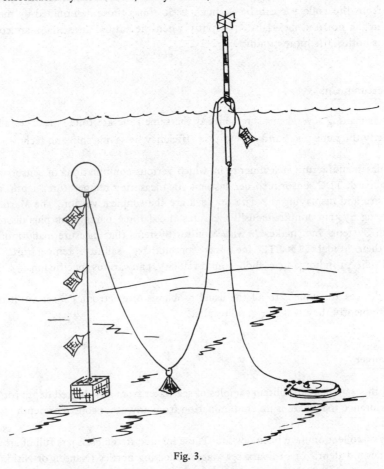

Fig. 3.

This collecting method makes use of animals as a filtering device combined with well known buoy techniques. The main marker is a large striped spar buoy slackly attached by a 1 cm diameter wire to a 200 kg clump anchor. The lines supporting the mussel sacks in the surface water are suspended from the buoy. An adjacent taut-moored buoy, entirely underwater, supports the deeper sacks; it is loosely connected to the main buoy by means of a weighted line so it can easily be found and retrieved.

The results obtained can be plotted as a family of smooth curves that have shown (for example) that over a period of time the pollution levels increase with depth and are much greater below the thermocline. Eventually the mussel's capacity is reached and a steady state condition exists. One problem is that the mussels may spawn and greatly decrease the pollutant level if the temperature changes abruptly; however, the level of their pollutant uptake soon returns to normal. In a similar but reverse experiment where the mussels were moved from polluted to clean waters, the half-life of the remaining pollutants was 14 days.

2.9. Television and 35 mm cameras

The objective of using television is to make it possible for the researcher to visualize the undersea world and thus understand it better. With a television camera on a bottom sled or other mount the watchers on shipboard can see the sizes and quantities of particulates in the water, the current direction and velocity, the composition and roughness of the bottom, the kinds of sea life, and the condition of outfall pipes.

Our Project has developed a light and camera package that features a light (left side of Fig. 4) that is peaked at 535 nm in the green sea water "window" and a television camera (below) that is most sensitive

Fig. 4.

to that part of the spectrum. This combination gives maximum light penetration with minimum power. The camera has an 8 mm focal length, automatic light control, and can be focused from the surface. It is connected to the ship by a cable 150 m long and observed on a monitor with a definition of 600 lines. All observations are recorded on video tape for replay later. An automatic take-up 35 mm camera (center of Fig. 4) and white strobe light (right) can be fired at will by the scientist watching the monitor. Thus, the television monitor becomes a view finder for the 35 mm camera. The latter has magazines of 36 or 250 exposures. The advantage of the still camera pictures over the video tape is better definition, color, and a more convenient record.

Researchers who are used to working blindly with remote instruments and samplers may have their ideas of bottom conditions changed abruptly by their first observations with television.

2.10. Trawls

A trawl is a bag-shaped fish net that is dragged along the sea bottom. When used for scientific purposes its objective is to sample demersal fish and large benthic invertebrates that live on the surface of the bottom.

Pollution investigators need to know something about the varieties, quantities and health of these bottom-dwelling animals. The common method of sampling is to tow a trawl of specified characteristics along the bottom for a specific distance. Otter trawls are commonly used (Fig. 5) in which the net mouth is kept open and the net is held against the bottom by hydrodynamic pressure against otter boards. The otter boards are made of wood and they slide along the bottom on steel "shoes"; this combination of densities assures that the boards remain on edge. The main part of the net is generally a large mesh that tapers to a small "cod end" lined with small mesh. The footrope at the net mouth is held against the bottom by a series of weights (often a chain) and the upper headline is held up by a series of small incompressible net floats.

Fig. 5.

We have examined a number of varieties named after the various net-makers and believe that the preferred characteristics and dimensions are as follows: Nylon net webbing with a mesh size in the main panels of about 4 cm; the cod end webbing is slightly smaller and cod end liner mesh openings are about 1.5 cm. Headrope is 8 m long, footrope is 10 m long (so headrope moves ahead of footrope) and leg lines that connect net to boards are about 1.2 m. The trawl boards are 1 m long, 0.5 m high and have a shoe width of 5 cm.

In operation these trawls are towed by a two-line polypropylene bridle 25 m long attached to the trawling wire by a swivel. In water depths to 100 m the net is towed with a line scope of about 3:1 at a velocity of 2.5 knots (1.3 m/s) for a duration of 10 min on the bottom. This covers an area about 5 m wide (the actual opening of the net when it is underway) times the distance traveled (780 m). Upon retrieval the cod end is opened and marine biologists aboard identify, measure and count the various creatures, being particularly watchful for any signs of disease or deformity.

2.11. Baited cine camera

Trawl nets do not always give a complete picture of the near-bottom fishes and large invertebrates. Some of the fish are able to swim out of the trawls and often there are notable species changes from day to night. In rocky areas it is not possible to trawl. Therefore the purpose of the baited cine camera is to obtain additional information on the fauna of the area being studied.

The device described here is an automatic camera developed by John Isaacs and Richard Shutts (Fig. 6). It operates as follows: The camera, with its self-contained battery power, lights and bait is lowered to the bottom and buoyed. At regular intervals (usually 10 min) the lights go on and the camera takes color motion pictures of water around the bait for 15 s. This equipment is left on the bottom for 24 h, during which time all the film is exposed.

The camera is a 16 mm Bell and Howell with 130 m film spools; film is Ektachrome 7242. Lens aperture is $f2.8$ and critical focal distance 2.5 m. Three 250 W lamps and the camera are powered by Yardley nickel-cadmium batteries. Various baits including whole local fish, punctured cans of sardines, and baited cages containing fish fragments were used. All seemed to attract fish.

The films obtained were all of high quality and were reviewed by a group of biologists who identified a number of species of fish not previously known to live in the area. Spiny dogfish and sablefish were found to be more abundant than indicated by the trawls, and various kinds of crabs, shrimp and gastropods were seen. The change in species abundance from day to night was evident. In addition, the direction

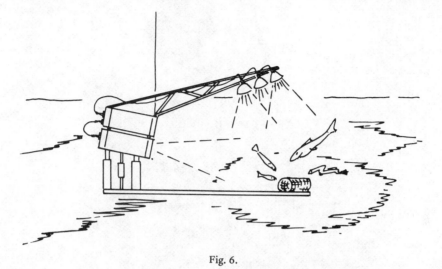

Fig. 6.

and velocity of water motion was made apparent by the movement of great numbers of plankton and particles in the water which was otherwise clear. Visibility was good to the limits of the lighted are (about 5 m).

2.12. Grab sampler

The objective of this device is to obtain a sample of the bottom for biological, physical and chemical analysis with minimal effort. The material retrieved is somewhat disturbed but reliable and repeatable samples can be obtained. Often from three to ten replicates are taken at one station and studied independently to determine the variability of animals in the area.

Our Project has systematically studied the principal kinds of grab samplers including the Van Veen, Smith-McIntyre, Shipek, Petersen, Orange peel, Clam-shell, and others. We were concerned with (1) the ability of each device to take a good sample and retain it (2) the convenience of rigging, retrieving and discharging it (3) reliability for taking many samples without loss or failure and (4) the cost of the device and the equipment required to use it properly.

As a result of our comparisons we believe that a well-built Van Veen type grab optimizes the criteria above. Figure 7 shows it closed upon retrieval from the bottom.

This sampler is held open by a chain that is released when it touches the bottom and tension goes off the lowering cable. At that moment lead weights on the outside of the bucket halves push the cutting edges into the bottom. As the cable is hauled in the extended lever arms force the edges together and hold them there. Screened flap valves in the top of the bucket permit water to flow through on descent but prevents loss of sample on retrieval.

2.13. Interstitial water sampler

The objective of this device is to obtain samples of the water in sediments for analysis by squeezing sediment samples. Levels of heavy metals and H_2S in the interstitial water are believed to be the most biologically available fraction of the pollutants.

Fig. 7.

Samples of ocean sludge, dark muds and other sediments believed to be polluted are taken with a grab sampler. The sediments are quickly transferred to the pressure chamber of the squeezer and capped with a piston. When pressures of about 100 kg/cm^2 are applied, the water is forced out through a series of millipore filters and tubing into closed sample jars.

The essential features of this device are a cylindrical polyvinylchloride pressure chamber 15 × 20 cm and a rugged steel framework that holds both the chamber and a hydraulic cylinder that is pressurized by a manually operated pump. A PVC piston sealed with "o" rings transmits pressure to the sediments in the cylinder after trapped air is bled out though a small valve in the piston.

This device was built by the Coastal Water Research Project to aid in the investigation of sediment quality, particularly to determine whether benthic infauna are exposed to high levels of pollutants. Generally, we have found lower levels of metals in this water than expected, presumably because they are firmly attached to particulates. The size of the chamber derives from the quantity of water required for metals analysis. For H_2S alone a much smaller chamber gives equally satisfactory results.

2.14. Corer (geological)

The objective of this corer is to obtain samples of the bottom sediment for chemical and physical analysis.

Corers have been standard tools of the oceanographer for many years; there are several kinds with dozens of variations, each intended to serve specific needs. Basically the corer is a pipe that is pushed vertically into the bottom to obtain a sample that may be from 1 to 10 m long. For the purpose of making surveys of recent sedimentary deposits in shallow water related to sewage discharges we have a version that is effective and reliable.

Our tool consists of a tube 8 cm in diameter and about 1 m long, plus another ½ m of weights and valve (Fig. 8). The tube is lined with a removable hard plastic liner held in place by a detachable tip with a

Fig. 8.

cutting edge. The overall weight of the tool is about 50 kg, the bulk of which is made up of lead shot in a chamber at the top. The essential feature of this device is a rubber stopper at the upper end that slides freely on a stainless steel rod. As the corer penetrates the bottom, flowing water raises the stopper but when the tool is retracted it drops to form a water-tight seal that holds the cored material in the tube.

It is good practice to lower this corer until it is about 2 m above the bottom and hold it there for a short while before allowing it to free fall into the bottom. We have rarely found it necessary to use a core catcher (only in sand or very soft muds). Generally within the first meter of sediment the tip encounters clay which seals the end nicely and holds the sediments above it. However, simple spring-metal fingers or a flap-valve that fits the inside of the pipe can be used as a core catching seal — with a loss of the lower 15 cm of core. Once the tool is retracted the plastic liner is removed, the ends are plugged and marked, and the sample location is written on the outside. If the objective is to determine the thickness of dark sediment or the depth to some strata this can usually be seen through the clear liner.

2.15. Box corer (biological)

The objective of this tool is to obtain samples of the upper 20 cm of muddy sea floor so that undisturbed biological communities can be examined. On occasion, chemists interested in details of the change in depth of certain pollutants take subcores of the box core samples to determine changes in the chemical content of sediments with age.

This device is intended to take a slow deliberate sample of an area of sea floor without blowing away any very light flocculates or animals that may be present on the surface (Fig. 9).

Fig. 9.

The box of this corer is made of thin stainless steel about 20 cm square and 1 m long, open at top and bottom. Approximately 100 kg of lead are attached to the upper part of the box which is mounted on a frame that holds it vertically just above the bottom. After the frame has touched down on the sea floor the box slowly settles the final few centimeters into the sediment. As the cable from the ship is hauled in it first acts on a lever system that swings a blade-like plate down through the sediments to seal the bottom of the box. After the box is closed the device lifts from the bottom and brings up an essentially undisturbed section of soft bottom. This broad shallow sample that extends below the depth that benthic animals live is preferred by biologists. The advantages of the deliberate box corer are evident; the disadvantages are that it is heavy, bulky, and requires a more substantial winch and handling gear than other sampling tools.

3.   CONCLUSIONS

The instruments described above are the most effective available to the pollution investigator but the mere possession of such equipment does not insure useful results. Good operational technique is of equal importance. Even greater significance in the overall investigation attaches to the plan of the experiment, the selection of measurement points, and the analysis of the data. Without a carefully thought out plan the best instruments will be of little value.

PLANNING AND EXECUTION OF COMPREHENSIVE ENVIRONMENTAL POLLUTION CONTROL PROJECTS

Alexander Gilad

World Health Organization, Regional Office for Europe

Summary — Environmental pollution is a complex phenomenon, having multiple causes and effects. Its prevention and control require a broad range of scientific, technical and economic tools and depends on active collaboration of professionals from many disciplines.

In this paper the contents, range and scope of environmental pollution abatement and control projects are discussed and the various components of these projects are defined. These include design of environmental information systems, manpower development, transfer and application of technology, planning and decision-making, private and governmental commitment, public support and feed-back and adjustment systems.

To be effective, comprehensive environmental pollution control projects must deal to a certain degree with each of those subjects at local, regional, national and/or international levels.

1. INTRODUCTION

The effective pollution control will ultimately depend on modification, reduction and dispersion of the wastes discharged into the sea. It is widely recognized that land-based sources are among the most important contributors to the pollution of the sea. It will, therefore, be necessary to institute and execute a series of local and regional pollution abatement projects. These will cover major population and industrial centres and will aim at selective elimination or reduction of the various categories of wastes and at the proper discharge and dispersion of the remainder.

The purpose of this paper is to discuss methodology of design of such projects and to describe methods of their execution.

2. DEFINITION

Identification and definition of the components of an environmental pollution control process will clarify subsequent discussion of planning methodology.

Figure 1 shows the relationship between various factors and activities involved in the generation of pollution and its control. All human activity, whether production or consumption, results in generation of residuals. Some of these residuals can be readily re-used or re-cycled back into the production process.

Note. The views expressed in this paper are those of the author and do not necessarily reflect the views or the policy of the Organization.

This is shown by arrows Nos. 1 to 3, Fig. 1. Those residuals which cannot be readily re-used or re-cycled become wastes. The wastes may be discharged into the environment, either in their original form or after treatment (arrows 7, 8 and 9). Our environment processes certain capacity for harmless assimilation of some of the wastes discharged into it (arrow 10). Those wastes which cannot be harmlessly assimilated into the environment result in pollution (arrow 11). Environmental pollution has certain detrimental effects. Some of these are known, others are suspected. Both the known and the suspected effects of pollution evoke public concern as shown by arrows 12 and 13 of Fig. 1. When public concern becomes

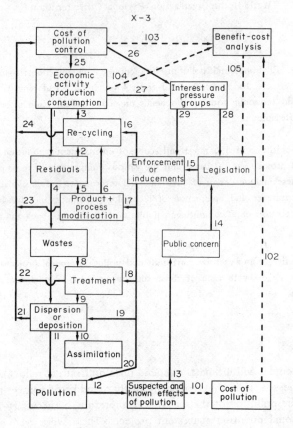

Fig. 1. Pollution control model.

strong enough, it usually results in environmental legislation (arrow 14). This legislation is enforced to a greater or lesser degree (arrow 15) and it influences the various elements of the pollution generation process. Some legislation is aimed at facilitating or encouraging re-cycling of residuals before they become wastes. This may be done by differential taxation, by introduction of charges, by subsidies or by other inducement for re-cycling.

It is frequently possible to modify the production processes, or even the products, to reduce the amount of wastes generated, or to change their composition so that they can be more easily re-cycled into the productive process (arrows 16 and 17). Other types of environmental legislation may prescribe mandatory treatment, or provide inducements for treatment, either through imposition of selective charges or granting of subsidies (arrow 18).

It must be borne in mind, however, that treatment does not reduce the amount of wastes but only changes their composition or physical state to facilitate their dispersion or assimilation. Some types of environmental legislation attempt to influence the dispersion process by requiring high stacks for dispersion of gaseous emissions or outfalls and diffusers for dispersion of liquid wastes (arrow 19).

Other types of environmental legislation attempt to regulate the permitted degree of pollution, leaving to the polluters or the enforcement agencies the choice of methods by which the prescribed environmental standards are to be achieved (arrow 20).

With the possible exception of re-cycling, all other pollution abatement methods cost money. The accumulated cost of all measures aimed at reduction of pollution, as shown by arrows 21—24 of Fig. 1, represents the total cost of pollution control. These costs, when charged either directly or indirectly to the products, are bound to effect the composition and mix of the economic activity (arrow 25). This, in turn, results in generation of interest and pressure groups whose purpose is to minimize the costs of pollution control and their possible detrimental influence on economic activity (arrow 26).

The interest groups exert pressure both on the legislative bodies and the enforcement agencies to avoid or to modify the legislation and the minimize its enforcement.

It is worth noting that the system is kept in equilibrium by the conflicting pressures of the public on the one hand, and the special interest groups on the other, thus depending essentially on the political process for resolution of conflicts and maintenance of equilibrium.

Adequate data for rational decision-making are not available. The present and future effects of environmental pollution are not precisely known and therefore the cost of pollution is very difficult to estimate. On the other hand, it is also extremely difficult to forecast with any degree of accuracy the effects of pollution control costs on the economy. For these reasons it is difficult to perform valid and dependable benefit-cost analyses and to arrive at rational decisions, as shown by the dotted arrows 101, 102, 103, 104 and 105 of Fig. 1.

3. PROJECT SCOPE

3.1. Environmental sectors

Narrowly conceived environmental pollution control projects are not likely to produce optimal solutions. Because the various causes and effects of pollution and its control are intimately inter-related, a comprehensive pollution control project must deal to some extent with all of them, taking into consideration the influence of each of these elements on the entire system. It is a multi-disciplinary activity involving close collaboration of scientists, engineers, public health professionals, lawyers, economists and other specialists.

The causes, effects and methods of abatement of air and water pollution, solid wastes disposal problems and other sectorial environmental activities are mutually interdependent. Many examples of interdependence of one environmental sector on others can be quoted. Certain solutions of the water pollution problem may aggravate the air pollution and the solid wastes problems in the same locality. Treatment of liquid wastes results in generation of large quantities of solids. These solids, if burned, may result in considerable air pollution. Disposal of solid wastes on land may contaminate drinking water supply, and incineration of garbage may contaminate the atmosphere. For these reasons, rational project design demands that human environment should be treated as a system and not as a conglomerate of separate and independent sectors. Total body burden and the cumulative effect of environmental insults on man and on the quality of life are the determinants of pollution abatement action.

3.2. The geographical area

The second important consideration in determination of the scope of pollution control activity is the area that is to be covered by the project. Local projects covering very limited areas circumscribed by administrative or political jurisdictional boundaries are not likely to produce optimal results. A solution which may be advantageous for a city may prove to be disastrous for a metropolitan area of which this city forms a part, because the effects of pollution are often external to the area in which the pollution is generated. To minimize the effect of these externalities on the planning process the environmental pollution control plans should cover the entire area within which emissions discharged at any point have significant environmental effects in other locations.

3.3. The planning horizon

Another parameter of project scope is the planning horizon. Protective and remedial measures which are effective and economical for the immediate future may become counter-productive in the long run. On the other hand, our capacity to forecast the conditions and to develop appropriate plans diminishes rapidly with the length of the planning horizon.

Figure 2 contains a three-dimensional matrix showing the spectra of environmental sectors, geographical areas and planning horizons that form the parameters of project scope.

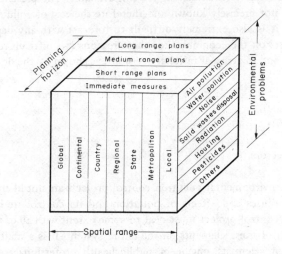

Fig. 2. Contents, range and scope of environmental pollution control plans.

Ideally, environmental pollution control projects should develop long-term plans, cover very large geographical areas and include all sectors of the environment. It is obvious, however, that such plans would require excessive amounts of resources and time and would be extremely difficult to execute, both from the conceptual and practical points of view. Within definite constraints of resources and time, the short-term impact of the projects tends to vary inversely with their scope.

One possible way of solving this difficulty is to institute a hierarchy of projects. Comprehensive, long-term projects country-wide, continental or even global in scope would aim at monitoring of general trends and establishment of broad guidelines and criteria. At the same time, regional or local projects would develop short and medium range plans aimed at solution of the local environmental problems and at the control of local sources, within the guidelines developed by the broad global projects.

Thus, a hierarchy of projects dealing basically with the same problem but ranging in scope, depth and detail is being established, and this multi-level approach assures effective and prompt action at the local level, providing at the same time, a broad framework for coordination and assuring consistency of approach.

4. PROJECT COMPONENTS

Figure 3 shows the main components of a typical environmental pollution abatement activity. These can be classified under the headings of information, manpower, plans and programmes, technology, public support and governmental commitment and action.

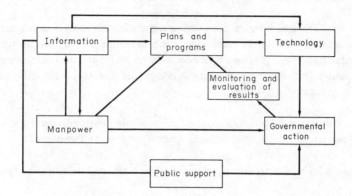

Fig. 3. Environmental pollution control activity (project) components.

4.1. Information

The first and foremost requirement for any rational activity is information, which serves to identify the problem and to define its magnitude and importance, and permits development and evaluation of policies and programmes for corrective actions. Although collection of environmental data is one of the most popular activities among environmental professionals, much of this effort is uncoordinated and disjointed and the results are less useful than they could be.

Often data are being collected just because they are collectable or because instrumentation or budgets are available, without any regard being paid to the ultimate purpose for which the information is being generated. Such procedures are not only wasteful in material and human resources but, most important of all, they frequently result in unnecessary delays of corrective actions, since protracted monitoring may serve as an excuse to delay decisions.

The World Health Organization, recognizing the importance of this activity, organized in January 1973, a Symposium on the Design of Environmental Information Systems in Katowice, Poland. Papers presented at this Symposium, edited by Professor Deininger of the University of Michigan have been published in book form [2].

The following information is usually needed for the purpose of formulation of marine water pollution control programmes:

(a) types and concentration of pollutants in the recipient waters;

(b) temporal and spacial distribution of the pollutants;

(c) the fate of the pollutants in the recipient waters, their decomposition, dispersion, absorption in food chains, etc.;

(d) the effects of the pollutants;

(e) sources, quantities and composition of discharges;

(f) forecasts and trends;

(g) dispersion and assimilation characteristic of the recipient waters.

It is necessary to establish and operate monitoring networks and analytical laboratories, and to collect and interpret large amounts of data. To obtain information on the present and future quantities and types of domestic and industrial discharges it is often necessary to perform industrial wastes surveys, to establish forecasts of population growth and industrial development and to obtain information on water consumption, industrial processes and so forth.

It is obvious that, unless the information system is carefully designed, the environmental information activity can take years and consume an inordinate amount of time and resources.

Two basic questions which have to be considered before a rational information system can be designed are:

(a) for what purposes will the information be used?;

(b) who will be the principal consumers of the output of the information system?

Obviously, the type and the form of the information necessary to diagnose the degree of pollution and to establish long-term trends will be different, both in contents and in form, from the information needed to develop and to select alternative abatement strategies, or to design treatment and dispersion facilities.

Considering the time needed for collection and processing of data and the cost, in terms of money and scarce manpower, the operating objective of information systems should be to collect, process and deliver the minimum amount of information that is necessary, rather than the maximum that is possible.

Given several alternative possible courses of action, it is necessary to test the sensitivity of the decision-making process to the information input. The result of such sensitivity analysis permits more rational design of the information system, by defining the type, amount and form of the information required.

As shown in Fig. 4, the information system can be subdivided into three basic components, viz: data collection, data processing and information delivery. Too often, at the outset of the programme, the entire effort is devoted to data collection, producing a mass of raw, unrelated and undigested data, of limited usefullness.

Early attention to the data processing activity is necessary for several reasons. First of all, it permits early evaluation of the data collection system and makes it possible to introduce the changes and adjustments needed to produce a coherent set of data. Secondly, it is inducive to early and parallel development of statistical and analytical skills, without which it is often impossible to make sense out of the collected data. Thirdly, it permits early activation of the information delivery system.

This last point merits special attention. The nature of the data collection activity is such that it is never finished and it is always possible, and frequently necessary, to keep improving the output by increasing the variety, the accuracy and the reliability of the data. It is, therefore, advisable to weigh carefully the risks of publishing imperfect information against the risk of delaying decisions of preventive or corrective actions until better quality data are available.

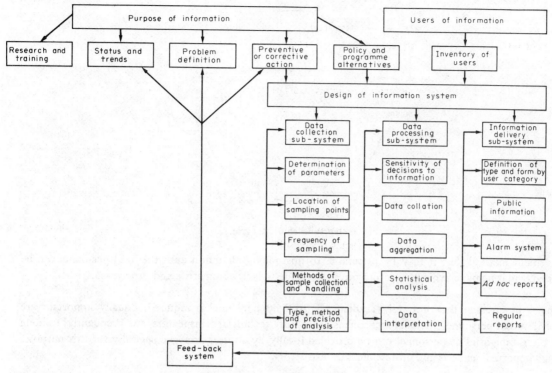

Fig. 4. Environmental information system.

All bureaucratic systems have an insatiable appetite for information and, not unlike other insatiable appetites this one also often results in indigestion. Therefore, as indicated in the last block of Fig. 4, it is necessary to design carefully the information delivery sub-system, to avoid waste, prevent overloading, and assure that the various information needs of the public and the agencies concerned are satisfied to the fullest extent possible.

4.2. Manpower

Meaningful programmes cannot be planned or executed without adequately trained manpower. Lack of scientists, engineers and technicians familiar with environmental pollution problems and having a knowledge of control and abatement methodology is often the greatest obstacle to successful action. Figure 5 illustrates the vicious cycle which develops frequently before the onset of environmental activities and which forms a major obstacle to environmental manpower development. Since no major environmental enhancement programmes are in operation, there are no meaningful career opportunities for environmental professionals, and, consequently, there is very little interest and few candidates for training. The consequent lack of training facilities and programmes results in inadequacy of environmental manpower, without which viable environmental programmes cannot be conceived, planned and put into operation.

A conscious, planned intervention is needed to break out of this vicious cycle to permit orderly development of environmental manpower at all levels. Since training of personnel normally takes a long time, it is important to begin this activity well in advance of the commencement of other project operations.

To break out of the manpower shortage cycle it is sometimes necessary to "import" personnel from other localities or countries for the initial stages of project operation. This is usually a costly, and often an in-

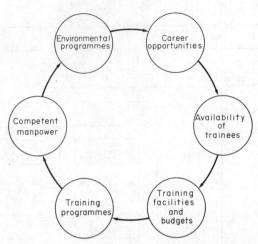

Fig. 5. Manpower shortage cycle.

convenient method, but it may be preferable to postponing all action until the local personnel can be trained, and it helps to avoid commiting grave errors due to lack of expertise and experience.

It must be stressed that not only academically trained personnel is required. Equally important are other levels of personnel like field technicians, laboratory technicians, inspectors, etc. Frequently, training for these categories of personnel can be provided locally by courses arranged specially for this purpose, using "imported" experts and professionals as instructors.

4.3. Plans and programmes

Planning has been defined as a process of identifying and defining the unmet needs and demands that constitute the problem, establishing realistic and feasible objectives, deciding on their priority, identifying the resources needed to achieve the objectives, and projecting action based on a choice of alternative strategies for solving the problem. Development of plans and programmes is undoubtedly the crucial part of project operation.

Various planning models have been proposed, discussed in literature and used by environmental planners. Most of the models using the systems approach contain the following elements.

4.3.1. *Identification of goals.* Goals are normally expressed in general terms, describing the value judgements of decision-makers. More than one major public goal relates to environmental quality. Among the most significant public goals relating to environment are: the health goal, conservation goal, the ecological goal and the economic development goal. Some of these goals are concurrent, while others are mutually conflicting [8].

Given the diversity of goals related to environmental protection, it is necessary to develop a policy which would reconcile the various goals or assign to them priorities. In some countries such environmental policies have been explicitly formulated and are available for environmental planners. In other countries lack of explicit environmental policies makes the planning problem vastly more difficult.

In such cases the planners must develop their own interpretation of public policy and submit this for the review of the decision-makers, before they can proceed with development of plans.

4.3.2. *Definition of objectives.* The term "objective" is used to denote desired ends that can be stated

more specifically than goals in quantitative terms. Objectives so defined can serve as milestones against which progress can be measured. Planners must take steps to assure consistency of operational objectives among various environmental sectors.

4.3.3. *Establishment of planning criteria.* One of the most important planning criteria is definition of the scope of the project. Environmental sectors to be dealt with must be precisely defined, uniform planning horizon must be established, and area to be covered by the project must be agreed upon. This permits establishment of a common data base for all environmental planning activities in the region. The data base includes such parameters as topography, hydrology, climatology, meteorology, demography, economy, etc. Based on this common data base, projections of physical, demographic and economic development can be made, serving as common planning criteria for all environmental sectors [5].

The above steps may be considered as preliminary to the main task of planning which is to devise the most feasible, satisfactory, efficient and economical method for solving the given problem within the established constraints. Several basic planning models are available to assist the planner in this task. Some of the most important of these are the planning, programming and budgeting system, linear programming, benefit-cost analyses and other variations of these. In using these techniques the following conditions must be met [7] :

(a) objectives must be clearly defined in quantitative terms prior to the consideration of the alternative courses of action;

(b) all possible courses of action are considered, independently from their proximity to, or divergence from, existing policies and procedures;

(c) the analysis is quantitative, requiring quantitative determination or estimation of costs and benefits. The benefits are quantified in monetary terms;

(d) the optimal alternative is the one that meets the predefined criteria at the minimum cost, or the one that produces maximum benefits at a predefined cost.

As can be seen from the above, the first requirement for application of these tools in any decision process is a clear definition of goals and objectives. However, given the divergence of public goals related to environment, it is not always possible to arrive at such clear and widely acceptable definitions.

The gaps in our knowledge on the effects of pollution are formidable. Quantitative analysis, which forms an integral part of the optimization techniques, requires quantification of benefits and costs. In the absence of sufficient data, planners and analysts are forced to make assumptions which may seriously jeopardize the validity of the results. Yet, since systems analysis or operations research has come into fashion, planners show a persistent devotion to the optimization methods. By definition, solutions arrived by these methods, should be the best available, and yet they often prove to be unacceptable to the decision-makers. This may be due to several reasons. One of them may be the difficulty in defining the objectives in quantitative terms as mentioned above. Another reason, and perhaps the most important one, is the failure of the optimizing method to rank the feasibility of the various solutions. In this method all alternatives being considered feasible are ranked in the order of their efficiency in achieving the objectives. However, not all feasible solutions are equally feasible and in practical life feasibility may be a more important criterion than economic efficiency.

Another important criterion frequently ignored in the optimizing process is "robustness" which has been defined as a measure of the useful flexibility maintained by decision [9].

Usually plans are implemented by stages spread over a considerable length of time. Considering the con-

stantly changing situations and the uncertainties involved in every planning process, and having in mind the continuously evolving and changing technology, it is obviously preferable to adopt solutions which leave us the greatest amount of residual flexibility at each stage of implementation.

Attempts have been made to assign a monetary value to the residual flexibility, or "robustness", so that it can be included in the optimization process. It seems, however, that such heroic attempts to quantify in monetary terms all the unquantifiable characteristics of various courses of action tend to obscure the issues and do not add significantly to the validity and acceptability of the resulting plans.

The shortcomings described above do not preclude the use of systems analysis in general, or of the optimization techniques in particular, in environmental planning.

On the contrary, experience has shown that these techniques are very useful at the lower level of analysis, dealing with sub-system optimization.

Given a definite quantity and composition of liquid wastes that are to be discharged into the sea and a number of feasible alternatives of accomplishing this task, consisting of the various points of discharge, and the various combinations of outfall, diffuser and treatment facilities, it is possible to optimize the discharge sub-system to arrive at the most economical method of reaching or preserving a pre-determined recipient water quality.

But, when the liquid wastes problem is posed in a broader context, including re-location of industries, changes of manufacturing processes, major modifications of land-use, etc., the amounts of data needed for optimization become so vast that the analysts are forced to make various estimates and assumptions, often of doubtful validity, thus reducing greatly the value of the output.

When the question is posed still more broadly, in the context of total environmental system, the problem of optimization analysis becomes so complicated that it approaches the impossible [4].

Broad policy decisions are seldom, if ever, arrived at through optimization analysis, but they are usually the result of what has been called the incremental decision process [3].

The incremental planning process assumes that future policies and programmes will evolve from the present ones and will differ from them only incrementally. The basic criteria for choice are feasibility (technical, political, economic) and the degree to which the result is likely to satisfy the various (and sometimes conflicting) objectives. The "best practical means" approach to environmental pollution abatement, as practised in several European countries, is an example of an incremental planning process.

It goes without saying that environmental planning cannot be done in isolation from other governmental agencies operating in the project area. First of all, the existing agencies possess a wealth of information which is extremely useful to the environmental planners. Secondly, there is a need to coordinate, to the fullest extent possible, the environmental plans with the plans of other governmental agencies. Thirdly, since the implementation of the environmental plans will depend to a large extent on the cooperation of the other agencies, there is a need to obtain full collaboration and cooperation as early in the planning process as possible. This may involve establishment of formal inter-jurisdictional coordinating committees as well as development of informal communication and collaboration links with the various departments.

4.3.4. *The output of the planning process.* It may be useful to consider the various types of plans which evolve in an environmental project planning process. In projects supported by the World Health Organization we normally divide the planning output into the following components:

(a) master plan;

(b) preliminary engineering;

(c) feasibility studies;

(d) detailed design and contract documents.

The definition and contents of the various types of plans are given below:

(a) *Master plan.* This is a phased long-range programme covering the Project Area and the period of 25 years during which construction of facilities is proposed to be carried out in stages defined by the Plan. It consists of a number of interrelated studies on technical, economic, organizational, managerial and financial aspects involved in the programme, and includes:

  a complete inventory and appraisal of existing conditions and facilities and of present short-comings and their causes;

  a review of past, evaluation of present, and forecast of future population growth and socioeconomic development with due consideration to existing and proposed economic and physical planning;

  a review of past, appraisal of present, and forecast of future loads and requirements;

  engineering, operational, economic and financial studies and alternate long-range stage-by-stage programmes for meeting these requirements with due consideration to the physical, technical, political and economic aspects involved.

On this basis, the Master Plan constitutes a justified (technical, financial and economic) programme which identifies how present and future loads and requirements should be met, including most suitable successive stages of implementation, through the development and construction of the required facilities at certain well-defined intervals during the planning period.

The Master Plan identifies and defines the first priority stage, covering facilities to be constructed within the first 4—5 years (from the estimated beginning of construction).

(b) *Preliminary engineering studies.* These studies comprise detailed investigations, surveys and technical analysis of alternative plans and schemes for the various elements of the system, which are proposed for a first stage implementation under the Master plan. Such studies are usually carried out for works which are to be constructed within the immediate 4—5 years. The preliminary designs are presented in a report form with comparative analysis, recommendations, justifications, preliminary specifications and cost estimates of the various alternatives, including cost of construction, operation and maintenance. The information presented should enable the preparation of the final engineering designs and contract documents with the minimum of additional field investigations and studies. The purpose of preliminary engineering studies is to provide guidance for selecting the most favourable alternative with respect to layout design criteria and capacity of the system.

(c) *Feasibility studies.* In addition to the preliminary engineering studies described above, these studies include pertinent legal, managerial, economic and financial matters, including a complete financial plan with estimates of capital and recurrent costs and the total income needs to cover the cost of operation and maintenance, interest and amortization of the completed works, based on studies and recommendations of the tariff structure.

5. TECHNOLOGY

5.1. Types of technology needed for pollution control

Every action aimed at reduction and control of pollution depends on facilities, equipment and processes, backed by expertise in their applications, use and maintenance. A broad range of various technologies is involved, depending on the nature of the intervention which is to be applied in each case.

The various technical means needed for pollution abatement fall into the following broad categories.

5.1.1. *Industrial technology.* Modification of certain industrial processes often results in marked quantitative and qualitative changes in the production of concomitant residuals, facilitating their re-use, reducing the amounts of wastes, or changing their characteristics to improve their treatability.

5.1.2. *Wastes treatment technology.* The purpose of wastes treatment is to change their chemical and biological composition and physical state so that they can be discharged into the environment without detrimental effects. A broad range of treatment processes is available for this purpose, depending on the wastes to be treated, the desired characteristics of the treatment plant effluent and many other factors.

5.1.3. *Wastes dispersion technology.* The environment possesses a finite capacity for absorbing and assimilating municipal sewage and many industrial wastes (after treatment, if necessary). To prevent overtaxing this absorptive and assimilative capacity of the environment, the wastes have to be adequately dispersed. Sea outfalls and diffuser systems are classic examples of dispersion facilities.

5.1.4. *Monitoring technology.* Effective environmental enhancement action depends on reliable data derived from monitoring of sources, types, levels and effects of the various pollutants. Monitoring systems and instrumentation vary from the most simple to the most sophisticated and their choice must depend on a variety of local conditions.

5.2. Development, adaption and utilization of technology

Given the great variety of technological requirements of every pollution control project and considering the unequal distribution of technological know-how around the globe, a question often arises whether it is more economical to develop the required technology locally or to import it from elsewhere.

The question is seldom considered on purely rational grounds, since such factors like national and local pride, professional ambitions and personal and group interests often play an important role.

Much too often, a great amount of time, resources and manpower is devoted to local development of

technology which has already been developed and tested elsewhere. Since lack of technological expertise often goes hand-in-hand with scarcity of trained manpower and resources, it is wasteful to devote the limited means to an activity of which the outcome is often doubtful and the product of which could be readily obtained or purchased elsewhere.

It is often argued that local research and development activities, regardless of the success of their outcome, contribute significantly to the training and development of local expertise. This is undoubtedly true, but the question remains whether the training objective could not be achieved by more efficient means than as a by-product of non-viable research activities.

The mixture of foreign and local contributions in transfer and adaptation of technology varies over a wide spectrum.

Import of technology can take many forms, ranging from purchase of complete installations under "turn-key" contracts, to acquisition of rights for local manufacture and use of equipment developed abroad.

"Turn-key" contracts, in spite of their obvious advantages of speed and efficiency, possess some very significant drawbacks in so far as they contribute almost nothing to the development of local technical capability, which is important for the subsequent operation and maintenance of the installations.

An optimum combination of the indigenous input and the imported technology should be aimed at in each case. The imported component should be the minimum that is necessary to assure the satisfactory product. The local contribution should be the maximum consistent with the existing constraints of time, trained personnel and physical facilities, taking into consideration their alternative uses.

While it seems certain that, in the short run, the greatest improvement in environmental quality can be achieved by more intensive application of existing technology, it is equally evident that the long-term solutions of the present and future environmental problems will require marked improvement of the existing technology and development of new technology.

The needed research activities can be broadly classified into the following categories:

- (a) identification and measurement of pollutants;

- (b) effect of pollutants on water quality, on aquatic eco-systems and on man;

- (c) origins of wastes;

- (d) control of wastes at their sources;

- (e) treatment of wastes;

- (f) disposal of wastes into the environment;

- (g) direct and indirect costs of pollution;

- (h) benefits derived from pollution control.

Best possible coordination of the research activities on the regional, national and international scales is needed to avoid unnecessary duplication of effort and to derive maximum benefit from the considerable investment of manpower and resources.

The World Health Organization assisted in establishment and operation of a system of national, regional and global reference centres, which serve as clearing houses for information on research in the various environmental sectors like air pollution, water pollution, solid wastes, etc.

Closer and more intensive collaboration of local and national research institutions with these reference centres would contribute towards increased efficiency of the total effort.

6. PRIVATE AND GOVERNMENTAL COMMITMENT

The preceding components of environmental pollution control project can be regarded as preparatory steps for action. Information, manpower, plans and programmes are meaningful only to the extent that they facilitate commitment of public and private resources and result in remedial actions taken by individuals, firms and governments.

In this context, governmental actions consist of legislation, budgetary allocation, taxation, subsidies, etc. An important element of governmental action is the establishment and maintenance of an enforcement system, without which all the other activities cannot be fully effective.

Recent experience in many countries has shown that a central environmental administrative body is essential for effective functioning of the system. Various countries established different forms of central environmental agencies, ranging from interdepartmental committees with coordinative and consultative functions all the way to full-fledged ministries of environment with policy setting, legislative and executive powers.

It is perhaps too early to evaluate the performance of the various types of environmental administrations, since most of them have been operating for a relatively short time.

In any case, the environmental administrative systems will vary from country to country, depending on local conditions, traditions and governmental systems. It seems clear, however, that in most cases, the functions and responsibilities needed for effective functioning of an environmental administrative system are:

  (a) formulation of environmental policy;

  (b) development of legislation;

  (c) coordination of those activities of other governmental bodies which have an impact on environment;

  (d) assessment of environmental impact of land use planning, major public works and industrial development;

  (e) supervision of enforcement activities (if those are entrusted to other departments);

  (f) responsibility for monitoring (either directly or through other governmental bodies);

  (g) participation in the process of allocation of funds for construction and operation of facilities;

(h) coordination of research and promotion of manpower development.

Time and space limitations do not permit full discussion of the entire range of fiscal measures that can be taken by government to reduce environmental pollution. Since generally the cost of pollution is external to the polluter, the various types of taxes and charges are aimed at internalizing the cost of pollution and providing incentives for preventive and corrective action by firms and individuals.

Individual action to prevent pollution can take many forms like change of consumption patterns, refraining from careless acts causing pollution, favouring products of those firms which show concern for environment, etc.

In response to public opinion, to the market mechanism and to the governmental actions, firms can modify their products to make the wastes more acceptable or suitable for re-use, they can modify their production processes and raw materials to minimize generation of wastes or they may install treatment facilities to reduce the quantity and change of composition of wastes.

Environmental pollution is a complex phenomenon, having various causes and producing many effects. No single activity is likely to make a significant dent in the problem. Well-prepared and energetically pursued action by all concerned is needed to stop the continuing deterioration of environmental quality and to start us on the road towards its improvement.

7. PUBLIC SUPPORT

Significant private or governmental action is unlikely without public support. Therefore, generation of public support is an integral part of any comprehensive pollution control programme.

This can be achieved through a combination of the following means:

(a) Media campaigns in the press, radio and television to produce public awareness of the problem and of the action contemplated, proposed or taken by the various levels of the government to solve it. This involves developing information in a form suitable for the media, making this information readily available to the media and encouraging special programmes and coverage.

(b) Development and distribution of educational material in a form suitable for use in the school systems encourages the teachers to discuss the subjects in class. Organizing excursions to points of interest (like treatment plans) contributes to the understanding of the issues and helps to educate a generation of environmentally oriented citizens.

(c) Assistance in the establishment of environmentally-oriented organizations, clubs and associations which may exercise significant influence on legislative and enforcement actions.

(d) Establishment of close cooperation with universities and professional organizations, helps to assure their support in the public discussions which may precede governmental decisions and allocation of funds.

Many worthwhile environmental programmes failed to materialize because of lack of public interest and support. This, in turn, was caused by reluctance of environmental scientists and professionals to get involved in public information campaigns and other related activities. In the long run, political decision-makers must respond to the will of the electorate and, therefore, public information activities are crucial to the success of environmental pollution control projects.

8. MONITORING, FEEDBACK AND ADJUSTMENT MECHANISMS

Environmental pollution control plans are conceived and developed in dynamic situations, they are based on imperfect information and they involve predictions and forecasts which contain various uncertainties.

Given the unavoidable time lap between planning and execution, it would be unreasonable to expect that the systems will be constructed and operated exactly as planned, that the results of their operation will be precisely as foreseen and that these results will remain optimal under the constantly changing conditions.

It is therefore imperative that the projects should contain a mechanism for modifications and adjustments. This mechanism consists of the following basic elements:

(a) *Monitoring.* The continuous selective monitoring of the significant parameters which formed the basis of the planning and design criteria, assures that the data and their extrapolations are still valid and that the programme addresses itself to the type and magnitude of the problem as it exists during the implementation stage.

The second, equally important purpose of continuous monitoring is to check and verify the results of the intervention, to assure that the system is functioning as planned, or to permit timely discovery of shortcomings and introduction of adjustments.

(b) *Feedback.* Information resulting from the continuous monitoring must be collated, interpreted and supplied to the decision-makers to enable them to draw appropriate conclusions.

(c) *Adjustment.* Adequate flexibility must be built into the system to permit its augmentation in response to new information or changing conditions.

Changes in plans, designs, laws and programmes do not necessarily result from faulty planning. As often as not, these changes are a sign that the system is viable and able to respond to the changing circumstances.

9. CONCLUSION

Time and space limitations do not permit full discussion of all the subjects touched upon in this paper.

It is hoped, however, that this general overview of the environmental pollution control planning process will serve to illustrate the broad spectrum of the issues with which it deals and will assist those who are involved in this important activity to structure their effort and to address themselves adequately to the various elements of the system.

REFERENCES

1. Butrico, F.A., *The Application of Systems Analysis and Other Mathematical Techniques to Water Quality Management.* Technical Publication WHO/W. Poll/69.3, World Health Organization, Geneva (Switzerland), 1969.

2. Deininger, R.A., Design of environmental information systems. *Proceedings of WHO Symposium, Katowice (Poland), 1973*; Ann Arbor Science Publishers, Ann Arbor, Michigan, 1974.

3. Gilad, A., Incremental decision model in environmental health planning. Doctoral Dissertation, University of Pittsburgh (U.S.A.), 1971.

4. Hoss, I.R., *Systems Analysis in Public Policy.* University of California Press, Berkeley, 1972.

5. Hufschmidt, M.M., *Metropolitan Water Resource Planning Model.* University of North Carolina, Chapel Hill, 1971.

6. Laub, J.M., Development of strategies for environmental conservation and pollution control. Proceedings of Third World Congress of Engineers and Architects, Tel-Aviv, 1973.

7. Lindblom, C.E., The science of muddling through. *Pub. Admin. Rev.*, 1959.

8. Myrdal, G., Economics of an improved environment. *Proceedings of Third World Congress of Engineers and Architects,* Tel-Aviv, 1973.

9. Rosenhead, J., Elton, M. and Gupta, S.K., Robustness and optimality as criteria for strategic decisions. *Oper. Res. Q.,* 23, (1971).

10. Schaefer, M., *Administration of Environmental Health Programmes, A Systems View.* Public Health Paper 59; World Health Organization; Geneva, 1974.

11. Toebes, G.H., Natural resource systems models in decision making. *Proceedings of a Water Resources Seminar*; Water Resources Research Center, Purdue University, Lafayette, Indiana, 1969.

ENVIRONMENTAL POLLUTION CONTROL IN ATHENS AREA

Gregory Markantonatos
Sanitary Environmental Protection Division,
Ministry of Social Services, Athens, Greece

Summary — Environmental pollution control in Athens area became a necessity, due to the rapid increase in modern human activity during the last few years.

In order to permit the decision-makers to reach rational decisions, based on reliable data, a joint project has been set up for comprehensive environmental pollution control. The project includes programmes for air and water pollution control, solid wastes management and noise abatement.

The project is succesfully training and developing its own staff and is efficiently transferring the foreign technology through international experts.

The liquid wastes programme, which constitutes a component of the Athens project, deals with the domestic and industrial wastewater disposal into the Saronikos Gulf.

To be effective, the above programme deals not only with the treatment plan, but, based on oceanographic studies, it has to develop the most appropriate combination of treatment and outfall-dispersion systems to handle the problem of liquid wastes disposal, without undue adverse effects on the marine environment.

1. INTRODUCTION

The rapid increase in population, economic activity and living standards in the Athens metropolitan area during the last 15 years has resulted in corresponding increases of the generation of air, water, solid waste and noise pollution.

Realistic projections of growth in population, economic activity and *per capita* income for this area imply significant increases in the rate of pollution loading in the next decade. This would create adverse health effects and might lead to environmental crisis, unless effective preventive and remedial action is started before the pollution reaches the critical stage.

The goal of protecting and enhancing environmental quality in Greece is part of the established policy of the Greek Government. The goal is, in truth, related to several other public goals, notably those of economic welfare, social well-being, public health and tourism promotion.

Taking into consideration the above aspects, the Greek Government, with the technical assistance of the World Health Organization (WHO) and the financial contribution of the United Nations Development Programme (UNDP), has set up a Project for comprehensive environmental pollution control, in the Metropolitan Athens area, including a liquid wastes programme for the effective water pollution control.

2. GENERAL OUTLINE OF THE PROJECT

2.1. Objectives

The main objectives of the Project can be summarized as follows:

(a) development of sectoral programmes for control of air, water and noise pollution and for solid wastes management;

(b) integration of the component sector programmes, for the development of a comprehensive environmental protection strategy;

(c) creation and training of professional and sub-professional manpower as well as development of laboratory facilities and other means for environmental policy;

(d) development of environmental policy, provision of legislation and creation of administrative machinery for enforcement and effective pollution control.

2.2. Duration

The project duration is 4½ years. It commenced in July 1972 under "Preliminary Activities" phase, which lasted 1 year. The project document was signed and full implementation was authorised in September 1973.

2.3. Budget

a.   Greek Government contribution, equivalent to U.S. $ 3,593,300

b.   United Nations (UNDP) contribution U.S. $ 1,021,200

TOTAL U.S. $ 4,614,500

The Greek Government's contribution is provided to cover the costs associated with local personnel, offices, equipment and laboratory facilities, as well as the major contract for the liquid wastes disposal study.

The UNDP contribution is used to finance expert services (49%), training abroad of project staff (16.2%), supply of basic equipment (18.5%) and some contracts (15.7%) for those parts of the job which can be effectively subcontracted to foreign consulting firms.

2.4. Institutional framework

a.   *The tripartite venture.* The Athens pollution control project, a main component of which is the liquid wastes programme, is a joint venture of the Greek Government, the United Nations Development Programme (UNDP) and the World Health Organization (WHO).

The Greek Government is cooperating through the Ministry of Social Services.

The World Health Organization has been designated as the Executing Agency for environmental pollution control project.

During the past 7 years, WHO has executed a large number of such projects, ranging from sectoral projects dealing with one aspect of pollution, e.g. wastes pollution, to comprehensive environmental projects, like the Athens project, dealing with several environmental sectors in an integrated manner.

b.   *Collaboration with existing local institutions.* It has been the project policy from the outset to assist and strengthen those national institutions which are capable and willing to contribute to the pollution control effort.

Notably, the programme has extended assistance to the Government Institute of Oceanographic and Fishing Research, which is conducting oceanographic studies in the Gulf of Saronikos. This assistance consisted of provision of expert services, oceanographic equipment, funds and training for the institute's staff, both by organizing local training courses given by foreign experts, and by sending Greek professional staff abroad for academic and practical training.

The project has also close collaboration with the Athens Sewerage Organization and its supervising Ministry of Public Works, as well as with other Governmental agencies, like the Greek Meteorological Service, which collects meteorological data related to air pollution control and the Association of Municipalities and Communes in the Athens area, which have the responsibility for solid wastes disposal.

c.   *Manpower training and development.* Plans cannot be conceived, evaluated or executed without adequate professional and sub-professional manpower. It is of utmost importance, therefore, to develop professional cadres, who will be able to deal effectively with the full spectrum of the necessary activities, involving monitoring, data analyses, planning, design, legislation, enforcement etc.

The project assists in manpower training and development through:

> Fellowship programme for studies abroad,

> Short courses given locally, with the participation of both foreign lecturers and Greek scientific and technical institutions and

> In-service training augmented by professional contact with foreign experts.

The Greek Government has assigned high priority to this project and has put at its disposal adequate financial, material and human resources.

The project staff consists of sanitary engineers, chemical engineers, chemists, physicists, bacteriologists, biologists, oceanographers, public health doctors, lawyers, economists, business administrators, as well as laboratory and clerical personnel. They are engaged in such activities as air quality monitoring, microbiological and chemical laboratory analyses, compilation of emission inventories, community noise studies, as well as review of and proposals for environmental legislation and evaluation of economic implications of pollution and its abatement. The project epidemiologist conducts selected studies on relation of pollution to health.

The present professional and auxiliary staff on the project payroll exceeds 60 and will probably double during the next year. It is forseen that a part of the project staff will serve as a nucleus for future Greek environmental activities.

The World Health Organization, utilizing experience accumulated on other similar activities, has mobilized international experts on a world-wide basis to advise and assist the programme staff and has enabled the Greek personnel to obtain in-service and academic training and experience at the most suitable laboratories or technical and academic institutions in the world. The results of these cooperative efforts have been most gratifying and give us a sound reason to believe that the programme will attain the main objectives and will further stimulate and multiply similar activities elsewhere.

2.5 Rationale

Environmental pollution control requires committment of considerable resources. It also exerts significant influence on industrial development, tourism, recreation, land use planning and many other important social and economic sectors.

In order to permit the decision-makers to reach rational decisions based on reliable data, the various alternative goals and the corresponding plans of actions must be identified, their feasibility and their costs must be determined, and their probable socio-economic impact must be ascertained. Order of priorities, based on need and feasibility, must be established, to maximize the benefits from the investment of public resources.

The Athens environmental pollution control project is designed to provide the sectoral stage-by-stage programme needed for this purpose.

3.   THE ATHENS LIQUID WASTES PROGRAMME

One of the main components of the aforementioned Athens project is water pollution control. This goal is attained through the proper municipal and industrial wastewater disposal into the Saronikos Gulf.

Since this International Congress focuses attention on the specific subject of "Marine Municipal and Industrial Wastewater Disposal", it was considered worth-while to analyze, as an example, the Athens liquid wastes programme.

3.1. Background information

a.   *Greater Athens area.* The Greater Athens Area comprises mainly the peninsula of Attica, approximately in the centre of Greece. Although its area covers less than 1% of the country, it contains more than a quarter of its population. The heavily urbanized portion of the area comprises about one fourth of the land area. It consists of 56 municipalities and communes, with a population of 2.53 million (1971 census). The largest municipalities in the roughly triangular Athens Basin area are Athens (867,000) and Piraeus (187,500). Projection of future population for the Athens Basin area is 4,000,000 by 1985.

The climate is hot and dry for about 8 months of the year, with average relative humidity of 50% in July and average maximum temperatures in August of 31.1°C.

Annual average rainfall is only about 500 mm, most of which falls during the months of December—March. The area has inadequate water sources for urban development and water is accordingly being transported long distance from river basins to the north and northwest of the area.

b.   *Existing sewerage facilities.* The sewerage system in Athens basin is basically a separate one, except for some central sections with combined systems.

The surface runoff water in the Athens basin is naturally drained towards Phaleron Bay, by means of the two main rivers, the Kifissos and the Ilissos.

The sewage of the town is collected in a trunk sewer, which, through a tunnel of horse-shoe section 3.60×3.60 m^2 and of about 4 km length, discharges at the small promontory Keos near Keratsini, opposite Psitalia island, about 2 km northwest of the entrance to Piraeus Port.

The bad condition at the present outfall point is aggravated by the works under construction for the extention of Piraeus Port, which eventually will enclose this point into the greater port.

c. *Saronikos Gulf.* The Saronikos gulf (see Fig. 1 in Appendix I) is a rather closed sea region bounded on the South and South-west by the Argolis peninsula and on the North and North-east by Attica. This semi-enclosed water mass of 2900 km^2 communicates to the Aegean Sea through a 50 km opening at the South-eastern limits of the gulf.

The mean depth of the gulf is about 140 m, while the deepest point reaches 425 m.

The tidal action in the gulf is very low.

4. THE OBJECTIVE OF THE ATHENS LIQUID WASTES PROGRAMME

The objective of this programme is to develop a rational plan for liquid wastes disposal and marine pollution control in the Metropolitan area of Athens, based on oceanographic studies, which are intended:

> (a) to gather basic reliable data concerning the physical, chemical and biological characteristics of the marine environment, in order to assess the absorptive capacity and the dispersive characteristics of the recipient waters, both in the vicinity of the existing sewage outfall and in the greater Saronikos Gulf ecosystem;
>
> (b) to assess the environmental impact (quantitatively and qualitatively) of the new liquid sewage disposal system on the recipient waters and monitoring after the construction of the treatment and outfall facilities.

Apart from the above oceanographic studies, the programme will include, but not necessarily be limited to, the following items:

> (a) master plan, preliminary engineering and feasibility study for liquid wastes collection and disposal system;
>
> (b) industrial wastes control programme;
>
> (c) programme for re-use of part of the waste water;
>
> (d) design of treatment plant;
>
> (e) design of sea outfall and diffusers.

5. DESCRIPTION OF THE PROGRAMME ITEMS

5.1. Preliminary oceanographic study

This study is intended to provide information necessary for the design of the sewage outfall. The study will include, but not be limited to, the following:

(a) studies concerning the general circulation patterns in the Gulf variations due to seasonal or climatic conditions;

(b) salinity;

(c) temperature;

(d) density;

(e) dispersion studies (bacterial, dye-stuffs, radiotraces etc.);

(f) physical and geological conditions of the sea bed;

(g) wave and swell effects;

(h) The following parameters will be monitored at selected stations of Saronikos Gulf once per month: the phosphorus compounds, the nitrogen compounds, transparency, dissolved oxygen, phytoplankton and chlorophyll(s).

The above studies will be performed for at least 1 year.

5.2. Environmental impact survey

A study of the environmental impact of the bulk of the sewage discharged into the receiving waters will follow, for the establishment of base-line data, so that the effect of future development can be quantitatively and qualitatively assessed and the nedd for additional work determined. The study shall, therefore, include a marine ecosystem survey, during which the above mentioned parameters will be studied at a reduced frequency, along with the following:

(a) sediment studies, including compounds of carbon, sulphur, metals and selected organic micro-pollutants;

(b) marine biology, including general quantitative description;

(c) quantitative monitoring of phytoplankton and/or zooplankton at selected locations;

(d) systematic monitoring of benthic flora and fauna, including tissue analyses for metals and selected organic micropollutants;

(e) systematic monitoring of selected indices of primary productivity;

(f) monitoring of selected fish and shellfish;

(g) microbiology, selected indices.

5.3. Master plan, preliminary engineering and feasibility study for liquid wastes collection and disposal system for the project area

The definition and contents of the above plans are presented in Dr. A. Gilad's paper and it is superfluous to repeat them here.

5.4. Quantity and characteristics of industrial waste waters (present and predicted)

A survey of industrial waste water sources, quantities and composition is provided to be conducted within the context of this study. Projections and forecasts of future discharges shall be made, and from these, both the present and the foreseen future industrial waste water inventories shall be prepared. A general programme will be developed for the disposal of liquid wastes, including industrial wastes from Elefsis Bay.

Specific recommendations for reduction and/or treatment of industrial wastes will be developed within the context of this Study.

The following steps are envisaged:

(a) identification of the 100 most significant industrial polluters in the Greater Athens area;

(b) collection of data on their products, raw materials, water consumption and effluents;

(c) collection of information on the industrial processes used by these polluters and on the approximate type and quantity of the pollutants included in their effluents;

(d) performance of analyses of selected effluents for selected contaminants and toxic substances;

(e) design of an industrial effluents testing and monitoring system;

(f) collection of information on the industrial development plans for the Greater Athens area;

(g) establishment of forecasts of quantity and type of industrial effluents which may be expected from future industrial development of the area;

(h) design of the industrial wastes treatment and disposal programme including:

(i) establishment of criteria of permissible discharges from individual plants to sewers, rivers and to the sea;

(ii) preliminary design of local industrial treatment plants to treat collectively industrial wastes produced by groups of industries;

(iii) establishment of criteria for quality of effluents of these plants discharged to sewers, rivers and to the sea;

(i) development of recommendations on the administrative and legislative provisions needed for the functioning of the programme;

(j) development of recommendations for a permanent monitoring, inspection and enforcement system;

(k) development of recommendations for fiscal measures such as fees, charges, fines, taxes, grants and subsidies aimed at reduction of the quantity and improvement of the quality of industrial effluents.

5.5. Preliminary engineering study of the sea outfall

5.5.1. *Definition.* These are studies comprising detailed investigations, surveys, calculations and technical analyses of alternative plans and schemes for the various elements of the disposal system. The preliminary engineering studies shall result in designs accompanied by a report with comparative analyses, recommendations, justifications, preliminary specifications and cost estimate of the various alternatives, including cost of construction, operation and maintenance.

5.5.2. *Alternative proposals for sewage disposal.* These will include:

      (a) discharge to Upper Saronikos Gulf;

      (b) discharge to open sea;

      (c) other discharge points;

      (d) reclamation and re-use of water and materials;

      (e) surface and/or underground disposal;

      (f) any combination of the above alternatives.

5.5.3. *Evaluation of alternatives.* In order to determine the location of the outfall, and to carry out its design, the following matters will be taken into account:

      (a) *Quality criteria of the recipient waters.* These will be established taking into account ecological, health, aesthetic, economic, recreational and fishery considerations and interests.

      (b) *Studies necessary for the preliminary design of the main Athens sewage outfall.* In addition to investigations carried out for an outfall near the present main outfall, the possible alternative of discharging to the open sea will be taken into account, by conducting appropriate investigations.

Besides, it will be necessary to collate and interpret the information gathered as specified above. For each alternative involving an outfall sewer, the alignment, the dimensions (diameter, length, etc.), capacity and the type of outfall sewer and the depth of outfall as well as dilution effect, will be determined and a survey carried out of the topography and geological conditions along the proposed routes in order to provide the basis for determining construction difficulties and costs. Besides, for each alternative, the locations, type, capacity, dimensions, layout and other relevant characteristics of any necessary treatment plant, will be defined.

Cost estimates will be prepared for tunnels, pipelines, treatment plants, outfalls and diffuser systems (for sewage), necessary for each alternative scheme and technical and economic analyses will be carried out, leading to an optimum solution specified in sufficient detail.

6. PROGRESS OF THE PROGRAMME WORK

6.1. Oceanographic studies

These studies are normally carried out by the Greek Institute of Oceanographic and Fishing Research, in conjunction with the Programme.

Data on physical, chemical and biological aspects of the receiving waters are collected during bi-monthly cruises and collated properly.

Prior to this Programme, much of the relevant data had been collected by the Saronikos Systems Project (SSP). This project was established as a joint programme of the Institute of Oceanographic and Fishing Research (IOKAE), Greece, and the Department of Oceanography, University of Washington, U.S.A., to conduct a complete marine ecological investigation of the Saronikos Gulf. The project was undertaken in order to study the effects of urban waste disposal from the Athens Metropolitan area into the Saronikos Gulf.

Data collected during the period December 1972 to June 1974 are currently being analysed, interpreted and presented under proper form, to facilitate the engineering part of the Study (master plan, sea outfall etc.). A summary of some preliminary conclusions, based on the above work, being made under contract by the Department of Oceanography of the University of Washington, is given in Appendix I.

Up to now the bulk of the preliminary oceanographic studies have been carried out and are currently being supplemented by the Programme, through Institute of Oceanographic and Fishing Research. Further investigations on water circulation patterns in the Saronikos Gulf are being carried out, using recording current meters as well as isotope radiotracers.

6.2. Work of the microbiological laboratory set up by the project

6.2.1. *Sewage analyses.* Regular sewage sampling programme was carried out at the sewage outfall and at two points in the sewerage network. The samples were analyzed for BOD, COD, pH and suspended solids. The results of these tests are summarized in the following table.

BOD, COD, pH and sedimentation Rates

Stations	Parameter	Number of determinations	Arithmetical mean	
Sewage outfall	BOD	21	372	ppm
(Keratsini)	COD	19	853	"
	pH	20	6.7	
	Imhoff	8	11.3	mi/l
1st point of	BOD	22	283	ppm
sewerage network	COD	20	594	"
	pH	21	6.6	
(Kallithea)	Imhoff	12	8.5	mi/l
2nd point of	BOD	18	448	ppm
sewerage network	COD	17	1547	"
(Rentis)	pH	18	7.3	
	Imhoff	8	39.5	

The testing programme has been already extended using automatic samplers, which permit determination of variations of the strength of sewage with time.

6.2.2. *Microbiological examination of coastal waters at bathing beaches.* Four hundred and ninety five samples were taken during the months of August and September 1974 from bathing places along the coast of Attica. The samples can be grouped according to their location in three regions:

    (a) from Piraeus to Sounion

    (b) from Perama to Elefsis

    (c) from Porto Rafti to Rafina

The samples were taken from 43 stations which were visited either once or three times according to their importance as a recreational centre. During each sampling, five specimens were taken from each station.

The sampling was done so as to render the specimens representative of the conditions in which the swimmer contacts the sea-water environment. The samples were taken at 2-5 m from the shore and at 100 cm depth. A small amount of sediment was taken in the same bottle with the sea-water.

Less than 50 Coli/100 m/l
○ Acceptable for swimming
● Between 50 and 500 Coli/100 m/l
Acceptable for swimming with some reservations

Total coliforms in recreational water of the Greater Athens Region.

All samples were examined for coliform and *E. coli* by inocculation of 10 ml, 1 ml and 0.1 ml quantities of well mixed sea-water and sediment, in McConkey Broth. Positive tubes were inocculated in Brilliant Green Big Broth at 44°C.

The results are shown in the summary form on the previous map. The tests show that the recreational water meets the current Greek Government standards at all locations examined. They also show the influence of the sewage outfall on the adjacent locations.

6.3. Engineering studies

It is the intention of the Programme to retain a well experienced and fully qualified consulting firm to prepare the master plan, the preliminary engineering design and the feasibility study for the liquid wastes collection and disposal system and to carry out the preliminary engineering study of the sea outfall.

For the selection of the most competent consulting firm, a short list of pre-qualified firms has been prepared, with the assistance of a WHO advisory panel of experts.

In the meanwhile, preparatory work has been done and new relevant activities are undertaken by the Programme staff, e.g. measurement of flow at the trunk sewer and the main collectors, sampling for microbiological, physical and chemical analyses at the sea outfall and other selected points of the net, estimation of leakage and infiltration, survey of industrial waste water sources, field treatability studies in pilot plan scale etc.

During this preparatory phase, a hydraulic profile drawing is planned to be also produced showing the deviations of all sewers in relation to the trunk sewer. This drawing will show if and how an increase in level of flow of the trunk sewer will affect other parts of the sewer system.

6.4. Training of personnel

An important function of the project, of which this programme forms a part, is that of maximum utilization of Greek manpower, resources and skills, and that of gaining experience and training local personnel. Therefore provision has been made in the Prospectus, that the selected consulting firm shall arrange to cooperate with the Greek Institute of Oceanographic and Fishing Research for the completion of the oceanographic studies, as well as with a competent Greek engineering firm for carrying out the engineering studies.

7. CONCLUSIONS

Some of the conclusions, which can be drawn from the previous presentation, are summarized below:

a. For the identification and effective control of environmental pollution in large urban areas, like Athens, an efficient approach may be the development of a comprehensive project, covering the main sectors like air, water, solid wastes and noise. This way the dispersion of responsibilities is avoided and all efforts are focused to the main objective, which is the pollution control abatement, regardless of the medium or the source of pollution.

b. To be effective, such a project should deal with subjects such as planning, identification of objectives, manpower development, design of environmental information systems, creation of administrative framework, etc.

c. Manpower development is of utmost importance. The Athens project has succeeded in developing its own staff through fellowship programme for studies abroad, short courses organized locally and in-service training augmented by professional contact with foreign experts.

d. The optimal way of transfer and adaptation of technological know-how constitutes a difficult problem. Although import of technology can be efficiently made awarding the various sectoral programmes to foreign contractors, this is not recommended, because it contributes almost nothing to the development of local technical capability, which is indispensable for the follow-up activities.

The Athens project is successfully assisted in this field by international experts recruited through WHO on world-wide base.

e. Concerning the specific subject of marine pollution control under consideration, since land-based dis-
    charges constitute the most significant source of pollution, the effective control of land-generated
    pollution is the most appropriate approach to the problem of preservation of coastal water quality
    and marine eco-systems.

f. To reduce and control pollution from land-based sources, it will be necessary to institute in each large
    population and industrial centre a programme similar in scope to the aforementioned Athens liquid
    wastes programme.

    This programme should deal not only with the land part of the sewerage system and treatment plan,
    but, based on oceanographic studies, it must develop the most appropriate combination of treatment
    and outfall-dispersion systems to handle the problem of the present and future liquid wastes disposal,
    without undue adverse effects on the marine environment.

g. Since all these programmes will deal essentially with similar problems, co-operation at all levels is of
    utmost importance. This could involve, i.e.:

        Exchange of information and data.

        Design of co-ordinated monitoring systems.

        Exchange of scientific personnel for training purposes.

        Establishment of guidelines for determination of permissible ecological and other changes
        in the recipient waters due to discharge of wastes.

        Designation of "model programmes" which could serve as models for the design and
        execution of similar ventures elsewhere.

        Establishment of a Model Code of Practice for the disposal of liquid wastes into the sea.

It is of utmost importance that each Government and each International Agency concerned should put
at the disposal of this joint endeavour their expertise and their accumulated experience as well as indispensable financial resources.

APPENDIX I

GENERAL CONCLUSIONS
ON WATER MASSES, CIRCULATION PATTERNS, NUTRIENT CIRCULATION
AND PRIMARY PRODUCTION IN THE SARONIKOS GULF

(Data from the Saronikos Systems Project
December 1972 through June 1974)

I.   WATER MASSES

1.   Definition

The temperature and salinity data from all cruises were used to define the basic water masses and their distributions in Saronikos Gulf. The basic water masses are (see Fig. 1):

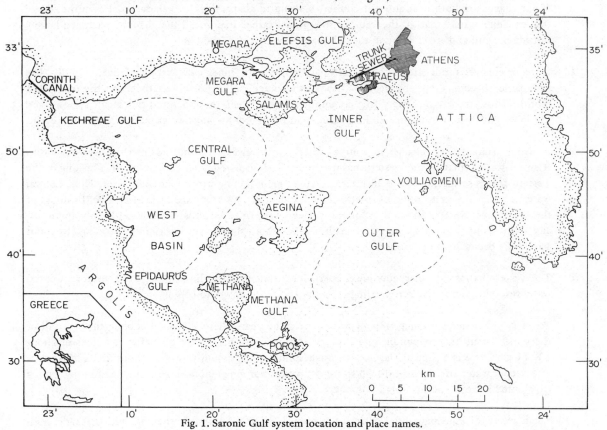

Fig. 1. Saronic Gulf system location and place names.

a.   *Outer Gulf* water occupies the deeper region southeast of Aegina Island and represents the Aegean Sea source water for the gulf system.

b.   *Inner Gulf* water occupies the head of the gulf adjacent to Athens and the major input, and extends to a varying degree south over the eastern section of the gulf.

c.   *Central* water is that characteristic of the western and Epidavros basins and to a varying degree enters the eastern gulf between Salamis and Aegina.

d.   *Elefsis* water is characteristic of Elefsis Bay. Restricted communication with the gulf precludes significant interaction with the gulf water masses, and Elefsis water therefore exhibits its own characteristic patterns and is not considered here.

Temperature and salinity are quasi-conservative properties. They are altered by heat and mass exchange across the sea surface, but with the water are modified only through advection and mixing. Diffusion (mixing) processes require significant energy inputs, and therefore property distributions are mainly due to advection. The degree of interaction (mixing) can be estimated from T–S (temperature-salinity) analysis because of the conservative nature of heat and mass. In the following sections we analyze the cruise data sequentially, for each cruise defining the water mass characteristics and their distributions, and then estimating the circulation patterns and mixing that were responsible for the observed distributions.

2.   Conclusions

From the cycling histories we reach the following generalizations about the alterations in water mass characteristics of Saronikos Gulf water:

a.   Major changes in both heat and salt content are due to changes in these characteristics in the source (Outer Gulf) water, that is, the causes are regional rather than local, the changes occurring largely outside the gulf and are effected in the gulf waters mainly by advection.

b.   A major change is reflected in the eastern part of the gulf (Inner Gulf water) relatively rapidly. The time scale to completely change (or "flush") the inner gulf region looks to be on the order of 1 month. Thus, there is relatively free communication between the inner and outer gulf regions, and the inner gulf region water has no long-term memory of any previous water mass modification event.

c.   A major characteristic change in source water is not always transmitted into the western basins (Central water), that is, the western basins are relatively isolated from the outer gulf. Flushing of the western basins seems to require an intermediate stage of mixing with Inner Gulf water. Thus, Central water is much slower to respond to characteristic changes, and the time scale of change (flushing) of the upper layer western basins is > 2 months. Deep western (Epidavros) basin water is only aperiodically flushed out through spring–summer, but much more slowly, and maximum warming (to about $15-16°C$) occurs in fall (October).

d.   The western basins are a markedly lower energy environment than the eastern gulf. Therefore, Central water responds more rapidly to heating in spring and more slowly to cooling in fall.

e.   There is one general cycling of temperature over the year. There is general cooling through fall of the upper layers to a minimum in Feburary–April, mid-winter, with temperature values of about $13-14°C$. There is a general warming through spring to maximum values of about $25°C$ in August. The deeper water also warms through spring–summer, but much more slowly and maximum warming (to about $15-16°C$) occurs in fall (October).

f.   There are two freshening-salting cycles each year, a fall and a spring freshening, with salting in mid-winter and summer. The degree of freshening (which is regional and not strictly local and presumably depends on the amount and timing of regional run-off) controls through a feed-back system the timing of the warming – cooling cycling. The freshening stabilizes the water column, and the partitioning of heat fluxed downward through vertical mixing and/or retained in the upper layer depends on the strength of the pycnocline between upper and lower layers developed by the addition of freshwater to the upper layer. The stronger the pycnocline, the more rapid the upper layer warming but the slower the effects penetrate to depth, and vice versa.

g. The base salinities of the system, i.e. the deep salinities, change only on some long time scale and are presumably the consequence of some major outside event or disturbance. Another way of saying this is that the freshwater annual variations are small quantitatively compared with the total volumes involved, and water mass modifications are only decreases and increases of upper layer salinities.

One major such event was documented by SSP data, a salting of the system of about 0.2°/oo. The salting first appeared in summer 1973, and persisted through the following summer, reflecting the long-term memory of the system to these events. One consequence of this event was the flushing of the deep western (Epidavros) basin.

II. CIRCULATION PATTERNS IN THE SARONIKOS GULF IN RELATION TO THE WINDS

1. General Conclusions

a. The flow field in Saronikos Gulf is wind-driven. There are no appreciable tidal effects. Baroclinicity (pressure gradient caused by internal mass differences) is only significant in summer (April–October) and then has only secondary effect on the motion. It amplifies to some extent flow through the Aegina–Vouliagmeni and Salamis–Aegina passages. It frequently causes small shear flow in the Aegina–Methana passage which is sheltered from much of the prevailing wind.

b. The winds develop two patterns of circulation in the gulf, cyclonic and anticyclonic. The cyclonic mode can be generated either by north (NW to E) or southeast (SE to S) winds, while anticyclonic circulation results from winds SW to NW. The characteristics of these modes are summarized in Table 1.

Table 1. Summary of circulation modes

		Anticyclonic	Cyclonic	Cyclonic (southeast)
Winds		SW–NW	NW–E	SE–S
Flow along Salamis		northwest	southeast	southeast
Flow between	) east	south	north	north
Aegina–	) center	south	–	north
Vouliagmeni	) west	south	south	south
Flow between	) north	east	east	west
Aegina–Salamis	) south	east	east	east
Flow through Aegina–Methana		west	–	east

c. The frequency of winds is such that cyclonic (north) mode will develop about 50% of the time, and the other two modes about 25% each.

d. About 1 day (or less with strong winds) is required to establish a circulation mode. Water speeds for winds > 5 m/s average 10–20 cm/s, but are frequently faster for short periods. It requires about 3 days to fully develop the distribution of water masses associated with a particular mode, and thus to establish the distribution pattern of other materials carried by the water.

e. It requires about 1 month to flush, or completely renew, the water in the eastern gulf and about 2 months or longer for the western basin water. Exceptions are that unusually strong (15 m/s) S to SE winds can flush the inner gulf with Outer Gulf water within a week; and unusually strong SW to W winds can flush the inner gulf with Central water in approximately the same, or slightly longer, time.

f. The turning point for flow just south of Psitalia from westerly to easterly comes with S winds, and from easterly to westerly with winds from W of NW.

g. In general, light to moderate winds produce less well-coupled circulations in the various topographic subregions of the gulf and less well-coupled flow through the different channels. Strong winds increase the gyres, often merging two adjacent similar ones or obliterating one of two adjacent opposing ones, and demand strong continuity coupling between the channels, with Salamis-Aegina or Aegina—Vouliagmeni dominating over Aegina—Methana.

III. NUTRIENT CIRCULATION AND PRIMARY PRODUCTION IN THE SARONIKOS GULF

a. The Western Gulf is the major nutrient storage area of the Saronikos Gulf, concentrating especially nitrate and silicate in excess of that expected from its volume. The region has the only significant areas in the gulf with depths greater than 100 m and these depths do not communicate freely with the Eastern Gulf. Oxygen may go to zero in the deepest waters in some years and the lowest value observed was 1.3 ml/l. Some mixing occurs in all years, restoring oxygen that is reduced again by the fallout of surface primary productivity. No external source of B.O.D. is required to account for the observed oxygen levels. In some years the Western Gulf is completely flushed with Outer Gulf, Aegean water.

b. The Western Gulf exported nutrient during the period studied, December 1972 to June 1974. More nutrient was exported during the stratified period of the year, i.e. in summer and fall, than during the unstratified winter period. For steady state, there must exist a local source of nutrients since the Keratsini influenced Inner Gulf Water rarely intrudes into the Western Gulf.

c. The source of local nutrient is apparently the Bay of Elefsis which has a thermohaline circulation driven by the presence of sills at both the eastern and western ends. The western sill water should be found at the top of the thermocline during stratified periods, enhancing both productivity in the Western Gulf and export to the Eastern Gulf. The sill water can be expected to find its way to the bottom of the gulf or to be completely mixed in the water column during the unstratified period.

d. The Bay of Elefsis turns over quite rapidly, on the order of 1-2 months according to estimates of the thermohaline circulation rates. The Bay acts as a pump, exporting both nutrients discharged into it directly and those nutrients imported from Keratsini in surface inflow. Low oxygen water originating in Elefsis Bay can be traced along the bottom of the Eastern Gulf. The Bay of Elefsis turnover rate is sufficient to replace the Western Basin water in a surprisingly short time, less than 10 years, and the effect on deeper waters by direct injection is much faster. The amounts of nutrient required to balance Western Gulf export can easily be provided through the Bay.

e. The amounts of nutrients stored in various parts of the Saronikos Gulf system and involved in Western Gulf export annually are of the same order as the discharge of nutrient from the present outfall annually (at least for ammonium).

f. The Keratsini—Elefsis Western Gulf system dominates and determines the nutrient regime in the

Saronikos system at the present time. High nutrients observed at any point in the system originate within the system. No significant source of high nutrients from outside (from the Aegean) has been observed.

g. The nutrient additions in the gulf determine the primary production there to a great extent. However, due to the very low back ground nutrient levels, the resulting production is not high considered on a world-wide basis — comparing well to natural production in richer parts of the tropical Pacific Ocean.

The contribution of phytoplankton particles to loss of transparency in the gulf (even in areas out of the immediate outfall area) is probably no more than half, the remaining loss due to non-biological suspended matter from the outfall.

SANITARY ASPECTS OF MARINE SEWAGE DISPOSAL

Alfredo Paoletti

Department of Hygiene, Faculty of Sciences,

University of Naples, Italy

Summary — Of the three types of marine pollution (radioactive, chemical and fecal microbiological), the last certainly affects public health more than the other two together; this is particularly true in the Mediterranean. Such a situation is quite anachronistic, since microbiological pollution is historically the oldest, the best known and the most easily controllable by the available techniques. Damages connected with bathing, although possible, are not unanimously admitted, or else they should not be considered serious on the basis of our present knowledge; instead, damages connected with pollution of edible molluscs are far more extensively proved.

1. INTRODUCTION

Before examining the wide problem in its different features five general points will be stressed.

(a) In the past, hygiene had to maintain our natural state of health. Now, hygiene not only aims at the maintenance of this state of health, but also at guaranteeing "welfare and joy of living". Hence marine pollution abatement must be such that not only will public health hazards be avoided, but our sight and senses are not shocked by disturbances of our environment, brought about by pollution (Fig. 1).

In this figure, solid wastes caused concern at a northern beach affected by a large river flowing at a distance; strangely enough, when surveys were done (July 1969), such waters complied with the standards for bathing, but aesthetically they were not at all satisfactory.

Furthermore, the problem of pollution must now consider the right — and so far neglected — ecological needs.

(b) Leaving out *thermal pollution,* which does not cause direct problems for public health, the situation, with regard to the other aspects of the problem, is anachronistic, both historically and technically. As a matter of fact: *Radioactive pollution,* which is the most recent and potentially the most dangerous, is better controlled, as both public opinion and responsible bodies were so impressed by the painful images of Hiroshima and Nagasaki that the necessary measures were taken. Therefore, the problem was studied and solved by convenient criteria, a fact which should be adopted as a guide for the study and solution of pollution problems of different types [1–4]; *Chemical pollution,* which evidently appeared with the industrial Revolution (1750) has dreadfully increased with time; it has already caused illness and death in the Minamata Bay, where methylmercury polluted fish were eaten [5, 6]. *Fecal microbic pollution* of the sea, which is as old as the world, might be easily abated by available technology; instead, surprisingly and anachronistically, it disgracefully affects public health in many countries to a greater extent than the other two types of pollution taken together. Now, everybody is aware of this sad situation, which sanitary experts and sanitary engineers have denounced for decades.

149

(c) Everybody knows that the sea is endowed with considerable *self-purification power*, and this fact will be further proved. However, in my opinion, this reality might appear false, were it not dimensioned within the right limits and corrected as follows: (a) the self-purification power of the sea is similar to that of any other natural environment with various types of microbial life (fresh water, soil, biological purification plants). (b) The sea, like the other mentioned environments, can destroy what is degradable (and these are the *positive aspects of self-purification*). However, the substances that are not degradable obviously cannot be destroyed neither in fresh waters, nor in purification plants, nor in the sea; on the contrary, they may enter into the various levels of the food chain and reach man in concentrated form (and these are the *negative aspects of self-purification*) [7, 8].

Fig. 1. Aesthetic pollution in estuarine area.

(d) By definition, domestic sludges almost exclusively contain (or should contain) natural and degradable pollutants (organic matter, pathogenic microorganisms and fecal pollution indicators) since industrial effluent discharges should be accepted only after reduction of the pollutants to tolerable levels for the environment or for the possible purification plant. Therefore municipal sewage discharges into the sea should not arouse public health problems with regard to chemical and radioactive pollutants. Only detergents, which are always present, might be concerned in the matter; they will be briefly considered when dealing with skin diseases.

(e) The pollutants present in industrial effluents are of two types: (a) *natural* (metals and in particular organic matter), which — when discharged into the sea — have not so far aroused public health hazards, because they are generally abundant in the earth's crust; (b) *artificial* (DDT, PCB, pesticides, methylmercury and other organic compounds), which rouse justified suspicions especially if they may accumulate. Furthermore, the well known damage produced by methylmercury is a warning that similar events might occur even by the action of other pollutants of this group. No doubt many of them must be controlled before discharge to within limits to be established (even if no public health hazard has been detected), if they cause ecological upset or if they arouse problems concerned with the depletion of the available earth resources [8].

It is convenient to stress the most debated and least clear question of marine pollution, i.e. pollution in relation to bathing.

2. MICROBIAL POLLUTION OF THE SEA

This problem was dealt with in Naples 12 years ago and the subject will be updated [8–18].

In theory, municipal waters convey to the sea all pathogens eliminated by sick persons and carriers; their amount is an index of the epidemiological conditions of the place under examination. They are:

(1) Metazoans: taenia eggs, ascarids, trichocephalos, hookworms, etc.

(2) Protozoans: cysts of dysenteric amoeba, lamblia, trichomonas, etc.

(3) Bacteria: of typhus, paratyphus, cholera; dysenteric, tubercular, staphylococci, etc.

(4) Pathogenic moulds

(5) Pathogenic Leptospira

(6) Viruses: viral heptatitis, poliomyelitis, other enteric viruses.

In practice, only few of them may return to man and cause diseases by two mechanisms: (a) ingestion of contaminated fish and molluscs; (b) bathing in polluted waters.

However, before examining these two features, it is convenient to remember that the occurrence of an infectious and parasitic disease involves the more or less optimal co-existence of three factors: (a) *The pathogenic agent load,* i.e. the introduction of a given number of microorganisms responsible for the disease is an important factor: the larger their number, the greater the possibility that infection is followed by disease. However, opinions on the infecting load required to cause a disease are neither clear nor consistent, because, on determining the disease, it is often difficult to distinguish what is due to this factor and what to other factors. It is enough to remember that a few volunteers acquired infection by ingesting less than ten *Salmonella typhosa*; in other tests, twenty volunteers did not contract the disease by ingesting nearly 1000 of them; in others, sixteen out of thirty-two people acquired infection by ingestion 10,000,000, and infection was always contracted by ingesting 100–1000 million such enterobacteria. As to bacillary dysentery a far lower infecting load seems enough. With regard to the salmonellae of paratyphus, pathogenic effects were detected with 100,000 ingested bacilli, etc. [17–20]. (b) *The virulence of the pathogenic agent,* i.e. its capability to cause disease more easily, the other two conditions being the same, makes it certainly more aggressive.

As a general rule, it may be stated that pathogens are the more dangerous to man the lower is the time elapsed from elimination by another man, since they are highly "humanized", i.e. adjusted to live in man. In fact, almost certainly their virulence is gradually reduced with permanence in an unsuitable external environment, probably because of the gradual loss of some well known virulence factors present in their cellular structures (antigen Vi, various capsular antigens, S–R modification, etc.). Furthermore, the percent reduction of entorobacteria in sea waters (and in other plurimicrobial environments) gradually proceeds with time as if they were placed into a mild disinfecting solution [9, 10, 21], but a small percentage of them survive for longer times (*resistant forms or exceptional Kruse cells*), which seem endowed with a particular pathogenic virulence owing to the reasons mentioned above (Fig. 2). (c) *The organic predisposition,* i.e. the greater ease with which disease is contracted by the least immunized subjects, is well known and does not require further explanations. For example, the other two conditions being the same, the Italian less easily contracts typhus or viral hepatitis than any other tourist coming from countries where such diseases are less spread and antityphus vaccination is less practiced.

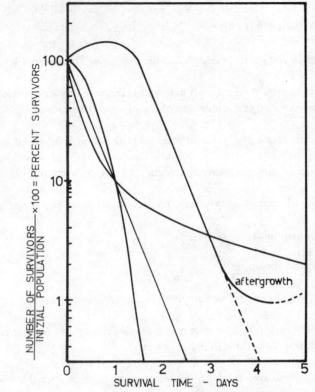

Fig. 2. Survival curves of *S. tiphi* on sea-water (from Beard and Meadocroft, 1935).

2.1. Bathing and public health

This is the most controversial feature of the problem of marine pollution, about which public opinion has not been thoroughly informed; and this is obvious since among epidemiological technicians, opinions are not unanimous. In an attempt to throw some light on the debated question, we must distinguish the sanitary aspects of beach pollution from the ecological. Though in fact they frequently co-exist, sometimes they are not associated, as Fig. 1 demonstrates; however, if all the sanitary requirements of marine pollution were fulfilled, the ecological, aesthetic and psychological requirements would also be adequately tackled.

It is axiomatic that the domestic sewage of urban centres contains pathogens; consequently, the receiving waters are likewise polluted.

It is equally true that an infectious disease occurs only if pathogens reach the patient; but it is also necessary that the three conditions discussed above must be fulfilled, which does not occur so easily. Therefore, the well known report by the British Committee of P.H.L.S., 1959, stated the general conclusions: the presence of pathogenic organisms in sewage contaminated sea waters more soundly proves the existence of the disease among people producing such sewages than a further risk of infection for bathers [11].

It has also been stated that the isolation of pathogens in sewages or in the receiving sea waters (which is possible by available technologies) has the same meaning as the recording of earthquakes by a seismograph:

they may be detected at a distance of thousands of km, but the damage they cause is limited to their epicentre [15]. Therefore, the presence of a few pathogens in the sea does not necessarily mean hazard to the bather. Certainly, coastal waters polluted by municipal discharges receive on average 100,000,000 fecal coli bacteria/100 ml of discharged sewage, and pathogens, such as salmonellae and enteric viruses, are present in variable amounts depending on the local epidemiological conditions [9, 10, 22]. However, it is also true that the number of enteric microorganisms gradually decreases on increasing the distance from the outfall; this is due to self-purification factors discussed at the 4th International Symposium of Medical Oceanography in Naples in 1969 [8, 23] and supported by other authors [24, 25] (Table 1).

Table 1. Self-purification factors of the sea

(1) Physical factors

Adsorption and sedimentation, dispersion and dilution, sun light, temperature, pH

(2) Chemical factors

Salts, oxygen, nutrients, growth factors

(3) Biological factors

(a) bactericidal power

(b) antibiotics

(c) lithic phenomena: *Bdellovibrio batteriovorus*, bateriolithic bacteria, bacteriophages

(d) predatory phenomena (filterfeeders, such as molluscs, tunicates, sponges, etc.; slime-eating animals, etc.)

Domestic sludges of Mediterranean communities contain a larger amount of intestinal pathogens than those in other countries, owing to the different epidemiological situation; that is why it seems right to impose more severe standards. However, the survival of such bacteria in the Mediterranean sea (expressed in terms of *Tl* 90) is much shorter than in the North Sea and elsewhere [24, 25]. In this way, such differences should disappear.

The occasional detection of pathogens in the open sea [13, 14, 26–29] does not invalidate the general rule that they gradually decrease and disappear quite rapidly.

In my opinion, pathogens may be passively conveyed and excreted by fish contaminated near the shore, or else they may be spread by polluted marine aerosols.

Whether the intervention of all or of some self-purification factors reported in Table 1 is accepted, one fact is certain and may be easily demonstrated by simple and routine laboratory techniques: on increasing the distance from the sewage outfall in the sea, all enteric bacteria and enteric viruses (as well as degradable organic substances) gradually and constantly decrease, until they disappear completely, at a far higher rate than would occur by dilution only. This is unquestionable and has been proved by all experts who have studied the topic by all available techniques. As mentioned at the Sanremo Meeting, we further proved and quantified this by up-to-date continuous monitoring techniques.

An example will better clarify this phenomenon: when looking at this wonderful Gulf of Naples, remember that it receives domestic sludges of about 4 million inhabitants both directly and indirectly, through water courses. This means that it receives nearly 690×10^{15} coliforms every day and a number of salmonellae of 12–13 figures (instead of 18), if we accept with the due updatings the generally accepted coliforms/salmonellae ratio [22, 60]. At a distance of a few hundred metres or of a few km from the outfalls, such enteric organisms are no longer detected, and only exceptionally, due to the reasons mentioned. Hence, for decades, all the microorganisms that have been discharged every day into this Gulf die; otherwise the situation would not have been the same for these 20 years, during which I have studied the problem. This is self-purification, which may be easily controlled here and elsewhere!

On the basis of these considerations, the much debated topic of the public health hazard connected with bathing may be considered only on a statistical basis.

With regard to *intestinal diseases*, the following information is available:

The P.H.L.S. Committee, on the basis of an inquiry set up from 1952 to 1956 in eighty coastal places, demonstrated that the incidence of enteric diseases is not higher there than elsewhere. It was surprisingly found that it is lower in those subjects belonging to the age group that bathe more frequently; this fact may be easily explained on considering that the various ages have different food habits, whereas bathing is not a determining factor.

However, the Committee also reported four cases of paratyphus B contracted by people who bathed in highly polluted waters, four further analogous cases possibly connected with bathing and two cases of typhus of people who drank sea water [11].

Another inquiry [11, 12, 16] concerned 300 children, of age below 15, who contracted poliomyelitis; they lived at seaside resorts, belonged to the same social class and attended the same school. Only 150 of them had practised sea-bathing during the 3 weeks before the disease. Hence the conclusion was reached that bathing had been an irrelevant factor for causing the disease.

In Australia, after leakage of a submarine outfall, coastal pollution of the Perth beach was thought to be the cause of ten cases of typhus [15, 16, 30]; however, no convincing proof was found that they were actually due to bathing [15, 16] because the required microbiological and epidemiological supports were lacking.

In Sydney, the eight submarine outfalls built a long time before the boom of sea-bathing, have not supplied a satisfactory protection of beaches, in compliance with the bacteriological standards presently in force (far less severe than the Italian ones). This notwithstanding, and though in the presence of pathogens in discharged waters, a questionnaire-inquiry set up on a large scale led to the conclusion that no increase in enteric diseases occurred among people who had practised sea bathing. Instead, it transpired that those who had drunk such waters ($\sim$ 4000 people) did not fall ill, whereas all cases of typhus that occurred in the last decade were not connected with sea bathing. However, local Authorities detected that considerable hazards for lung, eye and ear infections existed, although they recognized that the data collected were not above questioning and that the colifecal content in such waters ranged from maxima of 167,000 to minima of 37/100 ml; this notwithstanding such waters were satisfactory from the aesthetic point of view [30].

At Yalta and Eupatoria, Crimea, the statistical data have always shown that enteric diseases constantly increase during the bathing season; at Yalta it was decreased considerably after the installment of reclamation works of coastal waters [15, 16, 31]. Apart from the fact that such works consist of submarine outfalls (and not of purification plants (which is not always pointed out by Italian and foreign authors), the inquiries set up in the two towns do not demonstrate that the enteric diseases are connected with bathing [15. 16], but may be more easily attributed to the different food habits or sanitary conditions.

In France, a large-scale questionnaire-inquiry would have shown that coastal populations more frequently contract viral hepatitis and typhoid fever than other populations [16, 32]; however, not even such inquiry proves that such infections are connected with sea bathing [16].

With regard to *skin and mucous membrane infections* (bronchitis, vaginitis, angina, medium otitis, sinusitis, lung abscess, eczema, suppurated dermatosis, etc.) the following information is available:

A survey on 5000 people living on Lake Michigan has evidenced a higher number of such infections among those who bathed in bacteriologically acceptable waters (from 91 to 180 coliforms/100 ml) than among those who did not bathe; however, infections were even more frequent among those who bathed in more polluted waters (average 2300 coliforms/100 ml); in both cases, however, the influence of the atmospheric conditions was not sufficiently evaluated [16, 33, 34].

An inquiry was set up in Kentucky on 7500 bathers in both pure (3 coliforms/100 ml) and slightly polluted water (2700 coliforms/100 ml). They contracted such diseases more frequently than controls. Strangely, eye, nose, ear and bronchial infections were predominant in those who bathed in waters with a large amount of chlorine; in the second group, instead, intestinal infections predominated. However, on the whole, the largest number of cases of disease were found among those who bathed in acceptable waters [16, 33, 35].

An inquiry at Long Island on 5000 people did not evidence any correlation between disease and bathing in two differently polluted waters (averagely 610 and 253 coliforms/100 ml) [16, 33, 36, 37].

In France, an inquiry among coastal and inland populations showed that: bronchitis, angina and rhinitis are more frequent among inland populations; otitis, sinusitis and lung abscesses are equally frequent in both groups, viral hepatitis, typhoid fever, botulism, urticaria and suppurated dermatitis are more frequent on the coast [32].

From all these data drawn from literature, which we have reported with absolute objectivity, the following considerations and conclusions are drawn:

(1) Both those believing that bathing in polluted sea waters is risky [13, 14, 26] and those having an opposite opinion [11, 12, 15, 16, 30], agree that only accurate epidemiological inquiries, the difficulty of which must be stressed, may tell us whether bathing in non-acceptable waters is dangerous. They also agree that infection hazards increase on increasing water pollution and individual receptivity; however, no agreement has been found about the bacteriological limits within which any detrimental effect becomes possible.

Pollution of coastal waters may exert a direct influence on the bathers' health; however, such risks cannot be quantified. Health hazards due to bathing in polluted waters may be practically ignored, if such waters are not evidently and aesthetically contaminated by sewage outfalls.

This last hypothesis seems the most reliable on the basis of the available data, even if most inquiries do not refer to highly polluted areas. However, all admit that reasonable prudence should be more justified, without any excess in one sense or in the other.

(2) No doubt that the purer are waters with regard to pathogenic bacteria and fecal pollution indicators, the fitter they are for bathing; this is generally accepted. However, it being inevitable that coastal waters are more or less subject to pollution by enteric microorganisms depending on the population density, it is necessary to establish which is the tolerable microbial load that is not detrimental to the bather; on this matter opinions are quite different. In theory it should be admitted that even a quite polluted sea water cannot provoke intestinal diseases since we do not drink it; therefore, it is

impossible to introduce such an amount of pathogens as to exceed the infectious dose; this is the optimists' opinion. In practice, small amounts of water are accidentally ingested, especially by children; this seems to support the pessimists' opinion.

On the basis of unpublished research by myself, we may state that we ingest a larger number of coliforms whenever we eat raw vegetables (though well washed) than when we bathe; this notwithstanding, at least due to psychological reasons, it is not possible to accept the thesis according to which "bathing hazard may be ignored, unless the aesthetic conditions of bathing waters are repulsive" [11, 12, 15, 16]. But, if the few accurate inquiries available lead us to believe that no correlation exists between intestinal diseases and bathing, some studies accomplished at the beginning of this century lead to opposite conclusions [38, 39]. In my opinion, however, these were not well documented or concerned massively polluted waters or water ingested by boat overturning or by drowning accidents [11, 39].

On the basis of these facts standard limits for bathing waters can hardly be established: this explains the different criteria followed for standards in the various countries. Without quoting the permissible limits in the various nations and States of the same country (ranging from quite severe standards of 100 fecal coli/100 ml water to quite tolerant ones that impose an adequate distance of discharges or rely upon beach aesthetics), I think it convenient to underline the discrepancy existing between the permissible fecal coli limits for bathing and for mussels. As a matter of fact, standards in almost all countries generally admit that no hazard is involved in consuming molluscs containing up to 5 fecal coli/ml and even more, which means 500 or more fecal coli in ten big molluscs 10 g each; standards for bathing in some countries tend to impose a maximum of 100 fecal coli/100 ml water. The obvious conclusion is drawn, i.e. either *the former limits are too tolerant, or else the latter are far too severe*. As a matter of fact, it is enough to eat two molluscs of 10 g each with bacterial contents within safety limits, to introduce the same number of bacteria as when drinking 100 ml of acceptable sea water, which we never do. In my opinion, this remark is the most obvious proof that at least with regard to intestinal diseases, the scientifically most realistic fecal coli limits for bathing are those in force in the most tolerant countries. This must be said, although waters with the highest possible purity must be obviously preferred both for bathing and for breeding edible mussels.

(3) With regard to the infections of the upper airways and lungs, the available data would equally suggest that no hazard is involved with bathing. The nine cases reported in the literature [9, 10] concern children who accidentally fell into waters contaminated by the discharges of a sanatorium. Therefore, this infection was exceptional; its pathogenesis was analogous to that of *ab ingestis* pneumonia, which may frequently occur after accidental inhalation of an extraneous material, both on drowning and vomiting, etc.

(4) As revealed by some inquiries, rhinitis, sinusitis, pharyngitis, tonsillitis and otitis were more frequent on the coast; however, opposite conclusions have also been drawn. The other epidemiological conditions being the same, such infections of the mucous membranes were generally found to be more frequent in bathers than in control people, as well as in subjects who bathed in acceptable waters. Furthermore, they are also contracted by professional swimmers who practise this sport in swimming-pools where the water is acceptable owing to effective chlorination. Therefore, such infections are almost certainly connected with germs that dwell in a large number in everybody's upper airways (staphylococci, pneumococci, streptococci, etc.): due to diving or swimming, they are mobilized in paraoral cavities and their virulence is increased or their growth is favoured by the decrease in organic defences due to cooling (*a frigore infections*).

(5) According to the statistical data available, skin and genitals infections (dermatitis, vaginitis, eczema, urticaria, furuncolosis in general) have been more frequently found in bathers than among control population. However, such statistical data do not prove any certain connection with the bacteriological conditions of waters.

Apart from the fact that the responsible organisms (staphylococci, candidas, pyocians, etc.) have always teemed on human cutis and mucous membranes, such infections are now more frequent than in the past due — in the dermatologists' opinion — to detergents. In fact, they decrease the protective capabilities of the acid layer of human skin and consequently increase normal flora. Furthermore, sunlight favours such infections, since it increases the allergy-causing chemical action of these substances (photochemical action of the actinic rays) especially in those subjects that are particularly sensitive to some products. It is enough to think of the very many people who suffer from allergic solar dermatosis as soon as they sprinkle their skin with a given perfume, or wash themselves with a given soap or detergent, or use certain tanning lotions. The increase in the normal skin microbic flora easily takes place especially in such subjects and worsens the symptoms that, caused by a photochemical effect, finally attain a specifically microbial character. Similar considerations may be done for vulvovaginitis, which is now quite frequent especially among young people both in bathing and inland areas, both in summer and in winter. The underwear, if not thoroughly rinsed, causes allergic local reactions from detergents; then sand, though bacteriologically pure, mechanically irritates the injured part and favours the growth of local microbic flora. Staphylococci should be the main cause of this condition for it probably is favoured by sea salinity (haloresistant or better halophilic). However, if such a hypothesis is accepted, we should remember that salt is so abundant in perspiration to cause skin and mucous infections of this type even among non-bathers.

(6) The sea-bathing boom causes overcrowding, concentrated both in space and in time; furthermore, particularly for the poor, it gives rise to an almost inevitable deficiency of hygiene. These basic epidemiological principles cannot be denied. This notwithstanding, the above statistical investigations do not supply sufficient data to prove a different incidence of diseases among those who bathe in differently polluted waters. Furthermore, overcrowded waters, though bacteriologically convenient, are contaminated by the swimmers' microbic flora, which is always abundant on their skin and mucous membranes.

Such remarks drawn from the quoted literature and from the well known principles of the epidemiology of infectious diseases, would lead us to conclude either that bathing does not cause public health problems or that such a hazard is not statistically demonstrated, provided that the degree of microbic pollution is not exceptional or particular factors do not intervene. I personally think that, until better and more accurate statistical data are attained, we should take up a prudent position, for psychological reasons. With regard to the limits for bathing, some believe that they should be made more lenient especially in those countries where they are extremely rigid, since sea water is hardly drunk. In my opinion, this is scientifically acceptable: as a matter of fact, the criteria chosen for edible molluscs, which were established on the basis of far better documented epidemiological data should be otherwise considered unsafe.

However, the better the water is from the microbiological point of view the better it is "to comply with that highly social principle of hygiene, according to which life must be nobler and not only healthier" [40].

2.2. Fish and microbic pollution

As already reported [41, 42, 43, 44] fish and especially edible molluscs can be detrimental to human health through three types of factors:

(1) *Toxic substances*, either derived from the unicellular organisms especially taken up by molluscs or originated by bad storage. Such problems are not considered here being not connected with microbic sea pollution.

(2) *Pathogenic enteric bacteria*, which molluscs accumulate by permanence or breeding in polluted

waters, are often responsible for toxic food infections. Intoxications from *Clostridium botulinum* and *Vibrio parahaemoliticus* are not considered here, since such bacteria are spontaneously present in all oceanic waters and therefore are not due to the marine pollution caused by human activity [14, 45, 46]. Instead, we briefly mention the frequent infections from salmonellae of typhus and paratyphus, which find an important epidemiological factor especially in edible molluscs [14, 16, 39, 41, 42, 43, 44, etc.]. However, instead of listing data, we wish to explain why they are un-animously considered detrimental, in contrast with the conflicting opinions about bathing. The relative data may be summarized as follows [41, 42, 43, 44]: (a) molluscs concentrate the bacteria present in the waters where they live by particular and well known filtration mechanisms [8, 23]. Although such a filtration power varies depending on age, species and environmental conditions, the average concentration factor is 30—40 times, but it may even reach values of 100. This means that on eating ten raw molluscs of an average size, one ingests the bacteria that are present in 4—10 l. sea water where they lived. (b) Therefore, molluscs purify the sea (Fig. 3), but contaminate themselves, keeping the bacteria in their branchiae for a short time (Fig. 4) or in their intestine for 1—2 days at least. (c) It is not enough to guarantee satisfactory hygienic conditions in the breeding waters, but it is also necessary to take care of the various market stages. As a matter of fact, when stored in vessels containing polluted water, they purify it in a few minutes by taking up a large amount of the micro-organisms present. (d) Cooking of molluscs is enough to avoid infections from salmonellae. However, acute gastroenteritis is not always avoided: it appears 8—16 h after the meal, being connected with the enterotoxin they produce, which is thermostable. (e) Salmonellosis from infected molluscs is generally more serious, both owing to the higher microbic load taken up and to the aggressive or proinfectious action of enterotoxins [8, 23, 41, 42]. By analogy, *choleric infection* may be equally produced by polluted molluscs; this appears from the references. The cholera epidemic that recently took place in Italy had such an origin [51—53]; however, the cholerigenic vibrio was isolated in molluscs only in Sardinia.

(3) *Pathogenic enteric viruses* may be equally conveyed by sewage-contaminated molluscs. Such a possi-bility was demonstrated with regard to the viruses that may be cultivated in the lab (poliomyelitis, ECHO, coxsackie, etc.), but it has also been ascertained with regard to the viral hepatitis virus that cannot be cultivated yet [54—64]. In spite of the impossibility of a virological documentation, the epidemiological link between molluscs and viral hepatitis evidently appeared in the two American epidemics reported in the literature [56, 57]; they were caused by pollution due to temporary insufficiency of the existing purification plants, connected with the seasonal increase in bathing population.

3.   CHEMICAL POLLUTION OF THE SEA

In spite of its level and of its gradual increase, the chemical pollution of the marine environment has caused few thoroughly ascertained cases of damage to human health. Instead, its ecological aspects un-doubtedly are more important, though they considerably differ depending on the pollutant [8]; some-times, the environment is upset to the point of causing offensive conditions.

3.1. Chemical pollution and bathing

The literature does not report any damage to human health due to bathing in waters contaminated by chemical substances. However, it seems convenient to briefly point out the potential hazards involved.

A Swedish silver-plating industry in 1969 [65] accidentally disposed a large amount of silver cyanide into the Oslofjord. From the reaction that followed with marine salt, a silver chloride precipitate was formed with evolution of CN^- ions, which are highly toxic and which caused a severe fish kill. The characteristic

(a)

(b)

Fig. 3. (a) Very polluted sea water, with sandy bottom containing shellfish *Tellina*. (b) After 50–60 min the water cleared due to absorption of bacteria and seston by *Tellina*, through filter feeding mechanisms with inhalant and exhalant siphons. Sea water is purified but shellfish become highly polluted.

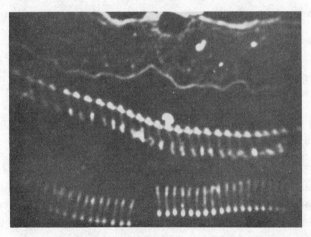

Fig. 4. Slice of *Mytilus* with a microscopic clump of non-pathogenic mycobacteria near the gills (fluorescent microscopy). Radioactive-labelled bacteria and autoradiography, flouresceing-antibody techniques and other means are very useful for evaluating the capture activity of shellfish on bacteria introduced in a laboratory aquarium.

smell of the poisonous gas frightened two experts, who were bathing in such waters. In this case ingestion of a small amount of water might have been fatal.

Detergents may increase skin and mucous membrane inflammations; but, as already mentioned, statistics are of no help in this respect, and other mechanisms may be invoked. Oils and other chemical pollutants certainly worsen the conditions of several waters, although health is not directly compromised.

3.2. Chemical pollution and fish

The problem was brought up by the well-known event of the Minamata disease from methyl mercury. It has become even more topical when it was found that considerable amounts of DDT are present in the fats of whales and of other animals (man inclusive). Hence all chemical pollutants have been incriminated or regarded suspiciously, because all or almost all concentrate along the various levels of the food chain and thus may return to man. In this respect, the situation may be summarized as follows [8] :

Natural pollutants, such as metals, are largely *necessary, essential* to the vegetable and animal life of the globe, since they are present in the earth's crust and have vital functions in particular metal enzymes. They are taken from the surrounding environment in variable amounts, depending on their concentration and on environmental and individual factors. With few exceptions, which will be mentioned hereinafter, diseases may be contracted for want of such oligoelements (their treatment consists in their additional administration in food). They may be quite safely tolerated within a wide range when their presence in the environment is in excess. Finally, they are seriously toxic if ingested in amounts exceeding more or less known limits (Table 2). Diseases in animals are known both for want and excess of some elements in

Table 2. Mineral elements that may be regarded as essential to man

In large amounts	In traces
Calcium	Iron (Selenium)
Phosphorus	Copper (Chromium)
Sodium	Zinc (Fluorine)
Potassium	Manganese (Nickel)
Sulphur	Cobalt (Vanadium)
Chlorine	Iodine (Silica)
Magnesium	Molybdenum (Antimony)
	Cadmium (Tin)

pastures; diseases from want of oligoelements are also possible for men (iodine, fluorine, iron, etc.), whereas diseases consequent to excess of them are only contracted in particular working environments. It has not been shown that diets based on fish have ever caused them.

Other natural elements may be considered as *inert* or *indifferent* since, though being present in the average human diet, do not cause either phenomena of want (if they are absent or in low amounts) or toxic effects (if their concentrations are quite high).

Finally, natural *detrimental* elements exist (mercury, lead, etc.), which, though always present in the human diet, have no ascertained useful functions, but cause harmful effects when taken up at concentrations exceeding certain levels. As to these, it may be said that the less taken up the better.

No record is found in literature concerning detrimental effects to human health caused by an excessive ingestion of such natural elements through fish, with the exception of mercury and perhaps of arsenic in the known intoxication from mussels, which took place at Wilhelmshaven in 1885. It is enough to think that not even the fish caught in the Red Sea are toxic to humans, though the concentrations of some elements (iron, copper, manganese and lead) in its waters are from 1000 to 50,000 times higher than in other seas [8, 66].

As mentioned above, *mercury* has already aroused problems of public health (see Minamata disease). As may be seen from literature (Table 3), the highest concentrations are found in the rivers of the Italian mercuriferous areas; that is why concentrations in the Mediterranean are 2—3 times higher than in other seas. However, the activity of man during the technological era contributes to markedly increase its content in some well determined coastal waters [68].

Table 3. Mercury concentrations in waters [67] — enlarged [8]

Type of water	Concentrations, ppb or μ/l
Mediterranean Sea	0.10 − 0.12
Oceans	0.05 − (0.03)
Minamata Bay (Japan)	1.6 − 3.6
Not polluted rivers (Italy)	0.01 − 0.05
Rivers in mercuriferous areas, Italy	5 − 140
Underground waters	0.02 − 0.07
Hydrothermal waters	1.5 − 2
Rainfalls	0.2
Greenland ices	0.06 − 0.10
Polluted rivers	0.1−0.3−3−2

Mercury is naturally deposited on the sea bottoms (0.02 − 1 ppb); however, higher values are found in the slimes in proximity of submarine outfalls in California (40 ppb) and exceptional values in the natural manganese sediments and volcanic submarine areas 800 ppb). It is inevitably concentrated along the various levels of the food chain and therefore appears in the usual human diet, where it should not exceed determined levels. According to the standards in force in some countries (U.S., Israel, Switzerland, Spain, Belgium, Canada), mercury content in fish should not exceed 0.5 ppm; in other countries (Italy and France), such a limit has been brought to 0.7 ppm; in others (Denmark, Japan, Sweden, Germany) up to 1 ppm mercury is tolerated in edible fish, provided that diet is not predominantly based on fish. In other cases, since the average Italian eats 6.5 kg/year of fish, the limits established by the Ministry of Health are not to exceed the prudent limit of *0.3 mg of mercury a week* tentatively fixed by the FAO—OMS experts.

Investigations made on the inhabitants of the San Pietro Isle (Sardinia) and on the food of the mercuri-ferous areas of Monte Amiata have not revealed toxic effects from mercury [69], although the content of such element exceeds the usual average values. Mercury, being a natural compound, hardly arouses problems of public health with the diet in general and with the diet based on fish in particular. However, that is not so if the metal is part of organic compounds (in particular methyl mercury), which are far more toxic and may even be formed in the environment and not only discharged as such from industries.

Artificial chemical pollutants (DDT, PCB, pesticides, methyl mercury, different organic compounds, etc.), being extraneous to the sea and earth biosphere, arouse further problems and, at least in theory, involve public health hazards, which vary depending on whether they are biodegradable or not, on whether they can be concentrated along the various levels of the food chain, on their acute or long-term toxicity.

It has been established that such pollutants return to humans through fish, and their amount varies from place to place. However, only *methyl mercury* ingested with fish has been so far detrimental to health. No doubt that humans' fats already contain fairly high amounts of DDT (and perhaps of PCB, which is far more toxic), but no inconveniences — at least ascertained — have been brought about. This must be certainly attributed more to transportation by air and earth food than to fish [8]. The Minamata event (146 people contracted the disease, of whom 48 died) was attributed to the acetaldehyde industries that discharged into the bay traces of metal as an organic compound. Less serious toxic effects also occurred in Scandinavia, where that compound had been used in pesticides for agriculture.

By proving that the environmental bacteria or hepatic enzymes of fish may transform metal mercury into an organic compound, the ion itself is under suspicion, as it would undergo such a transformation in the sea. However, no proof has been found so far that such a transformation has occurred in the mercuriferous areas to such an extent to make it toxic or even lethal in the usual daily diet. Only industrial effluents or agricultural products containing the organic product are really dangerous.

4. POLLUTION FROM RADIOACTIVE WASTES

This topic involves well defined areas under control by well qualified organizations. The knowledge acquired in radioecology should be taken as an example for the study and solution of problems concerned with pollution of different types, because natural and artificial pollution by radioactive substances are well quantified and because solutions are based on the ability of the total environment to receive, retain and transmit radioactive pollutants through whole ecological chains [8].

REFERENCES

1.   Ravera, O., Sono validi i concetti adottati dagli esperti di inquinamenti radioattivi per lo studio e il controllo degli inquinamenti convenzionali? Incontro tra gli esperti dell'inquinamento radioattivo e quelli del l'inquinamento convenzionale; F.A.S.T., Milano, 1970.

2.   Paoletti, A., Les enseignements apportés par la radio-écologie à l'étude de la pollution chimique du milieu. Atti Simposio Internazionale della Commissione della Comunità Europea su "Radioecologia applicata alla protezione dell'uomo e del suo ambiente". Roma, 1971.

3.   Antonelli, A., Possibilità di applicazione dei criteri e dei metodi della radioecologia al controllow degli inquinamenti convenzionali. Idem, C.E.E., Roma, 1971.

4.   IAEA — 144, *Pesticide Residues and Radioactive Substances in Food: a Comparative Study of the Problems.* Technical report of the International Atomic Energy Agency, Vienna, 1972.

5.   Fumiaki, Kuwabara, Minamata disease. *Sanichi Shobo* (1968).

6.   Paoletti, A., L'aggressione chimica del mare per opera dell'uomo. *Nuovi Ann. Igiene Microbiol.* **21**, 1970.

7.   Paoletti, A., Sources de pollution et autoépuration de la mer dans ses aspects positifs et négatifs. *Revue Int. Océan. Méd.*24, (1971). Atti 5° Colloquio Intern. di Oceanografia, Medica, 1971; Ellebi Edit., Messina, 1973.

8.   Paoletti, A., *Oceanografia Medica ed Inquinamento.* Liguori, Napoli, 1975.

9.   Paoletti, A., Microviventi patogeni nell'ambiente marino. Atti Convegno Naz. su "Problemi attuali di Igiene e Medicina Sociale", Napoli, 1963.

10. Paoletti, A., Microorganismes pathogènes dans le milieu marin. Atti Simposio Intern. su "Pollution marine par les microorganismes et les produits petroliers; CIESM, Monaco, 1964.

11. Committee on Bathing Beach Contamination, P.H.L.S., Sewage contamination of coastal bathing waters in England and Wales. *J. Hyg. Camb.* **57**, 1959.

12. Medical Research Council. *Sewage Contamination of Bathing Beaches in England and Wales*; Memorandum No. 37; H.M.S.O., London, 1959.

13. Brisou, J., La pollution microbienne, virale et parasitaire des eaux littorales et ses conséquences pour la santé publique. *Bull. Org. Mond. Santé* **38** 1968.

14. Brisou, J., Les Aspects Médicaux de la Pollution des Mers. *Bordeaux Méd.* **4** (1968).

15. Moore, B., The present status of diseases connected with marine pollution. *Rév. Int. Océan. Méd.* **18-19** (1970).

16. Pike, E.B. and Gameson, L.H., Effects of marine sewage disposal. Annual Conference of The Institute of Water Pollution Control; Douglas, Isle of Man, Sept. 1969. Conference paper No. 4.

17. McCoy, J.H., Sewage pollution of rivers, estuaries and beaches. *Pub. Hlth. Inspect.* **74** (1965).

18. Kehr, R.W. and Butterfield, C.T., Notes on the relation between coliforms and enteric pathogens. *Pub. Hlth. Rep. Wash.* **58** (1943).

19. Wilson, G.H. and Miles, A.A., *Topley and Wilson's Principles of Bacteriology and Immunity*. Vol. 2. Arnold, London, 1955.

20. Hornick, R.B. and Woodward, T.E., Appraisal of typhoid vaccine in experimentally infected human subjects. *Trans. Am. Clin. Clim. Ass.* **78** (1966).

21. Orlob, G.T., Viability of sewage bacteria in sea water. *Sewage Ind. Wastes.* **28** (1956).

22. Clarke, N.A., Berg, G., Kabler, P.W. and Chang, S.L., Human enteric virus in water: source, survival and removability. 2nd Int. Conf. on Water Poll. Res., London, Sept. 1962.

23. Paoletti, A., Facteurs biologiques d'autoépuration des eaux de mer: points clairs et points obscurs d'une question discutée. *Rév. Int. Océan. Méd.* 1970, **18-19**.

24. Aubert, M. and Aubert, J., Pollution marine et amenagement des rivages. *Rév. Int. Océan. Méd.* Suppl. (1973).

25. Aubert, M. and Aubert, J., *Océanographie Médicale*; Gauthier-Villars, Paris, 1969.

26. Brisou, J., Limites de l'autoépuration et de la biodégradation. *Rév. Hyg. Méd. soc.* **17**, 19 (1969).

27. Buttiaux, R., Pollution marines et Santé publique. *Rév. Int. Océan. Méd.* **11**, 157 (1968).

28. Bonde, G.J., Studies on the dispersion and disappearance of enteric bacteria in the marine environment. *Rév. Int. Océanog. Méd.* **9**, 17 (1968).

29. Shuval, H.I., Cohen, N. and Furer, Y.Y., The dispersion of bacterial pollution along the Tel-Aviv shore. *Rév. Int. Océanog. Méd.* **9**, 107 (1968).

30. Flynn, M.J. and Thistlethwayte, D.K.B., Sewage pollution and sea bathing. In *Proc. 2nd Int. Conf. Wat. Poll. Res.* Pergamon Press, London, 1965, Vol. 3.

31. Gorodetskij, A.S. and Raskin, B.M., *The Hygiene of Coastal Seas.* Medizina, Leningrad, 1964 (in Russian).

32. C.E.R.B.O.M., *Rapport d'activité,* 1970.

33. Stevenson, A.H., Studies on water quality and health. *Am. J. Publ. Hlth.* **43**, 529 (1953).

34. Smith, R.S., Woolsey, T.D. and Stevenosn, A.H., *Bathing Water Quality and Health. I. Great Lakes.* U.S. Federal Security Agency, Public Health Service, Environmental Health Center, Cincinnatti, Ohio, 1951.

35. Smith, R.S. and Woolsey, T.D., *Bathing Water Quality and Health. II. Inland River.* U.S. Federal Security Agency, Public Health Service, Environmental Health Center, Cincinnatti, Ohio, 1952.

36. Stevenson, A.H., *Bathin Water Quality and Health. III. Coastal Water.* U.S. Public Health Service, R.A.Taft Sanitary Engineering Center, Cincinnatti, Ohio, 1961.

37. Berger, B.B., Jensen, E.C., Ludwig, H., Romer, H., Shapiro, M.A. and Senn, C.L., Coliform standards for recreational waters. *J. San. Eng. Div., Am. Soc. Civ. Engrs.* **89**, SA, 57 (1963).

38. Reece, R.J., *38th Annual Report to Local Government Board* 1908-9, Suppl. with Report of Med. Officer for 1908-9, Appendix A, No. 6.

39. D'Arca, S.U., Problemi epidemiologici dell'inquinamento costiero. Atti Conv. Naz. su "Problemi attuali di Igiene e Medicina Sociale", Napoli, gennaio 1963.

40. Del Vecchio, V., *L'inquinamento costiero, problema di Igiene sociale.* Ist. Med. Sociale, Roma, 1962.

41. Paoletti, A., Les problèmes hygieniques des coquillages. Atti Simposio Int. su "Pollution marine par les microorganismes et produits pétroliers". CIESM, Monaco, 1964.

42. Paoletti, A., Velocità di contaminazione dei mitili e localizzazione dei germi nei loro tessuti. Impiego del microscopio a fluorescenza. *Rend. e Atti Accad. Sci., Napoli* **120** (1966).

43. Paoletti, A., Aspetti igienici della mitilocultura nel Golfo di Napoli. *Orizzonti Economici* **62**, 60 (1966).

44. Paoletti, A., Ferro, V. and De Simone, E., Problemi igienici dei mitili: moderni concetti di ecologia e fisiologia quali fondamenti di una efficace profilassi delle malattie da essi trasmesse. *Ann. San. Pubbl.* **29**, 1033 (1968).

45. Eklund, M.W. and Paysky, F., Incidence of Cl. botulinum type E from the Pacific coast of the United States. *Botulism 1966,* Chapman and Hall, London, 1967.

46. Kravchenko, A.T. and Shishulina, L.M., Distribution of *Cl. botulinum* in soil and water in the U.R.S.S. *Botulism 1966.* Chapman & Hall, London, 1967.

47. Aiso, K. and Matsuno, M., The outbreaks of enteritis-type food poisoning due to fish in Japan and its causative bacteria. *Jap. J. Microbiol.* **5**, 337 (1961).

48. Fujino, T., Taxonomic studies on the bacterial strains isolated from cases of "Shirasu" food-poisoning and related microorganisms. *Biken J. Osaka* **8**, 63 (1965).

49. Zen-Yoji, Epidemiology, entero-pathogenicity and classification of *Vibrio parahaemolyticus*. *J. Infect. Dis.* **115**, 436 (1965).

50. Dadisman, T.A., Nelson, R., Molenda, J.R. and Garber, H.J., *Vibrio parahaemolyticus* Gastroenteritis in Maryland; *Am. J. Epidem.* **96**, 414 (1973).

51. De Lorenzo, F., L'episodio di colera in Campania. *Giorn. Mal. Inf. Parass.* **26**, 193 (1974).

52. Schiraldi, O., Osservazioni sull'epidemia di colera in Puglia. *Giorn. Mal. Inf. Parass.* **26**, 297 (1974).

53. Angioni, G., L'episodio di colera in Sardegna. *Georn. Mal. Inf. Parass.* **26**, 215 (1974).

54. Ross, B., Hepatitis epidemic conveyed by oysters. *Svensk Lakartidningen* **53**, 980 (1956).

55. Crovari, P., Sulla depurazione dei mitilit nei riguardi del virus poliomielitico. *Igiene Moderna* **51**, 22 (1958).

56. Dougherty, W.J. and Altman, R., Viral hepatitis in New Jersey 1960-61. *Am. J. Med.* **32**, 704 (1962).

57. Mason, J.O. and McLean, W.R., Infectious hepatitis, traced by the consumption of raw oysters. *Am. J. Hyg.* **75**, 90 (1962).

58. Heldstrom, C.E. and Lycke, E., An experimental study on oysters as virus carriers. *Acta Path. Microbiol. Scand.* **48**, 153 (1963); *Am. J. Hyg.* **79**, 134 (1964).

59. Mosley, J.W., Shellfish associated with infectious hepatitis. Hepatitis Surveillance Report, Communicable Disease Center, U.S.P.H.S., 1964, No. 19, 30.

60. Slanetz, L.W., Bartley, C.H. and Metcalf, T.G., Correlations of coliform and fecal streptococcal indices with the presence of *Salmonellae* and enteric risues in sea water and shellfish. *Proc. Int. Conf. in Water Poll. Res., Tokio 1964.* Vol. 3, Pergamon Press, 1964.

61. Cioglia, L. and Scarpa, B., Attivita predatoria e contaminazione dei molluschi eduli. *Nuovi Ann. Ig. Microbiol.* **16**, 28 (1965).

62. Metcalf, T.G. and Stiles, W.C., The accumulation of enteric viruses by the oysters. *J. Infect. Dis.* **115**, 68 (1965).

63. Paoletti, A., Problemi igienici dei mitili; capacita di concentrazione nei riguardi di *E. coli* e *S. Typhi* I; del batteriofago II; del virus poliomielitico III. *Boll. Soc. Ital. Biol. Sper.* **26**, 906, 910, 1580 (1965).

64. Mitchell, J.R., Presnell, M.W., Akim, E.W., Cummins, J.M. and Liu, O.C., Accumulation and elimination of poliovirus by the eastern oyster. *Am. J. Epidem.* **84**, 40 (1966).

65. Foyn, E., Disposal, distribution and effects of organic and inorganic chemical waste in the marine environment; *Rev. Int. Océanog. Méd.* **17**, 51 (1970).

66. SCEP, *Man's Impact on the Global Environment.* MIT Press, 1970.

67. Vigliani, P., Oligoelementi nelle acque di superficie. Sep. Pollution '74 Giornate di Studio.

68. Aubert, M., Le problème du mercure en Méditerannee. *Rev. Int. Océanog. Méd.* **37-38**, 215 (1975).

69. Loreto, G., Aspetti sanitari relativi alla presenza di metilmercurio nei prodotti ittici; *Atti IV° Conv. sui problemi della pesca.* Cesenatico (Forli), 1973, Camera di Commercio, Ind. Art. e Agric. Edit.

APPLICATION OF FIELD DATA TO SYSTEM DESIGN

H. Roy Oakley

J.D. & D.M. Watson High Wycombe, Buckinghamshire, England

Summary – The disposal of municipal and industrial waste to a marine environment is acceptable only if a suitable system is designed to collect, treat and discharge the wastes in a controlled manner. Before such a system can be designed it is necessary to state the problem and the objectives and to establish relevant facts. Field data is often required for the latter purpose, and experience in applying field data to system design in different situations shows that a preliminary assessment of the problem should precede any field investigation so that investigation is related to specified needs and only relevant data acquired. The data should be continually analysed and assessed and action to implement a control system should not be deferred until complete solutions are available. Close and sympathetic collaboration between engineers and scientists of other disciplines should be exercised from the outset.

1. INTRODUCTION

The acquisition of data is a means to an end. In the present context the objectives are identification and quantification of factors affecting the quality of the marine environment, and formulation of the measures that should be taken to protect or improve the environment. The latter may necessitate the collection, treatment and disposal of pollutants, and the purpose of this paper is to show how the engineering task involved depends on the data available, and influences the data collection scheme. It will be useful to define the terms used.

System Design is used in the sense of the planning and engineering design of a system for collection of pollutants, and the treatment and disposal of pollutants into a marine environment.

Pollutants are consequently defined as waste substances which modify the environment to an undesirable or harmful degree, typically municipal sewage or water-borne industrial waste, but not excluding solid wastes.

A system may embrace intercepting sewers, pumping stations, land-based treatment facilities, marine outfalls and diffusers, and occasionally marine based treatment facilities such as artificial aerators. All these components must be integrated into an effective whole if an efficient design is to result.

Disposal of pollutants into the sea is seen as part of the natural cycle, but safeguards are necessary to limit the total quantity of persistent pollutants and to reduce the concentration of degradable matter to within tolerable limits.

Field data is taken to mean facts which have been acquired by investigation and observation which are relevant to the design of disposal systems. They may relate to the design of sewers or other conveyer systems or treatment facilities, but in this paper emphasis is given to those relevant to the design and construction of marine outfalls. The data required will include:

 (a) The characteristics of the waste including quantity and rates of discharge, nature and composition.

(b) Topography, geology and land utilization of the relevant part of the coastline.

(c) The physical, chemical and biological characteristics of the relevant part of the sea, with particular emphasis to water movement.

(d) The characteristics of the sea bed and underlying sub-soil.

(e) The use made or desired of the sea for recreation, food or industrial purposes.

2. OBJECTIVES

Engineering is an applied science, and the design of waste disposal systems cannot proceed without definition of the desired result. This may not be difficult to state in qualitative terms, but considerable difficulty has been experienced in defining tangible and measurable objectives to which design can be related and the success of the system evaluated. In the present context three sets of objectives can be recognized:

(a) Those relating to land use including consideration of land values, amenity protection, air and noise pollution. It should be kept in mind that the establishment and construction of land-based treatment works may solve a problem of marine pollution at the expense of creating problems different in character but of no less magnitude on shore.

(b) Those relating to marine disposal of waste matter. This has been discussed extensively, if inconclusively, elsewhere [1–3].

(c) General economy. Whilst this is often conveniently measured in financial terms, recent concern as to the use of energy, and in particular fossil fuels, serves as a reminder that true economy is concerned with the efficient use of human and material resources, and that alternative systems should be evaluated in terms of energy, material, skilled and unskilled labour, as well as in financial terms.

3. DESIGN APPROACH

The design of systems in engineering terms is thus seen as directed to satisfying the established objectives at least cost in terms of effort, energy, use of material, money or any other relevant criteria. The generic term for a design procedure of this nature is *optimization*, and the collection and use of field data should be applied to this end.

It is not possible in this short paper to attempt a full discussion of engineering aspects of optimizing collection and disposal systems, but brief reference may be made to the specific comparison of alternative combinations of treatment and outfall length, noting that a fuller discussion is contained in previous papers [3–4].

The optimization of treatment systems has not been fully established. Models have been proposed [5] and a recent development in the U.K. [6] is promising, but even for conventional systems for treatment to normal standards, data is generally inadequate for complex models and their development for the particular requirement of marine disposal is probably not likely in the immediate future.

The cost and effect of treatment methods are often little understood by laymen, and not always by

Table 1. Performance of alternative treatment systems

Treatment method	Effluent Quality						Sludge yield (tonnes dry solids per 1000 capita per annum)
	Suspended solids, mg/l	Biological oxygen demand, mg/l	Total nitrogen, mg/l	Total phosphate, mg/l	Synthetic detergent mg/l	Faecal coliform bacteria, per 100 ml	
None (typical municipal sewage)	300	250	40	20	20	10×10^6	
Sedimentation	100	150	35	15	15	2×10^6	25
Chemical precipitation	55	60	ND	4	ND	1×10^6	39
Flotation	90	130	ND	8	ND	1.5×10^6	27
'Partial' biological (to 50:50 standard)	45	45	30	10	10	1×10^6	32
'Full' biological (to 30:20 standard)	20	15	20	8	5	4×10^5	35
'Full' biological with anoxic zone	20	15	8	6	5	4×10^5	35
Electrolytic	85	100	ND	1	ND	4×10^3	31

Note: (1) ND = No reliable data available.

(2) The above figures are approximate values for works of normal design in typical sewage and should be taken as 'order of magnitude' figures for comparative purposes only.

Table 2. Comparative cost of alternative treatment systems

Treatment method	Comparative capital cost	Comparative operating cost	Energy requirements	Labour requirements
Sedimentation	1	1	Low	Low
Chemical precipitation	1.4	1.6	Low	Medium
Flotation	1.2	1.3	Medium	Medium
Partial biological	2.5	2.6	High	High
Full biological	3	2.9	High	High
Full biological with anoxic zone	3.1	2.9	High	High
Electrolytic	1.5	1.9	High	Medium

N o t e s: (1) These figures are approximate comparative values only taking sedimentation as unity. Considerable variation must be expected in varying circumstances and each case must be taken on its merits.

(2) Only 'direct' energy used in operation is considered.

(3) The cost of sludge treatment and disposal is *not* included: see Table 1 for quantities. Sludge disposal costs may be up to 40% of the total cost of municipal sewage treatment.

scientists concerned with marine environment; relevant facts are therefore re-stated in Tables 1 and 2 which show the average effect and comparative cost of the principal alternative methods of treating municipal sewage.

It is emphasized that 'full' treatment is a misnomer and that treatment methods should be orientated towards the main objectives. If, for example, the principal concern is with nutrient discharge, then normal biological treatment methods are inefficient and may be inappropriate.

It is more difficult to make general comparisons on the cost and effect of constructing outfalls of varying size and different length and location, and one of the objectives of data accumulation will be to enable relevant factors to be assessed so that comparison can be made for particular locations. The data requirements for such comparisons may be summarized as follows:

(a) Feasibility and cost. This covers engineering considerations which will include sediment depth and configuration; bedrock characteristics; wave, wind and fog occurrence; the magnitude, direction and variation of currents; restrictions imposed by amenity or industrial considerations; and location of suitable land based construction sites and harbours.

(b) Behaviour of the discharge. This is concerned with factors which will determine the movement and dispersion of an effluent, and of both the settleable and floatable solid fractions. These should include current direction, strength and variation with depth; temperature, density and other physical characteristics; wind direction and magnitude; and horizontal and vertical diffusion coefficients.

(c) Effect on the environment. This will embrace those physical, chemical and biological considerations which will determine the changes in the effluent and the consequential changes on the environment. The latter, in particular, is beyond the scope of this paper, but is should not be forgotten that the environment reacts on the effluent and that there are consequential changes in particle size and behaviour and in chemical and biological characteristics which must be assessed. In other words if, as has been suggested, marine disposal is regarded as part of the natural biological cycle, then the changes in the effluent after discharge, and the consequential effects on the marine environment, should be seen as a process complementary to any treatment afforded before discharge, and as part of the design of the disposal system.

4. METHODOLOGY

When applying this general approach to a specific set of circumstances it is prudent to keep in mind the considerable cost and time of field work in a marine environment. Some figures have been quoted [7] which show that unplanned investigations may be expensive mistakes. Unfortunately there is a 'chicken and egg' situation in that until alternative solutions can be outlined, the data requirements for their evaluation and comparison cannot be fully specified; equally, alternative solutions cannot be devised until a minimum of data is available. It is therefore important that system design and consideration should run in parallel with data collection and that there should be a constant feedback of data and interplay between the investigation and design process.

Typically the problem should be examined in the following sequence: identify the problem – specify objectives – make initial examination of available data as to topography and geology, water movement and the environment – outline tentative alternative solutions – specify data needs to evaluate these alternatives and assess the cost and value of each set of data in relation to the available budget – commence field work – feed back initial results so as to make an early evaluation of alternative systems – modify the tentative schemes and data requirements – acquire and assess the data necessary for comparison of selected alternatives and confirmation of the preferred solution. This can be shown in tabular form as in the following sketch:

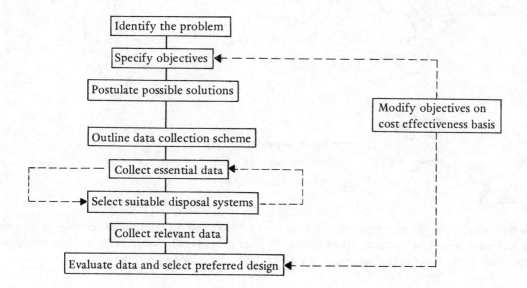

5. CASE HISTORIES

It is scarcely necessary to emphasize that particular investigations do not all follow the same pattern, and it may be of value to outline three recent examples with different features.

The investigation in the Bay of Naples was primarily concerned with evaluating and proposing remedies for an existing situation of unacceptable pollution.

The South Hampshire investigation concerned the particular needs for a proposed development of particular location and size, and is of interest because of the sensitive nature of the area, which is highly valued for amenity purposes.

In contrast the Hong Kong investigation was related to development plans of a more fluid nature and showed the way in which consideration of waste disposal to the marine environment, in comparison with the alternative of land treatment and disposal, influenced the development plans.

5.1. Bay of Naples

In the 110 km of coast of the Province of Naples between the Lido di Licola and Punto Germano (Plan No. 1) sewage and industrial waste from a population of 2,400,000 discharges with little or no treatment.

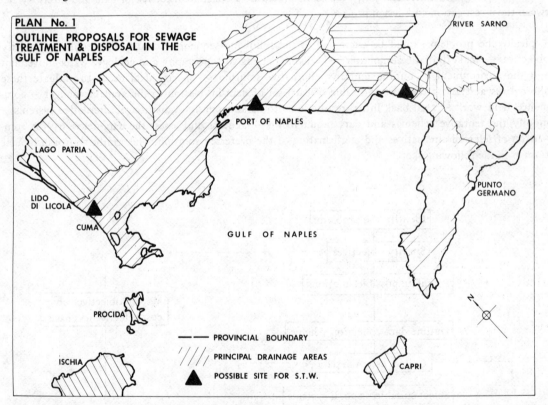

Concern with the public health situation is well known [8, 9] and the effect on the ecology of the water has also been the subject of study over a number of years [10, 11]. The unsatisfactory amenity situation, particularly with respect to the beaches, is less well publicized but evident to any visitor.

In 1971, the Provincia di Napoli commissioned a study with the following terms of reference:

(a) study the sources of civil and industrial pollution in the Gulf of Naples from Lago Patria to Punta Germano including the coastline of the isles of Capri, Ischia and Procida.

(b) the results of the study will be presented as follows:

   (i) a map of the existing pollution situation, highlighting the more serious situations.

   (ii) evaluation of the capacity for self-purification of the sea at varying distances from the coast and on the eventual dispersion of the wastes.

   (iii) economic evaluation of the various solutions.

   (iv) a proposed plan for the improvement of the situation.

   (v) a proposed system for controlling the discharge of domestic and industrial wastes to the coast.

The results of the investigation were set out in three volumes [12] :

   Volume I   *Metodologia,* February 1973

   Volume II   *Risultati delle Endagine,* May 1973

   Volume III   *Proposte,* June 1973

Evaluation of the present and future quantity and nature of the waste discharge was a major undertaking requiring visits to more than thirty local authorities and numerous major industrial undertakings. All the various official organizations involved in the control of pollution or representing industry were contacted and drawn into the study. The coastline was walked and every discharge mapped and the principal streams sampled.

Examination was made of the existing sewerage and sewage treatment facilities, topography and development so that drainage areas of each discharge could be delineated and the present and future pollution loads assessed. Using this as a basis the tentative lines of intercepting sewers and sites of treatment works were proposed.

The study indicated three main areas of pollution at Cuma, the Port of Naples and the mouth of the River Sarno. Investigation of water movement was thus concentrated at two points, off the coast at Cuma and Sarno. An outline examination was made of the present physical and chemical condition of the coastal waters and sufficient data was obtained to enable a simple mathematical model to be constructed from which the behaviour of a discharge to any point within the Study Areas could be predicted. This in turn, in combination with an assessment of the degree of waste treatment required, enabled the feasibility and cost of outfalls in the two locations to be assessed.

Broad proposals were thus put forward for the improvement of the existing situation and proposed a number of sewage treatment works and outfalls. Each of the proposed systems would require examination in detail, and following presentation of the Report and other evidence the interested authorities have proceeded to invite proposals for the detailed design and construction of treatment works and outfalls generally in accordance with the study recommendations.

The study showed that it is not completely satisfactory to study pollution problems within administrative boundaries, but rather within drainage basins. More than 80% of the pollution discharged to the Bay of Naples from the River Sarno came from outside the provincial boundary.

5.2. South Hampshire

It was proposed to develop a new city for upwards of 1½ million people in the area between Southampton and Portsmouth (Plan No. 2). Natural water courses are small and it was decided to examine alternative

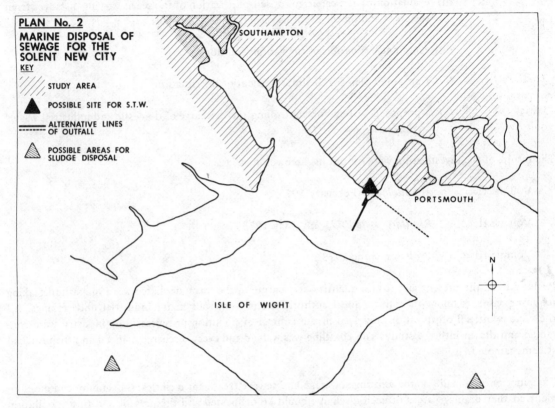

strategies of a treatment works on the coast designed to discharge effluent of high standard (30:20 BOD and SS) through a short outfall and the discharge of sewage, after the minimum of pretreatment, to deeper water.

The Solent is an area of great natural beauty, well known for its yachting and other recreational activities, and unusual in that it possesses some of the characteristics of an estuary and some of the open sea.

The terms of reference were to examine and report on the feasibility and probable acceptability of marine disposal of partially treated sewage arising from a specified area; to identify the most suitable points and methods of discharge from the outfall sewers; to examine suitable methods and probable acceptability of marine disposal of the sludge arising from the treatment of sewage; and to report, under appropriate heads, on the estimated costs involved in the foregoing.

A preliminary examination of the problem was undertaken using data already available and examining the quantity and quality of sewage likely to arise, the probable location of intercepting sewers, the principal characteristics of the adjacent waters and the environmental circumstances. It was concluded that marine discharge was likely to be feasible at a cost comparable with full treatment, and the data requirements for the delineation and assessment of a suitable system were outlined [13].

Work then proceeded in more detail. The principal areas of investigation were:

(a) An assessment of present conditions as a reference basis, including consideration of water contact sport, angling and fishing, navigation, existing outfalls, existing chemical and biological conditions of the water and oil pollution.

(b) A statement of objectives, considering health, amenity, commercial and other considerations.

(c) Collection and assessment of data as to mass water movement; local water movement; surface movement; dispersion rates; sediment movement; geology; navigation; feasibility and cost of sewage treatment and sludge disposal.

To allow flexibility in evaluation of alternative points of discharge, a mathematical model was constructed to permit prediction of water movement over an area large enough to cover all feasible points of discharge. Initial examination of geology and possible construction techniques and other constraints, suggested that only two or three alternative designs of outfall needed detailed consideration, and further examination of the sea bed profile and subsoil was restricted to these routes.

Similarly, consideration of the probable movement of sediments indicated areas for more detailed examination by sea bed drifters and other means.

A particular difficulty was the assessment of the present biological conditions which necessarily depended on such historical records as were available and such appreciation of the situation as could be obtained in the limited time of the survey. A systematic collection of data was initiated which it was hoped would extend over future years so that natural fluctuations could be assessed.

In the result, comparison of alternative trial designs led to the selection of the most favourable, which could be compared with the alternative of a full treatment works discharging through a short outfall [14]. In the event, the latter alternative was selected.

5.3. Hong Kong

The Hong Kong Study has some of the characteristics of each of the previous investigations described. The background circumstances and approach to the investigation have been described elsewhere [15] and it is sufficient here to say that the problem was to examine the situation resulting from the discharge of largely untreated sewage from a population of about 4 million into the harbour and adjacent coastal waters; and to consider the consequence of population growth and of development of new urban areas both within the existing development and in new centres elsewhere in the New Territories (Plan No. 3).

A Preliminary Report was made in which broad consideration of the problem of New Territories was given, and comparison made of the alternatives of treatment and discharge to inland water courses and disposal to coastal waters [16]. This recommended a more comprehensive marine survey which was initiated in 1970. The marine survey investigation covered a complex situation along some 70 km of coastline and within the large inlet known as Tolo Harbour. It compressed into a 12 month period a comprehensive sampling programme designed to determine the condition of the water in those areas of sea affected by the present sewage discharges. Examination was made of the background as to topography, geology, meteorology and physical factors, and current meter observations and float tracking exercises were carried out to give sufficient information on water movement for comparative designs to be evaluated. Bacterial tracing was used to examine the behaviour and dispersion of discharges from the principal existing outfalls. The Report [17] provided a reference document which is used in the consideration of alternative development strategies. When these had been determined, more detailed investigation was made of particular areas to facilitate the design of waste disposal systems. For North West Kowloon, a

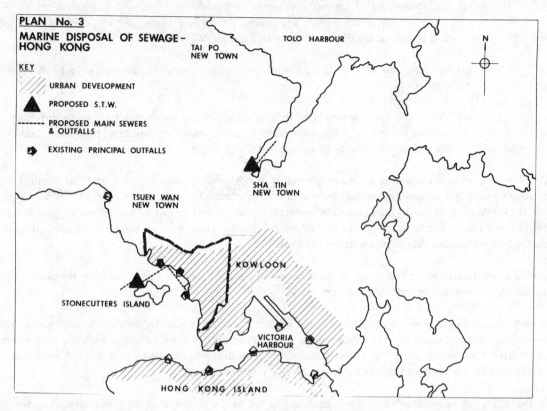

feasibility study was made of a system of intercepting sewers and treatment works to relieve the pollution from a drainage area serving 1 million people. The basic data on water movement from the previous Report was supplemented as alternative designs were developed so as to provide more specific information related to the particular locations, and data was also obtained relevant to the feasibility and cost of construction [18]. For Sha Tin, where a new town for 500 000 people is to be built, the previous data was supplemented by further investigation so that the treatment needs and comparative costs for alternative points of outfall within Tolo Harbour could be examined. A major consideration was the increase in nutrients which might result and the consequential effects on the ecology. The time scale did not permit a programme of experimental work relating to this complex problem, and the proposals consequently incorporate facilities for biological denitrification at the treatment works which can be applied should the need become apparent as the population and flow builds up [19].

6. CONCLUSIONS

It is hoped that this paper will have demonstrated that no set values can be applied and that it is necessary to be flexible in approaching problems of this nature. There are, however, some general conclusions that can be drawn:

(1) Time spent in reconaissance is seldom wasted. A careful preliminary examination of the problem, collection and evaluation of available data readily obtainable, and the development of tentative outline solutions, is an essential pre-requisite to any field investigation.

(2) More haste — less speed. To rush into field work without adequate preparation, and to collect data without thought to its analysis and use, may waste effort, money and time.

without thought to its analysis and use, may waste effort, money and time.

(3) Perfection is never attained. The danger of perpetual investigation, to the exclusion of constructive action, is self-evident. It is illusory to seek complete solutions to the complex problem posed by waste discharge to the sea, and the time will come when the information available should be judged as adequate to make a sound but flexible scheme which can be implemented and monitored.

(4) Teamwork is essential. Close and sympathetic collaboration between design engineers and the many other disciplines involved in the investigation of marine pollution problems is essential to a sound solution and should be practised from the outset. It has been shown that consideration of engineering design needs materially affects the programme of investigation; and that periodic evaluation of data in relation to system design should be made throughout the investigation, and interim appraisals made which may well modify the investigation and even the objectives. The paper will have served its purpose if it assists in developing an understanding of these needs.

REFERENCES

1.   Pike, E.B. and Gameson, A.L.H., Effects of marine sewage disposal. *Wat. Pollut. Cont.* **69**, 355 (1970).

2.   Key, A., Water pollution control in coastal areas: Where do we go from here? Proceedings of Institute of Water Pollution Control Symposium, 'Water Pollution Control in Coastal Areas', Bournemouth, May 1970. p. 111.

3.   Oakley, H.R. and Cripps, T., Investigation and design: Proceedings of Institute of Water Pollution Control Symposium, 'Water Pollution Control in Coastal Areas', Bournemouth, May 1970. p. 64.

4.   Cripps, T., Treatment prior to discharge — Cost Benefit considerations. Proceedings of 23rd Commission Internationale Pour L'Exploration Scientifique De La Mer Mediterranee Congress, Athens, Nov. 1972. p. 173.

5.   Shih, C.S. and De Filippi, J.A., System optimization of waste treatment plant process design; *J. Sanit. Engng. Div.; Am. Soc. Civil Engs.* **196**, SA2, 409 (1970).

6.   Construction Industry Research & Information Association. Cost-effective sewage treatment — the creation of an optimising model, CIRIA Report 46: May 1973.

7.   Oakley, H.R. and Staples, K.D., Interpretation of field data. *Discharge of Sewage from Sea Outfalls.* (Ed. A.L.H. Gameson), Pergamon Press, Oxford, 1975.

8.   Paoletti, A., Pollution fécale du littoral de cumes et considérations sur le pouvoir d'autoépuration du milieu marin. *Rev. Int. Océanogr. Méd.* (1966).

9.   Pierantoni, A., La miticoltura nel golfo di Napoli. *Boll. Soc. Nat. Napoli,* **76**, (1967).

10.   Puri, H.S., Bonaduce, G. and Malloy, J., Ecology of the Gulf of Naples. *Staz. Zool. Napoli,* **33**, Supplement 87-199 (1964).

11.   Bacci, G., *Staz. Zool. Napoli,* **37**, Supplement 87-199 (1964).

12.   Eurostaff S.p.A. & Dagh Watson S.p.A., Studio Sull'Inginnamento Del Golfo Di Napoli. 1973.

13.   J.D. and D.M. Watson., Report on Marine Disposal of Sewage for South Hampshire Plan Advisory Committee: July 1970.

14. J.D. and D.M. Watson, Marine Disposal of Sewage and Sewage Sludge for South Hampshire Plan Advisory Committee. May 1972.

15. Oakley, H.R. and Cripps, T., Marine Pollution Studies in Hong Kong and Singapore. *Marine Pollution and Sea Life*. Proceedings of Food and Agriculture Organization Conference on Marine Pollution and Its Effects on Living Resources and Fishing, Rome, Dec. 1970.

16. J.D. and D.M. Watson, Report on Sewerage and Sewage Disposal in the Northern New Territories for Government of Hong Kong, Nov. 1965.

17. J.D. and D.M. Watson, Report on Marine Investigation Into Sewage Discharges for Government of Hong Kong, July 1971.

18. J.D. and D.M. Watson, Report on Investigation Into Sewage Treatment and Disposal for North West Kowloon for Government of Hong Kong, April 1974.

19. J.D. and D.M. Watson, Report on Investigation Into Sewage Treatment and Disposal for Sha Tin New Town for Government of Hong Kong, Dec. 1973.

DESIGN AND CONSTRUCTION OF SUBMARINE PIPELINES AND DIFFUSERS

W.G. Gerald Snook

Parry, Froud and Snook — Consulting Civil Engineers,
Clevedon, Avon, U.K.

Summary — There is continuous development of new constructional techniques, materials and instrumentation not necessarily associated with public health engineering. The effect of these developments on overall design considerations, including public health aspects, is discussed. It is essential for the public health engineer to maintain efficient liaison and interchange of research and development data with various disciplines and fields of engineering, science and industry.

1. CONSTRUCTIONAL TECHNIQUES

To indicate advances made insofar as Europe is concerned it is surely sufficient to state that in 1950 it was doubtful if any submarine pipeline extended more than 1 km below low water mark. As a result of offshore oil and gas development, it is anticipated that by the end of 1975 the total length of submarine pipelines will run into many hundreds of kilometres. The world total length of submarine pipelines runs into several thousands of kilometres of which 95% were laid subsequent to 1950.

1.1. Submarine pipelines

The constructional techniques for installing the submarine pipeline section of a waste disposal scheme fall into the following main categories:

1.1.1. *Bottom tow method.* This method has probably been the most widely used for "near-shore" submarine pipelines.

Lengths or "strings" of pipe equivalent to the total length of submarine pipeline are constructed in their entirety on land. For instance for a 6 km submarine pipeline there may be 12 "strings" of 500 m. The number of "strings" depends upon the construction site and the area and length available. The first "string" is placed on rubber-wheeled conveyors and is pulled into the sea along the seabed by powerful winches mounted on barges anchored offshore. For joining the individual lengths of "strings" a "tie-in" area is established above the high water mark. When the tail-end of the first "string" reaches the "tie-in" area the pull is stopped while the second "string" is moved on to the conveyors and joined to the first. The pull recommences until the tail-end of the second "string" is in the "tie-in" area and the procedure is repeated until the total length of submarine pipeline is installed on the seabed or in a previously prepared trench in the seabed.

It is not necessary that the "strings" are in a straight line on the projected line of the submarine pipeline. Often, due to site restrictions, the "strings" have to be curved, but, of course, the radius of curvature has to be within the stress limits of the pipe and facilitate reasonable handling conditions on site. The seaward tangent point on the curve must obviously be well back from the "tie-in" area so that the jointing process takes place on a straight line.

Various stresses are imposed on the submarine pipeline during this type of construction. These include direct tensile stress due to the pulling force necessary to overcome friction on the sea bottom; bending stress resulting from curvature on land and undulations on the seabed, also as a result of being supported on the conveyors (continuous beam effect).

The foregoing are some of the stress conditions which can exist during a bottom tow operation and when equated the total stress usually dictates that steel must be used as the material for the submarine pipeline in order to contain these stresses.

For the particular utilization under consideration steel needs to be protected against corrosion, both internally and externally and these aspects are discussed under Section 3 below.

The author has successfully utilized non-ferrous metals and plastics for "bottom-tow" operations, but these have usually been in small diameters, comparatively short lengths or in sheltered waters with low current speeds.

1.1.2. *Pipe by pipe method.* This entails using divers to joint pipes under water — the pipes either laying on the seabed or in a trench previously excavated in the seabed.

The joints are inevitably mechanical joints in the form of flanges or preferably a flexible "push-in" type of joint. The author would always advocate burying a submarine pipeline and subsequently backfilling over the pipeline. In consequence the pipeline is not subjected to the forces of waves and currents or damage from external sources such as ships' anchors and fishing nets. Therefore, this method facilitates the use of the wide range of plastic pipe materials available.

The economics of this method depend upon the speed at which the divers can work, which in turn is severely affected by the weather and sea conditions. The method also relies upon being able to maintain a trench during the jointing of the pipes and in consequence the type and variation of seabed material is an important consideration when considering this method at the design stage. Generally the more cohesive the material the more attractive this particular method becomes.

In cases where local conditions allow this type of construction the author has found that this particular method shows considerable economies compared with other methods described in this section.

1.1.3. *Float and drop method.* This involves floating a pipeline or sections of pipeline on the surface of the sea, towing them on the surface to their final location and sinking them on to the seabed or into a previously prepared trench.

Depending upon circumstances the pipe may be fabricated in its entirety or in sections. If in sections jointing can either take place on the bottom or by bringing the previously laid seaward end to the surface and joining on to the next section before lowering to the seabed and thence repeating the process.

Obviously this method is greatly affected by weather and sea conditions. It is extremely difficult to control bending of the pipe and severe stresses due to radius of curvature will exist unless the selected pipe material is compatible with the radius of curvature.

1.1.4. *Lay barge method.* The lay barge method is used almost exclusively for laying long distance submarine pipelines for the oil and gas industry. They are extremely expensive but efficient items of equipment operated by highly skilled crews on a 24 h/day basis. The method is somewhat restrictive as to the diameter that can be handled and at present this appears to be a maximum of 1250 mm. Laying rates of 150 m/day are not uncommon.

The method is essentially a seaborne operation as all pipe jointing takes place on the barge. As pipes are jointed the assembled line passes down a ramp into the sea and on to the seabed. As each new joint is made the barge is winched forward on previously laid anchors over a distance equivalent to one pipe length. The normal material used is steel and jointing is almost inevitably achieved by welding. The jointing is carried out in five or six separate positions (known as stations) on the barge. A specialist team mans each station and the work of each team is programmed so that as far as is possible they complete their section of the jointing operation simultaneously with all others so that the barge can be winched forward one pipe length. The work carried out by the first team passes to the second team who will apply their particular process. The first station usually aligns the pipe and carries out the all important first runs of weld (root-bead); the second station usually completes the weld; the third station carries out the radiographic examination of the weld; the fourth station makes good and completes the protective coating on the outside of the steel pipe (usually bituminous enamel reinforced with fibreglass); the fifth and usually the last station makes good and completes any reinforced concrete weight coating which may be applied to the pipeline.

To commence operations the barge is normally brought into the shore and commences jointing pipes which are then pulled off the barge towards the shore by a landbased winch. Once the pipe has reached the shore it is secured and the barge commences the operations described above working seaward. Occasionally, when very long submarine pipelines have to be laid for offshore oil recovery several barges are used on the same line. Each barge is allocated a section and on completion of any two sections the ends are brought to the surface, joined and lowered back on to the seabed.

Stress considerations are all important when considering lay barging, particularly in nearshore coastal waters. Although the barges used are exceptionally large nevertheless they do, of course, roll and pitch much as any vessel. The configuration of the pipe between the barge and the seabed usually takes the form of a reversed curve catenary. In order to reduce curvature on the surface the jointing stations are placed on an inclined ramp so that the pipe does not have to be contorted to a horizontal position on the barge for jointing. In order to reduce bending moments and shear forces due to curvature in the vertical plane and transverse forces of currents on the horizontal plane the pipe is supported for some considerable distance below the surface on a semi-buoyant extension of the ramp (this is referred to as a ladder or stringer). In spite of the foregoing precautions it can be seen that excessive movement of the barge can cause considerable stressing of the pipeline in the immediate area between the barge and where it enters the sea. Should sea conditions become sufficiently bad so as to produce movement beyond a pre-determined amount the operation of jointing is discontinued and the pipe lowered on to the seabed. The barges are equipped with cranes and davits for this purpose.

Under normal circumstances for the short length of submarine pipeline involved the lay barge method is not economic due to the cost of mobilization and running costs of such barges. However, on the smaller diameter pipes it is possible to improvize equipment to produce a lay barge rather than employ purpose-made equipment. The author has found that with the increasing incidence of lay barge operations in European waters it has become attractively economic to negotiate with lay barge operators to carry out short submarine pipeline projects between major offshore contracts.

One difficulty of utilizing this method and steel pipe is that the pipes have to be welded individually on the barge and in consequence would burn off the internal lining material unless this lining was kept back out of the heat range of the weld. In this case the lining must be made good after welding and, of course, inspected thoroughly before the particular joint concerned entered the sea. In spite of the possible use of rapid setting lining materials for this making-good process nevertheless the operation does involve considerable delays to the normal output of a lay barge without the necessity for making good such linings. This problem can be overcome by utilizing longer lengths of pipeline with flanges or special double-walled joints (welding taking place on the outside wall) or by inserting prefabricated resin coated resiliant plastic sleeves after welding.

Lay barges can also be used to lay certain types of plastic pipe or aluminium or other non-ferrous pipes.

The author is particularly interested in the development of ductile iron as a material for submarine pipe-lines and discusses this material in Section 3 below.

It is usual when using the lay barge method for the pipe to be placed on the seabed and subsequently entrenched by high pressure jetting devices which are pulled along the pipeline displacing the seabed material and entrenching the pipe. In consequence the pipe must be heavy enough to punch its way through the mixture of seawater and seabed material which inevitably has a much higher density than seawater itself. The use of lightweight non-ferrous or plastic pipe materials under these circumstances presents problems in providing this necessary weight. Depending upon the type and variation of seabed material it may be possible to pre-trench prior to lay-barging which would overcome the problem of utilizing lightweight non-ferrous and plastic pipes. Trenching is discussed in more detail in Section 2 below.

1.1.5. *Reel-off barge.* This method is essentially similar to the lay barge method described in 1.1.4 above, but the pipe is reeled on to a large drum positioned on the barge. This operation is usually done at the place of manufacture of the pipe and the barge with the pipe aboard is towed to the construction site which can often be several thousand miles away. The barge is positioned near the shore and the end of the pipeline is unreeled, taken ashore and secured. The barge is then winched seaward on its anchor wires and the pipe unreeled from the barge on to the bottom of the sea or into a previously prepared trench.

Obviously the limitations of this method are in the flexibility of the pipe material concerned and in the economic size of reel and barge relative to the diameter of the pipe. To the author's knowledge the maximum pipe diameter attempted by this method has been 250 mm and the maximum length of this diameter taken on to the drum has been 12 km.

A variation of this method has been to instal on the barge a plastic pipe extruding machine and to extrude the pipe as the barge is winched seaward.

2.   TRENCHING – EXCAVATION AND BACKFILLING

Depending upon the type and variation of seabed strata it is the author's experience that excavation and backfill necessary to bury a submarine pipeline in near shore coastal waters can often involve the major factor of expenditure on a project or become the major criteria influencing design consideration of the whole scheme.

The main problem seems to be maintaining a trench for a period long enough to safely instal the submarine pipeline prior to backfilling. Obviously the more cohesive the seabed the easier it is to maintain a trench as the angle of repose of the material increases. A pre-dug trench in sand can quite easily disappear over one storm or even one strong Spring tide. Under these conditions it would, therefore, seem logical that a method involving the laying of the submarine pipeline prior to excavation or laying pipe by pipe might be the most economic solution on the basis that all work done is work permanently achieved. Unfortunately the decision-making parameters are not straightforward and involve application of experience to the results of a careful hydrographic survey and investigation programme. This situation often creates a problem for the consulting engineer in that of several contractors tendering, ones method of laying may be advantageous compared with others, but this method of excavation and backfill with the plant in his possession may be completely inferior to some of the other contractors. The author has found in recent years that it is sometimes advantageous to engage several contractors on a negotiated basis taking advantage of their particular specialization and equipment for various sections of the work. Several major projects have been most successfully and economically completed utilizing this type of contractual arrangement where the consulting engineer acts as co-ordinator on the project, not only designing the works, but also deciding on the particular method of construction and modus operandi for various sections of the work.

2.1. Jetting

During the last 10 years great advances have been made in the development of high pressure jetting equipment which enables a submarine pipeline to be entrenched into the seabed having previously been laid on the surface of the seabed. Water pressures in excess of 100 kg/cm^2 have been used and these devices have successfully cut through quite stiff clay. As discussed in Section 3 below it is essential that the pipe is strong, flexible and heavy in order that the curvature necessary to take the pipe from the seabed to the final bottom of trench level can be readily and safely achieved and that the pipe is heavy enough to punch through the increased density of seawater/seabed material mixture.

Utilization of this method of entrenching quite often requires that the backfilling is achieved by natural processes which is not always acceptable or, indeed, feasible.

2.2. Ploughing

Since 1960 developments in submarine ploughing have provided a means of pulling a gigantic underwater plough along the seabed utilizing winches as for the "bottom-tow" (1.1.1 above). Submarine ploughs come in all sizes up to an approximate maximum capable of providing a 3 m deep trench with sloping sides, thus producing a trapezoidal section of trench. The speed of ploughing depends upon the size of the plough, its strength and the horse power available for pulling relative to the type of material being excavated. Speeds vary from 2 to 6 m/min. The plough is usually of the "double-throw" type which deposits the excavated material either side of the trench. It is possible to backfill the trench by pulling a backfilling plough along in the same manner as the excavating plough. Often a pipe is attached behind the plough so that the trench and the pipe are installed in one operation. The ploughs are controlled hydraulically so that the depth of cut can be varied to accommodate undulations on the seabed should this be desirable.

2.3. Grabbing

A normal excavator equipped as a grab (clam shell) and mounted on a barge is a very commonly used piece of equipment. Whether or not to use this equipment depends upon the type and variation of seabed material. The method is most economic in cohesive materials and grabs are available of several cubic metres capacity capable of digging in stiff clay.

In deep water it is difficult to land the grab exactly in the trench and this difficulty is further aggravated if cross-currents prevail. However, it is always a source of amazement to the author that good excavator drivers, normally land-based, very quickly adapt to this sort of work and can manage to land a grab on the trench line with relentless accuracy even in the most adverse conditions.

One advantage of this method is that due to the jibbing radius of the excavator the backfill material can usually be placed well away from the side of the trench, thus avoiding surcharge. In the case of a reasonably good cohesive material the excavated material can be re-dug for backfilling. Excavation and backfilling is usually achieved by the use of accurate transit lines fixed ashore or by laser beam.

2.4. Bucket dredging

It is very seldom that bucket dredgers can be used in open sea conditions due to the fact that the bucket ladder is a rigid attachment to the vessel which in all but calm weather can cause severe shock loads to the ladder and the vessel itself as a result of pitching and rolling of the vessel. Where it is possible to use such equipment the outputs are very high and the excavation can be very accurately achieved.

2.5. Trailer suction hopper dredgers

These vessels are capable of removing seabed material by suction when under way. They are essentially a self-propelled bottom opening hopper barge equipped with suction pumps and a trailing suction pipe terminating with a "drag-head" which can vary in design for the particular material being excavated. Some of these vessels are large enough to take on board 8000 m^3 of material in periods ranging from 1 to 5 h, depending upon the material being excavated.

The same vessels can be used for backfilling and the accuracy of excavation and backfilling can be achieved either with shore-based transits sited visually, but normally such vessels would utilize either laser beams and/or electronic plotting devices enabling them to fix themselves within 1 m several miles from the shore.

2.6. Rock cutting suction dredgers

The rock cutting suction dredger involves a fixed "ladder" terminating in a revolving cutting head driven by a shaft on the ladder. Both the trailer suction hopper dredgers (2.5 above) and rock cutting suction dredgers are almost inevitably equipped with swell compensators to enable them to work in quite heavy swells without damage to the suction pipe or ladders.

2.7. Blasting

In certain types of seabed material, not necessarily hard rock, it may be economic to blast prior to excavation.

The development of shaped charges for underwater use enables a series of shaped charges to be placed on the line of the trench at a spacing depending upon the type of material to be blasted. Blasting of all charges takes place simultaneously and it is possible to create a complete trench in that the blasted material is deposited either side of the trench formed. This is achieved by a two-phase explosion separated by milliseconds. The first phase projects a conical jet into the seabed and the second phase fills this jet hole with explosive gases producing an upward and outward explosion forming a semi-elliptical trench. The greater the depth of water the more efficient is the process.

2.8. Underwater excavators

In recent years traditional land-based excavating equipment has been modified to work underwater. The type of equipment adapted in this way has included circular trench diggers, Back-acter (back-hoe), Bulldozers and Mole ploughs. This equipment is normally operated by divers, but remote control of such equipment is being developed. One of the problems with the use of this equipment is the lack of visibility underwater created by the excavation process itself.

The author understands that remote control, electronic alignment and depth of trench monitoring equipment is being developed. When proved, this type of excavation equipment should make a great impact on the economic aspects of this particular field of engineering.

3. MATERIALS

In the author's opinion the ideal material for submarine pipelines utilized for waste disposal in the marine environment would be taht which would provide a pipe which is very strong, flexible and heavy.

Weight is usually essential where a pipe may at any time, either during construction or in its working life, be subjected to the forces of waves, currents or moving bed conditions. Where this weight is not completely provided by the intrinsic properties of the pipeline material it has be be provided artificially and this is usually achieved by applying reinforced concrete coatings to the outside of the pipe. If the construction method involves placing the pipe whilst empty or if there is the danger of a pipe ever becoming empty during its working life it is essential that the total weight of the pipe is at least sufficient to provide a vertical component of weight to resist the upward force of buoyancy. Where pipes can be subjected to the force of currents and waves the weight required to provide a vertical component sufficient to resist these forces can be calculated.

In opening this section reference was made to strength, flexibility and weight and the degree of either of these properties is entirely dependent upon the inter-relationship between the local conditions, method of laying selected and the pipeline materials available. Obviously, therefore, each case must be examined on its merits and it is impossible to generalize when discussing this aspect of the subject.

3.1. Steel

Steel has for many years been the traditional material for submarine pipelines and as mentioned in Section 1.1.1 it is essential that it is lined internally and externally to resist corrosion. The internal lining must also be capable of resisting abrasion particularly where high velocities, together with the solids content of an effluent, can produce an abrasive environment.

In the author's experience the ever-increasing cost of lining materials and processes is gradually eroding the economic advantages of utilizing steel for this particular application of submarine pipelines.

Mention was made in Section 1.1.4 of the difficulty of making good the lining after welding on a lay barge. The same problem arises when making a joint in the "tie-in" with the bottom tow method mentioned in Section 1.1.1. This has been overcome by double-walled joints and also incorporating stainless steel stubs at the end of each "tie-in" joint, the weld being achieved with stainless steel electrodes, thus providing continuity of protection, if not lining.

3.2. Alloys and non-ferrous metals

Various alloys can, of course, be produced to resist the corrosive effects of seawater, but these are inevitably extremely expensive materials and often difficult to weld under the normal field conditions experienced in this type of work.

Of the non-ferrous metals various grades of aluminium appear to be the most attractive and developments in welding this material have now eliminated the problem of corrosion and cracking due to heat stress and stress corrosion. However, the question of providing weight to this material still arises. Development of constructional techniques utilizing the wide range of plastic materials diminishes the economic and technical advantage of using aluminium alloys for this type of work.

3.3. Ductile iron

The author is extremely interested in this relatively new material. The material has the advantage of possessing properties of both steel and cast iron in that it is anti-corrosive and yet ductile. In addition, of course, the material has the same weight characteristics as steel and cast iron, which is a further advantage. Unfortunately at present the material is usually produced in relatively short lengths and the joints provided are either a mechanical type of push-in flexible bolted joints or flanges. Research is continuing into the welding of ductile iron and the author considers that when this material can be welded satisfactorily under field conditions it could well have quite a drastic economic impact on this particular field of engineering.

3.4. Plastics

Of the numerous types of plastic material available for pipelines the most commonly used for submarine pipelines have been u.PVC, glass reinforced PVC, fibreglass (G.R.P.), high density polyethylene, rubber (natural and synthetic — reinforced and un-reinforced), together with various combinations of these materials and ferrous materials.

One disadvantage of this class of materials is that they are almost inevitably very light in weight and require additional weight to be provided artificially.

A most interesting combination of materials is a high density polyethylene liner reinforced with steel tape and armoured wiring which can be considered as a submarine electricity cable without the conductor. Unfortunately this pipe is only made to 170 mm i.d., but can be continuously manufactured in lengths of up to 12 km, facilitating laying by the "Reel-off barge" method (1.1.5). This particular type of pipe is indeed very strong, flexible and heavy.

4.   DIFFUSERS – PROTECTION, OPERATION AND MAINTENANCE

The diffuser section usually takes the form of small diameter vertical riser pipes fixed to branches on the soffit of the main bore of the submarine pipeline and terminating above bed level in a 90° bend and a nozzle facilitating ejection of the waste in the horizontal plane. In order to maintain velocity in the main bore the main bore section of the diffuser length is normally reduced in section, the terminal of which is brought on a gentle bend to the surface and finished with a flange into which is mounted a horizontal diffuser nozzle of lesser diameter. The terminal flange is usually hinged and secured with a single bolt to facilitate future maintenance. The author advocates the placing of access flanges periodically along the main bore of the diffuser. These access points consist of vertical branches of the same diameter as the main bore terminating in an easily removable flange. Often these can be incorporated as part of a particular diffuser port.

4.1. Protection from fishing nets and ships' anchors

As the diffusers project above the seabed level they must be protected against the forces of currents, waves and physical damage from external sources such as ships' anchors, fishing nets and sand waves.

This protection is usually best provided by domed reinforced concrete "igloos", the diffuser nozzle discharges through a slot in the concrete dome and is set back within the external extremities of the concrete dome. The dome itself sits on precast concrete rings which are in turn founded on concrete bagwork.

A manhole is placed in the top of the dome to provide access for maintenance of the diffusers.

Should a bottom trawl fishing net encounter the dome it would slide over the dome without causing damage either to the dome or the net.

4.2. Operation and maintenance

Obviously it is desirable to build into the design any facilities which reduce the necessity for maintenance to a minimum. In addition to careful selection of materials it is possible to incorporate various features in the design, all of which will reduce and ease the task of future maintenance. These are briefly discussed below:

4.2.1. *Types of discharge.*
Basically there are two types of discharge, either pumped or by gravity. In both cases the discharge can be continuous or intermittent depending upon the flow characteristics. In either case velocities should be self-cleansing and high enough not only to keep material in suspension during discharge, but also to re-suspend material which may have deposited itself on the invert of the pipe during times of no discharge.

4.2.2. *Grease build-up.*
In coastal resorts the gastronomic inclination of the holiday populus is inclined towards fried foods and consequently a much greater proportion of grease is contained within the sewage from such an area. It is, therefore, essential that velocities both in the main bore and through the diffuser ports are sufficiently high to prevent the build-up of grease.

4.2.3. *Blockages.*
In a properly designed scheme blockages should, in theory, not be possible. However, operation manuals and instructions are sometimes ignored and it is remotely possible under these circumstances that a partial blockage could occur. Provision must be made for such a remote possibility and this takes the form of regular access points along the length of the submarine pipeline and particularly in the diffuser area, all as described in 4 above.

Also as mentioned in 4 above the diffuser section should terminate in a hinged flange containing the terminal diffuser port. This flange enables the system to be flushed after all other diffuser ports have been temporarily blocked off with flanges bolted or clamped to the nozzles. This produces an extremely high velocity through the diffuser section.

4.2.4. *Air and gas release.*
In the case of an undulating seabed it may be possible to form high spots. It is remotely possible that if there is a forced period of no flow extending over several weeks gas may be released from the wastes and collect in these high spots. In consequence very small diameter diffuser ports are installed on these high spots, the diffuser terminating in a loosely bolted mesh screen which will trap solids, but release gases.

4.2.5. *Cathodic protection.*
In the case of steel pipes cathodic protection is inevitably provided. The method employed is usually the impressed current system which enables the monitoring of the feed current to be observed throughout the working life of the submarine pipeline. Occasionally sacrificial anode type of protection is specified in which case the anodes are inspected during the regular diving inspections throughout the working life of the submarine pipeline.

In the case of non-ferrous metals or plastics obviously this particular protection is not required.

4.2.6. *Regular inspections.* It is essential that regular diving inspections are carried out, particularly on the diffuser section and air release ports.

During construction it is prudent to lay some easily identifiable underwater markers on the route of the submarine pipeline which facilitates divers swimming along the route during the regular inspections. Alternatively, depending upon visibility, inspection can be made by underwater closed circuit television following transit lines and picking up the route markers on the seabed.

Diving inspection should include examination of all diffusers and preventative maintenance in the form of brushing inside the diffuser nozzles to remove any possible accumulation of grease etc. Spare diffuser bends and nozzles should be kept on shore, together with the temporary blank flanges for blanking off diffusers should flushing of the diffuser section become necessary (Section 4.2.3).

Diving inspection should take place annually during the first 3 years of operation and at least every 2 years thereafter.

5.   HYDROGRAPHIC SURVEY AND INVESTIGATIONS

Instrumentation and development of specialized techniques relative to hydrographic survey and investigation has advanced tremendously over the last 30 years and the rate of development and innovation is increasing continuously.

More use is made of remote control recording equipment for the measurement of all parameters at sea. This enables recording to continue during adverse sea and weather conditions which would prevent observations from surface craft.

Hydrographic survey and investigations of necessity provide data for design considerations relative to assessment of behavioural characteristics of sewage/sea mixtures and also the selection of the type of submarine pipeline and method of laying. The data required for these two basic design considerations are, from economic considerations, often collected simultaneously and cannot be divorced.

5.1. Dilution, dispersal and *E. coli* mortality parameters

For dilution and dispersal assessments we are interested in determining current patterns at all depths; tendency for the seawater to stratify relative to salinity and temperature; seasonal variations of current patterns over a reasonably wide area in the location of the outfall — its extent depending upon the magnitude of the project (on average an area 10 km either side of the outfall and 10 km seaward of the high water mark is typical); the effect of seasonal variation in these current patterns as they affect secondary currents in the immediate location of the outfall; the chemical, biological and bacteriological characteristics of the receiving waters on a seasonal basis; daylight attenuation measurements at various depths, general turbulence and wave data, wind data and assessment of wind induced surface currents etc., etc.

5.2. Engineering design and construction — parameters

For the construction and operation of the submarine pipeline and diffuser system, we need further information from the hydrographic investigations. This would include bottom topography (by echo sounding); nature and geology of the seabed; wave and wind patterns; current speeds at all depths, particularly near the bottom; any evidence of possible seabed movement (particularly sand waves); mass transportation of solids and siltation etc. Due to the increase in the number of submarine pipelines installed in nearshore waters for the oil and gas industry in recent years it has been discovered that the incidence of sand waves is far greater than ever anticipated. Much still remains to be discovered regarding this phenomenon which is capable of producing "marching" sand waves with near vertical and unstable faces as much as 3 m high. The effect of such waves encountering a submarine pipeline of comparatively lesser diameter could well be the creation of suspensions which rapidly develop into areas of scour causing suspensions of the pipe over large distances depending on the speed of the bottom current. In addition such sand waves have an obvious detrimental effect upon the diffusers and their efficiency. Providing the occurrence of sand waves can be forecast, their effect can be nullified in the design of the pipeline. Oblique asdic (side scanning sonar) assists in determining the presence of sand waves. The same equipment operating continuously from a fixed base provides a means of measuring the rate of movement of the waves and also an indication of their shape and magnitude by reference to calibration stakes of different heights installed on the seabed.

5.3. Instrumentation — techniques

During the last 20 years electronics, nuclear physics, bio-chemistry and fluorimentry has replaced the rowing boat, ballasted beer bottles, oranges and combinations of wood and metal.

Scientific development in the field of hydrographic investigations has enabled a much more rational and logical approach to be adopted at the investigation stage in simulating the effect of discharging wastes into the sea.

The use of radio-active isotopes, bio-chemical and fluorescent tracers, enables mixtures of similar specific gravity to the waste to be discharged either on the bottom or on the surface of the proposed receiving waters. Radiographic, bio-chemical, chemical and electronic analysis of samples taken will produce an accurate assessment of the behaviour of the wastes which may subsequently be disposed of in the same receiving waters. All of the foregoing techniques have been tried and truly tested and are available to the investigating engineer.

The author has investigated the use of infra-red aerial photography to observe the track of patches of hot water ejected on to the surface of the sea at a proposed diffuser location. The results of this investigation are most encouraging. Temperature differentials of fractions of a degree centigrade can be observed and recorded by day and night and irrespective of weather conditions providing it is possible to fly. Anchored hot or cold sources, floating on the sea surface, act as reference points.

Research and development in oceanographic instrumentation is continuous and increasingly ingenious. As such instrumentation becomes more and more sophisticated the investigating engineer becomes more dependent upon specialist interpretation of the results. In the author's experience this often leaves much to be desired, particularly in the case of seismic interpretation of the first 5 m depth of seabed — in spite of the use of sophisticated seismic instrumentation, coupled with adequate calibration or reference boreholes.

6. CONCLUSIONS

During the last 10 years there has been an intensification of the search for offshore oil and gas deposits accompanied by a general upsurge in offshore and underwater construction works.

As a result improved techniques of design and construction have been evolved which can be applied to submarine pipelines and diffusers utilized for waste disposal in the marine environment.

In order to take full advantage of this situation it is essential that public health engineers specializing in this field maintain close liaison with the offshore industry which, in turn, draws upon the resources of practically all known disciplines of engineering and science.

Developments in techniques for welding ductile iron pipes together with improved methods of under-water excavation must be encouraged and intensified. The author considers that the success of these two developments will provide a further major boost to the already well established economic viability of extended submarine pipelines and diffuser systems utilized for waste disposal in the marine environment.

THE USE OF ROTOR STRAINING AND CHLORINATION PRIOR TO MARINE DISCHARGE OF WASTEWATER

G. Shelef[1], Y. Yoshpe-Purer[2], Y. Sheinberg[3] and A. Fried[4]

Summary – The paper describes laboratory and pilot-plant experiments using rotor straining and chlorination of raw sewage of the Dan Region (Greater Tel-Aviv) aiming at reducing beach pollution caused by the 880 m long 1.5 m in diameter Reading Marine Outfall until a planned full sophisticated wastewater treatment and reclamation plant will become operational. Using a stainless steel rotor strainer with 0.75 mm screen openings followed by chlorination with 20 mg/l active chlorine and 15 min contact time reduced settleable solids and suspended solids by 41% and 38% respectively, reduced floatables to a very low level while fecal coliform concentrations were also substantially reduced. The expected improvement of effluent quality may allow the reopening of various beaches and may prolong the use of others.

The rotor strainer residue solid content is approximately 15% and its disposal by further dewatering and incineration, anaerobic digestion, composting or other means is now being evaluated.

1. INTRODUCTION

Pretreatment of wastewater prior to marine discharge with removal of floatables, part of the settleables and chlorination has been proposed recently in various wastewater marine discharge schemes for one or more of the following purposes:

(a) improving performance of existing outfalls;

(b) emergency measures when treatment plant is overloaded, bypassed or when sea currents situation is unfavourable;

(c) temporary solution until permanent treatment facilities will be constructed;

(d) treatment of combined sewer overflow prior to marine discharge; or

(e) part of the permanent design in combination with a marine outfall.

---

[1] Department of Environmental Engineering, Technion—Israel Institute of Technology, Haifa, Israel.

[2] Regional Laboratory; A. Felix Public Health Laboratories, The Ministry of Health, Tel-Aviv, Israel.

[3] The Dan Region Association of Towns (Sewerage), Tel-Aviv, Israel.

[4] The Dan Region Association of Towns (Sewerage), Tel-Aviv, Israel.

The city of Capetown, South Africa, relying on a usually most favourable off-shore sea current situation is about to build its Camps Bay Sea outfall preceded by particle size reduction to less than 3 mm in diameter by maceration as well as applying electrolytic chlorination (Morris and Prentice, 1974). The proposed marine outfall in Tijuca Beach, Rio de Janeiro, Brazil, includes the installation of rotor straining before marine discharge (Ludwig and Almeida, 1975).

The operation of 20 mesh screen and chlorination at the Joint Water Pollution Central Plant of the Los Angeles County Sanitation Districts (Parkhurst, 1975) aimed at improving effluent quality prior to marine discharge at times of high hydraulic loadings and/or at time of unfavourable ocean conditions, is another example.

This paper describes bench scale and pilot plant scale experiments in the use of rotor straining followed by chlorination aimed at improving the performance of the existing Reading wastewater marine outfall serving the Dan Region (Greater Tel-Aviv), Israel.

2.   THE DAN REGION READING MARINE OUTFALL

The Reading marine outfall today serves the major part of the Dan Region (Greater Tel-Aviv) area with a population of over 1 million. It discharges about 150,000 m^3 of raw domestic and industrial (mainly food and canning industries) sewage per day (approximately 40 mgd) with peak hourly flow of 15 m^3/h. The flow increase per annum is between 5 to 10%. The outfall is 880 m long, 1.5 m in diameter and terminates with a short diffuser 10 m below the surface of the Mediterranean Sea.

Partial comminution of the raw sewage prior to discharge was practised for the last 13 years.

The Dan Region sewerage Master Plan calls for diverting the sewage stream from the existing Reading pumping station into a proposed modern nitrification-denitrification activated sludge treatment plant at the Rishon Le Zion area 15 km south of the Reading outfall. This proposed plant will provide a high degree of treatment producing effluent to be recharged into a nearby aquifer for further reclamation and water reuse for agricultural and other purposes (Shelef, 1975).

The Rishon Le Zion Wastewater Reclamation Project will be completed in between 5 to 10 years depending on the availability of funds. After its completion, the Reading outfall will continue to serve for emergency and plant bypassing purposes.

The existing Rishon outfall is responsible for an increasing pollution of the nearby popular beaches of Tel-Aviv and some were closed to the public after the existing standard of 2400 coliform organisms per 100 ml in at least 80% of the time was surpassed (Shelef, 1973).

Two more popular beaches will reach an alarming pollution situation as the sewage flow at the Reading outfall increases.

Physical dilution at the Reading outfall was determined by radioisotopes and initial dilution ratio of approximately 50 was measured while total dilution of between 300 to 1000 was determined at the fringe of various beaches along the shoreline.

The various beaches affected by the Reading outfall provide the major summer recreation facilities to the public and to tourism with a bathing season extending for almost 8 months a year. The removal of identifiable floatables, part of the suspended and settleable solids and the reduction of bacteria, even before the proposed Rishon Le Zion treatment plant is to be operational, was sought for and among other alternatives, the combination of rotor straining and chlorination was selected for further evaluation. It should be noted that the "solid" floatables constitute the most objectionable floatables, while oil, grease and other hexane extractables were at relatively lower concentrations.

3. THE EFFECT OF ROTOR STRAINING FOLLOWED BY CHLORINATION PRIOR TO MARINE DISCHARGE

3.1. Materials and methods

Bench scale rotor straining with a capacity of 1 l/min using a rotor stainless steel screen with openings of 0.75 mm (0.03 in. or 22 mesh equiv.) was used for the preliminary experiments. A commercial unit produced by the Hydrocyclonics Corp. with screen openings of 0.75 mm served for the pilot-plant operation (Fig. 1).

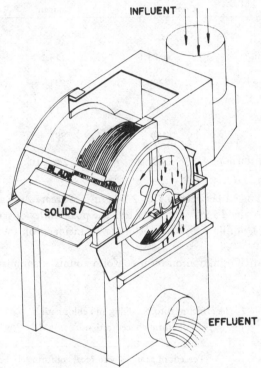

Fig. 1. Schematic view of the rotor strainers used in the Reading field studies.

This unit contains a screen cylinder of 45 cm in diameter and 60 cm in length. It was operated at approximately 5 rpm with application rate of approximately 150 m^3 of raw sewage per h. The sewage was diverted into the strainer after the main pumping station of the Reading outfall, thus the sewage particles were partially reduced in size. Following the rotor straining the effluent was chlorinated with 20 mg/l of active chlorine at 15 min contact time.

3.2. Results and discussion

The results summarized herein represent experiments with both the bench-scale and pilot-plant with screen openings of 0.75 mm.

The rotor strainer removed all particles of 0.75 mm and above, i.e. fecal matter, food scraps (such as bones, corn grains, pieces of tomatoes and orange peels and seeds), cigarette filters, fibers, pieces of wood and other identifiable and unsightly crude matter. Part of this material is floatable and causes pollution of the bathing water and the beach shoreline.

Table 1 describes the effectiveness of the rotor strainer in reducing some of the wastewater components. It should be noted however, that the experiments were carried out during the height of the orange canning and processing season (winter and spring) when the Dan Region sewage is particularly strong.

Table 1. Results of rotor straining (without chlorination)
of raw sewage at the Dan Region Reading outfall

Component	Comminuted raw sewage Concentration	Rotor strained raw sewage	
		Concen.	% Reduction
Settleable solids (1 h Imhoff Cone) ml/l	24.1	14.1	41
Suspended solids, mg/l	900	551	38
BOD_5 days, mg/l	940	718	24
COD mg/l	1380	895	35
Fecal coliforms (per 100 ml)	5.7×10^7	2.0×10^7	65

The amount of screen residue ranged between 0.5 to 2.5 kg/m^3 of treated sewage (wet basis), its solid content averaged 15% and it contained 95% volatile matter. The possibilities of screen residue disposal by anaerobic digestion, by sanitary landfilling, by composting or by incineration are now being investigated.

The effect of rotor straining and chlorination on fecal coliform counts is summarized in Table 2 and in Fig. 2.

Table 2. The effect of rotor straining and chlorination
of fecal coliform concentration

	Percent of samples with fecal coliforms concentration per 100 ml at given order of magnitude							
	10^1	10^2	10^3	10^4	10^5	10^6	10^7	10^8
Comminuted raw sewage (no chlorination)					4	16	72	8
Rotor strained raw sewage (no chlorination)					5	45	37	13
Comminuted and chlorinated raw sewage (20 mg/l Cl_2, 15 min contact)	35	25	5	10	10	10		5
Rotor strained and chlorinated raw sewage (20 mg/l Cl_2, 15 min contact)	60	20	12	4	4			

The rotor straining of the raw sewage, even without chlorination, resulted in a significant reduction of fecal coliforms by removing a considerable amount of fecal matter.

While only 20% of the samples contained below 10^7 fecal coliforms per 100 ml in the comminuted raw sewage, more than 50% of the samples were below fecal coliform concentration of 10^7 per 100 ml following rotor straining. Rotor straining proved to be even more useful in increasing the effectiveness of chlorination. While in chlorinated comminuted raw sewage only 65% of the samples contained less than 10^4 fecal coliforms per 100 ml as much as 92% of the samples of the rotor-strained chlorinated sewage contained less than 10^4 fecal coliforms per 100 ml.

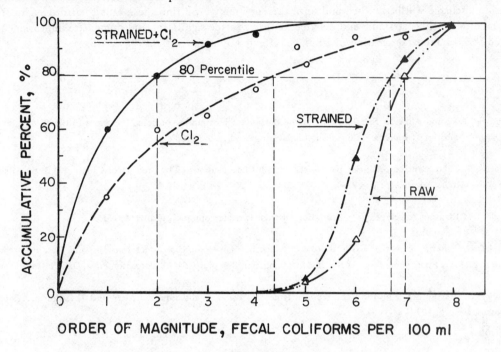

Fig. 2. Accumulated frequency percentage of samples at given order of magnitude
of coliforms concentration following various treatment combinations for
rotor straining and chlorination.

As seen from Fig. 2 the 80 percentile of coliform count of the strained and chlorinated wastewater fell at order of magnitude 2 (between 100 to 999 coliforms per 100 ml) as compared to orders of magnitude of between 2 to 5 with chlorinated wastewater without straining.

Thus, the removal of the coarse material by rotor straining not only physically removed a significant amount of bacteria but also rendered a more effective disinfection of the remaining sewage stream.

4.   SUMMARY AND CONCLUSIONS

Rotor straining alone can significantly reduce pollution at the beaches affected by the Reading outfall by removing the most objectionable floatables, reducing the settleable and suspended solids and reducing a significant part of fecal coliforms.

The combination of rotor straining and chlorination with 20 mg/l active chlorine at 15 min contact proved to be quite effective in reducing fecal coliforms to a point where 92% of the samples contained less than 10^4 fecal coliforms per 100 ml; thus any additional dilution between the outfall and the affected beaches of 150 or more would render compliance with existing bathing-water microbial standards.

The screen residue is considerably dewatered (approximately 15% solids). The method and economics involved in its final disposal requires further investigation.

Although rotor straining followed by chlorination seem to be economical when compared to conventional treatment, it is still arguable whether such treatment is sufficient when factors other than floatables and coliform concentration are of concern. It should be nevertheless noted that until a consensus between engineers, scientists, public-health officials and environmentalists regarding the optimal treatment prior to marine discharge will be reached and until the community will find the means to afford such a treatment, methods such as straining and disinfection can provide significant and immediate improvement to polluted beaches and to the marine environment affected by ocean outfalls.

REFERENCES

1.   Ludwig, G.R. and Almeida, S.A.S., Plans for the Barra de Tijuca marine outfalls. Presented to the Rio de Janeiro Health Authorities by Encibra — Consulting Engineering Co. Brazil. (1975).

2.   Morris, S.S. and Prentice, S.G., Camps Bay sea outfall. Report submitted by the City of Cape Town to the Ministry of Water Affairs, South Africa. (1974).

3.   Parkhurst, J.D., Los Angeles County Sanitation Districts (private communication). (1975).

4.   Shelef, G., Criteria for marine waste disposal in Israel. In: *Marine Pollution and Marine Disposal*. Eds. E.A. Pearson and E. de Fraja Frangipane, Proc. of the 2nd Int. Cong. San Remo, pp. 67-73, Pergamon Press, Oxford, 1975.

5.   Shelef, G., Wastewater renovation and reuse in Israel. In: *Water Renovation and Reuse*. Ed. H.I. Shuval, Academic Press New York, 1975.

DESIGN OF MAJOR REGIONAL SEWERAGE SYSTEMS WITH DISPOSAL AT SEA

W.G. Gerald Snook

Parry, Froud and Snook, Consulting Civil Engineers,

Clevedon, Avon, U.K.

Summary — The problems of sewerage and sewage disposal peculiar to coastal areas are discussed. Continuous improvements in design and construction techniques for waste disposal in the marine environment have contributed to the establishment of the economic viability of centralized disposal on a regional basis. The evolutionary tendency is for the sewerage reticulation area to become larger with each new regional scheme. Design considerations of such schemes within the U.K. are discussed.

1. POLLUTION OF COASTAL WATERS — THE BASIC PROBLEM

The generally accepted fact that the standard of living and affluence of most countries has risen drastically since 1945, leads to far more intensive use of beaches and coastal waters for recreational purposes. The sewage disposal systems (outfalls or land-based biological treatment works) of leisure resorts were often designed to cope with some increase in Summer population, but the "explosion" of Summer populations encountered today, almost inevitably cause the existing systems to be hopelessly overloaded. This problem is further aggravated by drastic fluctuations in flow and the increasing tendency for industrial development in coastal areas. This situation inevitably creates severe pollution problems of nearshore coastal waters.

The problem can, therefore, be very simply defined as the prevention or elimination of pollution of beaches and coastal waters by the selection of the most economic, safe and speedy method for the disposal of sewage and/or trade waste in coastal areas.

Regrettably the function of the large majority of new sewerage and sewage disposal schemes carried out in coastal areas is the elimination of existing pollution and not the prevention of possible future pollution.

2. BATHING WATER AND TREATMENT STANDARDS

The selection of method to solve any particular pollution problem obviously depends upon the standard of bathing waters to be achieved. Standards vary from country to country and often from district to district within a country. Many countries have no bathing water standards. Some countries, such as the U.K., have rigid standards for land-based biological treatment and subsequent dilution of the resulting effluent, but no categoric standard for bathing waters.

Most bathing water standards are modelled on that produced by the State Water Pollution Control Board, California (U.S.A.) which is based on the bacterial (*E. coli*) content of seawater samples and states that 80% of all samples shall contain not more than 10 *E. coli*/ml and the balance of 20% shall contain not more than 100 *E. coli*/ml.

In the U.K. the standard for full biological treatment of sewage is known as the "Royal Commission Standard" (R.C.S.) and requires that the effluent shall contain not more than 30 parts per million (ppm) of suspended solids (SS) and has a biochemical oxygen demand (BOD) of not more than 20 ppm. This standard, which is generally applied to effluent discharged into rivers, was intended for application to rivers which diluted the effluent by at least eight volumes of river water to one of effluent. Where this degree of dilution is not available, a higher degree of purification may be advisable. Where the degree of dilution is great, such as in tidal estuaries or the sea, less stringent standards are permitted. There is no stipulation as to any permitted bacterial content of the treated effluent. In this context the standard states that crude sewage may be discharged into a stream of sufficient size to dilute it by 500 volumes subject to the provision of screens or detritus tanks that might appear to be necessary. In cases where insufficient receiving waters are available to provide the minimum requirement of eight dilutions then a higher degree of purification is necessary. Standards of 10 BOD; 10 SS are not uncommon.

Although no bacteriological standard for bathing waters exists in the U.K. nevertheless the author has, since 1955, designed coastal sewerage and sewage disposal schemes on the basis of the "Californian" standard (10/ml). It is significant that, when designing on this basis, it has been found that, in the majority of schemes fully investigated, disposal at sea by long submarine pipelines and diffuser systems have shown considerable economic advantage over disposal of effluent from land-based biological treatment plants.

3.   ECONOMIC SIGNIFICANCE OF STANDARDS APPLIED TO ALTERNATIVE METHODS OF DISPOSAL

It would appear from published figures that a land-based biological treatment plant operating within the R.C.S. would achieve a 90% reduction in *E. coli* (measured in the unchlorinated final effluent). Assuming that raw sewage contains on average 10^6 *E. coli*/ml such a works would reduce the figure to 10^5/ml. Assuming that the R.C.S. required a further ten dilutions this would reduce the *E. coli* count to 10^4/ml. To achieve this dilution factor in the sea it would be necessary to discharge the effluent through a submarine pipeline and diffuser system.

In general, in the U.K., the inclination of the seabed for any distance seawards varies between 1:50 to 1:200. The tidal range varies between 4 m and 14 m. In consequence the length of foreshore between high water and low water varies considerably, but the inclination is usually less below low water than on the foreshore. In order to discharge "R.C.S." effluent it would seem logical that the minimum depth of water over the diffusers should be 3 m. In consequence the average length of submarine pipeline and diffusers for this purpose (including the foreshore section) would be approximately 700 m. If the "Californian" standard is to be achieved then the length of submarine pipeline and diffuser would have to be greatly increased.

Therefore, the economic criteria is the cost of the increased length of submarine pipeline. This increase depends upon any lesser degree of treatment provided.

For any particular location it is as well if the first economic exercise is to assume full secondary treatment on land and to determine the length of submarine pipeline and diffuser section required. This then produces a common factor which is the cost of constructing this length of submarine pipeline and diffuser which would, of course, include the cost of mobilization and setting up functional units for supply, fabrication and construction. From this basis any increase in the length of submarine pipeline is not reflected on a *pro rata* basis, but on increased cost of material and extension of plant/labour hire time. It is this non-*pro rata* increase in cost which has to be offset against reduced costs of lesser treatment provided. However, there are other economic and sociological aspects to be considered in establishing secondary treatment in heavily built up coastal areas and these are briefly discussed in Section 4.

4. SECONDARY TREATMENT IN COASTAL AREAS – SOME ECONOMIC AND SOCIOLOGICAL CONSIDERATIONS

In lay circles secondary treatment is normally referred to as "full" treatment and is widely considered to be the best solution. There are, however, in most cases the following engineering, aesthetic, economic and sociological problems to be overcome by municipalities and regions with existing sea outfalls.

4.1. Establishment of site

It is normal to site a new treatment works well away from existing towns to allow for future expansion and development of the towns. Large areas of valuable land and possible future development areas are taken up with a serious lowering of property values and restriction of usage in the immediate vicinity of the works (particularly to the leeward!) due to deterioration of existing amenities and aesthetic values.

4.2. Pumping station and rising mains

Usually combined drainage systems exist which fall by gravity to the lowest point prior to discharge to sea through a short outfall or outfalls. Building and general development has taken place landward of these points making it extremely difficult, costly and inconvenient to construct pumping mains from the collection points back through the built up areas and existing services to the site of any proposed treatment works.

The costs of pumping and rising mains are usually a common factor for any alternative considered.

4.3. Storm overflows

In the U.K. it is stipulated that treatment shall be given to a quantity of flow in accordance with a formula which approximates to six times the dry weather flow of sewage and in consequence stormwater overflows have to be provided along the coast and this entails screening of the stormflow before discharge to the sea through short outfalls taken well below the low water mark. As a result there is a commitment to a common cost factor relative to marine construction, the effects of which are similar to those described previously in Section 3.

4.4. Social and commercial disruption

There is generally a lengthy period from instigation to commissioning of a scheme involving land-based secondary treatment. This causes considerable inconvenience to the municipalities during construction. In coastal tourist resorts there is a very real danger of loss of revenue over several holiday seasons.

4.5. Capital and operating costs

The international tendency for persistently high rates of inflation drastically affects any solution which involves lengthy periods of construction. Machinery for sophisticated treatment often has extended delivery periods which, when combined with the time factor required for the accompanying detailed design, pre-contract documentation, administration and procurement, tends to render such schemes vulnerable to the ravages of inflation.

The operating costs of secondary treatment plants contain a high labour element which contributes to a persistent rise in such operating costs which, in turn, increase with the degree of treatment installed.

4.6. Fluctuations of flow

In most seaside resorts the increase of Summer population over Winter resident population varies between a multiple of five and ten. This increase in population is inevitably related to school holidays and is extremely abrupt, creating a sudden increase in the amount of sewage to be treated. By the very nature of modern bacteriological treatment processes this means that during the first 4 weeks of the holiday season there is a great reduction in the efficiency of biological treatment.

4.7. Sludge disposal

Secondary treatment works produce large quantities of sludge with a moisture content, depending upon the type of treatment, varying between 90 and 98%. The processing and disposal of this sludge can be an acute embarrassment particularly in areas of high amenity values, such as popular coastal holiday resorts.

5.   A METHOD OF TREATMENT AND DISPOSAL OF SUSPENDED SOLIDS

In the U.K. it is found that average domestic sewage contains approximately 250 ppm of SS in terms of dry matter. The average moisture content (mc) is 80–85% which approximates to 1500 ppm by volume of dry weather flow (DWF). As a comparison, an activated sludge process treating sewage with a BOD of 300 ppm produces an average quantity of sludge at 97.5% mc of 12,500 ppm by volume of DWF.

On the basis that secondary treatment merely removes 90% of E. coli, 90% of SS and 95% of BOD the author considered a high degree of removal of SS as an alternative to biological land-based treatment.

5.1. General principle of disposal

Sewage containing approximately 10^6 E. coli/ml would be discharged through a submarine pipeline terminating in a diffuser section located so as to achieve "Californian" standard of bathing waters along the adjacent coastline.

Raw sewage is subjected to very fine screening, the screened sewage being passed to a balancing tank with a capacity of at least half the volume of the submarine pipeline. The screens remove solids above a particle size of 2 mm and the dry screenings are conveyed to a separate solids sump.

It is preferred, if local topography permits, that the screens, solids sump and balancing tank are situated on high ground at sufficient elevation to provide gravity discharge through the submarine pipeline at relatively high velocities. If this is not possible the screens, solids sump and balancing tank can be situated under cover beneath a promenade or sea front when discharge takes place by pumping.

In the case of a gravity system, discharge is achieved by motorized valves controlled by the level in the balancing tank. These valves must be situated relative to the level of the low water mark so that when they are closed vaporization cannot take place downstream of the valve at times of low water.

The solids sump feeds into devices capable of disintegrating the solids into almost liquid state. When the screened sewage reaches a pre-determined top level in the balancing tank an adjustable timing device first-ly activates the disintegrators which pump liquified solids into the outlet chamber of the balancing tank. After a set period, starting/opening signals are sent to either the pumps or the control valves. Disintegration continues during the discharge period until either the solids sump is emptied or the lower level in the balancing tank is reached when a signal shuts off the disintegrators at a pre-determined period before shutting off the pumps or commencing to close the control valves.

By this method liquefied solids are discharged through the diffusers at the same rate and in the same quantity as they arrive in the raw sewage — 250 ppm.

It will be noted that this method ensures that each batch discharged achieves a continuous periphery on the surface of the sea, thus assisting in more efficient surface dispersal.

Provision was made in the design for both settlement and diversion of disintegrated solids should either of these precautions be considered desirable in the future.

5.1.1. *Theory of solids "treatment"*. As mentioned the finely screened solids are "liquefied" by dis-integration to very fine limits and returned to the screened sewage during discharge to sea through the submarine pipeline and diffuser.

"Liquefied" solids are, therefore, discharged through the diffuser in the same proportion as they arrive at the balancing tank and are subjected to the same initial dilution factors as the liquid faction.

This method of disposal, in effect, "treats" the solids in that they would be:

> more readily oxidized due to the very small particle size;

> more efficiently presented to the seawater, thus accelerating the mortality rate of
> *E. coli*;

> more vulnerable to grazing and predation;

> more likely to increase their specific gravity thereby creating slight negative buoyancy. This would induce a slow sedimentation process over a wide area and would, therefore, be less likely to cause aesthetic or psychological nuisance or concern.

5.2. Performance

During the last 6 years several such schemes have been installed and monitoring of the bathing waters has shown that the quality is well within the "Californian" standards in spite of several monitoring surveys being set up during severe on-shore gales coinciding with the flood period of Spring tides.

5.3. Maintenance

The main bore of the diffuser section is reduced in stages in order to maintain velocity and prevent settlement.

The balancing tank principle ensures that the designed velocity through the diffuser ports is maintained irrespective of the flow into the balancing tanks which, of course varies considerably depending upon the time of year. The diffuser port velocity is, of course, dependent upon the economic static or pumping head available, but generally varies from scheme to scheme between 3 m and 7 m/s. As a result of this relatively high velocity there is no visible evidence of any grease build up on the diffuser ports.

5.4. Preference for gravity discharge

On the basis that the normal economic pumping velocity is found to be approximately 1 m/s the diameter of submarine pipelines can become somewhat large in major schemes.

Almost inevitably, where local topography permits, it has been found more economic to pump the sewage through a comparatively shorter distance to balancing tanks situated at sufficient elevation to provide higher velocities through the submarine pipeline and in consequence greatly reduce diameter. Average velocities of 3 m/s are normal. A major scheme about to be commenced is described in Section 7 below where discharge will take place through a 1200 mm diameter pipe at velocities varying between 4 and 5 m/s. With velocities in excess of 2 m/s it is advisable to instal grit extraction prior to screening in order to avoid the possibility of abrasion of the internal wall of the submarine pipeline and the diffuser ports.

6. REGIONALIZATION

Regionalization is the principle of combining the sewerage reticulation schemes and centralizing the disposal facilities of areas which would normally support independent schemes.

Several regional schemes have been constructed in the U.K. incorporating disposal methods based on those outlined in Section 5 above.

The successful operation of these schemes has encouraged public health engineers to give consideration to progressively larger areas of regionalization. Typical is the scheme briefly outlined in Section 7 below sewering an area of approximately 250 km^2 with disposal through a single submarine pipeline and diffuser system providing a maximum rate of discharge of 3×10^5 m^3/day.

Various economic, technical and sociological aspects of regionalization are discussed below.

6.1. Economics

Reference was made previously in Sections 3 and 4.2 to the common cost factors applicable to any alternative pre-treatment method. These are the submarine pipeline and diffuser system, together with pumping and rising mains. It was further suggested in Section 3 that the economic criteria when considering the degree of pre-treatment to be provided is the cost of the additional length of submarine pipeline necessary compared with that required for the discharge of effluent from a secondary treatment plant.

Advances in constructional techniques and development of new material [1] have tended to reduce costs of actually laying a submarine pipeline after mobilization has been achieved.

Establishment of a gravity discharge system (Sections 5.1 and 5.4) facilitate reduction in the diameter of the submarine pipeline, thus achieving further reductions in cost. The increased velocity achieved by gravity systems affords a means of drastically increasing the capacity of the submarine pipeline and diffuser system for quite small increases in diameter. This extra capacity can usually be achieved most economically and is often the major contributory factor influencing the decision to extend the area of a sewerage reticulation scheme.

As the area of a sewerage reticulation scheme is increased it is usually found that further economies can be achieved by the replacement of existing sewage treatment works by pumping stations, thus reducing operating costs and facilitating the release of valuable land for further development.

By careful design of the trunk sewers maximum diameter can be installed compatible with gradients and minimum rates of flow. For comparatively little extra cost this philosophy provides potential for unforeseen development. In the author's experience the establishment of a gravity trunk sewer inevitably attracts intensive and unforeseen development and in many cases is responsible for the instigation of complete reappraisal of planning policies within the area.

6.2. Receiving waters

Effluent from any treatment process has to be passed to receiving waters which, in the case of inland towns, villages and industrial development consists of the most convenient river. As development increases so does the quantity of effluent produced. There is usually a stipulation as to the minimum amount of dilutions to be effected by the receiving waters, the quantity of which usually varies from season to season. This factor in itself often attracts industry to coastal areas so that the sea can be utilized for abstraction of cooling water and discharge of effluents. This is not always desirable, particularly in areas which are economically dependent upon tourism. Regionalization could lead to the policy of establishing industrial development inland, utilizing rivers for cooling water and discharging effluent into the regional reticulation scheme for ultimate discharge to sea. This policy would also assist to decrease and perhaps even eliminate pollution of fresh water streams and rivers.

6.3. Regional sewerage reticulation scheme

The following aspects have to be taken into account when considering the extent and the design of a large reticulation scheme.

6.3.1. *Septicity*. The extent of a reticulation scheme often depends upon the maximum time of travel of sewage within the scheme. There is a real danger, particularly in hot weather, that sewage entering the system at the furthermost point under low-flow conditions could become septic before it reaches the balancing tank. Apart from amenity and public health considerations there is also the danger of sulphide attack on pipes and concrete structures within the system. This problem can be overcome by oxygen injection into the system at appropriate times in selected locations. Nevertheless it is a wise precaution to monitor for sulphides at regular intervals during the first 2 years of operation in an attempt to establish any clearly defined pattern of sulphide formation.

6.3.2. *Pumping stations and rising mains*. Pumping stations which replace existing treatment works with consequent release of valuable land may eventually be surrounded by development. It is essential that these stations are made inconspicuous. Often such stations are placed underground at lower levels than is required from technical considerations in order to preserve local amenity.

It is essential that diesel standby generators are provided for each pumping station or perhaps groups of two or three pumping stations served by one diesel standby generator.

It is desirable that all pumping stations are designed to operate automatically and are monitored from a central control station.

As far as is possible all equipment should be standardized to economize on maintenance and such items as stock-holding of spare parts.

6.3.3. *Gravity and pumping mains.* The length of pumping mains should be kept to a minimum and at least duplicated with possible triplication of varying sizes of pipe for areas with large fluctuations of seasonal flow. Where long periods of low-flow can be anticipated a smaller diameter rising main can be utilized, the large diameter rising main being evacuated and filled with inhibited fresh water.

As discussed in Section 6.1 it has been found to be most cost effective to size up the trunk gravity sewers to the maximum possible compatible with gradients available and minimum flow conditions. This sizing up should, of course, be progressive from the balancing tank.

6.3.4. *Trade wastes.* In the case of biological treatment plants certain types of trade wastes are completely unacceptable without pre-treatment, otherwise they may kill off bacteria in filter beds etc. Providing such effluents are acceptable for marine discharge by virtue of the provision of sufficient dilutions there may well be a beneficial effect in accepting such effluents into a reticulation scheme in that the particular toxicity may well assist in reducing bacteria in sewage within the system prior to discharge to sea. However, it is, of course, essential that no trade waste is accepted without full analysis and laid down conditions governing variations in the contents of a particular trade waste.

6.4. Provision for future additional treatment

It is considered desirable to make provision for possible additional future treatment. This usually takes the form of pumping macerated solids to a convenient site for settlement after mixing with screened effluent drawn off from the nearest point in the reticulation system. Allowance can also be made for the provision of aeration to this process to assist in creating sludge and oxygenation of the sewage returned to the system.

Where space is available near the balancing tank provision is made for possible future installation of sedimentation tanks prior to discharge into the balancing tank and the initial hydraulic head is designed to allow for incorporation of this possible future treatment facility.

The advantage of residual chlorination of screened sewage containing "liquefied" solids is in some doubt. However, provision for future incorporation of this facility is a relatively inexpensive item in the initial expenditure.

7.   A TYPICAL REGIONAL SEWERAGE SCHEME

Figure 1 is a diagrammatic illustration of a regional scheme designed to collect and dispose of all sewage and trade wastes from South Pembrokeshire, South Wales, U.K. The scheme has received technical approval from the appropriate Government Departments. Construction is about to commence and will be phased into the most cost effective sequence.

The region lies within a "National Park" and is further designated as being "an area of exceptional natural beauty". A flourishing tourist industry exists and the Summer population is often ten times that of the resident Winter population. Superb sandy beaches abound in the area.

The region is bounded on the South by the Bristol Channel and on the North by Milford Haven, which has been claimed to be the world's second largest natural deep water harbour. Oil refineries are established on the North and South sides of Milford Haven and are serviced by ocean-going oil tankers up to 200,000 tons capacity.

Offshore oil exploration has commenced in the "Celtic Sea" to the West of the region and there is a fair degree of confidence and optimism as to the results. If oil or gas is found it will most probably be brought ashore to the West of the region.

The region is, therefore, classed as a "growth area" relative to industrial development and tourism.

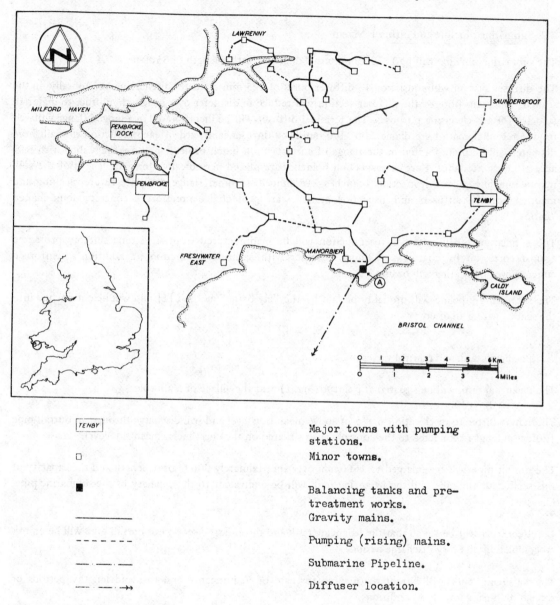

TENBY	Major towns with pumping stations.
▫	Minor towns.
■	Balancing tanks and pre-treatment works.
————	Gravity mains.
— — — —	Pumping (rising) mains.
—··—··—	Submarine Pipeline.
—··—··→	Diffuser location.

Fig. 1.

Prior to the establishment of the oil refineries the economics of the region was based on a thriving agricultural community and farming on an intensive basis continues throughout the whole area.

7.1. Flows

The scheme is designed on the basis of anticipated flow in the year 2020. The local planning authorities forecast of the population and industrial development for the year 2020 was equivalent to a maximum flow condition of 2×10^6 m^3/day. However, in order to allow for unforeseen circumstances the submarine pipeline, balancing tank and main trunk sewers have been designed for an ultimate maximum rate of flow of 3×10^6 m^3/day. It is estimated that the 50% increase in capacity can be provided for 20% increase in costs.

7.2. Submarine pipeline and diffuser system

The submarine pipeline will be 1120 mm internal diameter with a length of 3500 m.

The diffuser system will consist of 42 diffuser ports of 125 mm diameter discharging horizontally on the seabed. The main bore of the diffuser section will reduce in diameter over its length in order to maintain velocity for self-cleansing purposes. The terminal diffuser will be incorporated in a hinged flange situated on the terminal main bore diameter of 300 mm. The diffuser landward of each reduction in main bore diameter will be incorporated in the flange of an inspection hatch projecting above the seabed with easy access for maintenance. Further inspection branches are placed periodically along the length of the submarine pipeline. These inspection branches and hinged terminal flange facilitate future cleaning and maintenance. All diffusers and inspection branches are protected by reinforced concrete dome-shaped "igloos" [1].

The submarine pipeline and diffuser section will be manufactured from steel tube suitably protected against corrosion by coating and lining, together with cathodic protection. In addition a reinforced concrete weight coating will be applied.

The submarine pipeline will probably be laid by the "lay-barge" method [1] and will be entrenched into the seabed with 1 m of cover.

7.3. Balancing tank and control valves

The balancing tank will be situated at point A (Fig. 1) near the village of Manorbier.

The tank will be at an elevation of 46 m above mean tide level and will discharge through an outfall pipe 1200 m in length connected to the control valves situated on the coastline at mean tide level.

The outfall pipe will be enlarged to 1500 mm over approximately 900 m of its length and the capacity of this section of the pipe and the balancing tank will be equivalent to the capacity of the submarine pipeline.

Level controls will be set so that the balancing tank and the enlarged section of outfall pipe will be emptied completely on every discharge period.

The balancing tank will be compartmented for ease of maintenance and to avoid lengthy periods of stagnation during low-flow conditions.

Three sets of control valves will be installed, thus providing one duty valve, one standby valve and one emergency valve. The valves are controlled by three independent electrodes which facilitate bringing in the standby or emergency valve in the case of malfunction of the duty valve. All valves are signalled to close when a pre-determined low level in the tank has been reached.

7.4. Pre-treatment

Sewage arrives at the pre-treatment works at an elevation of 3 m above the top level of the balancing tanks in order to provide sufficient hydraulic head to operate the pre-treatment processes on a gravity system. Allowance is made in this hydraulic head for the provision of future sedimentation tanks in the unlikely event of these being considered necessary in the future. Generally the principle of pre-treatment is as described in Sections 5.1 and 5.1.1 above. All plant is duplicated and provision is made for inclusion of additional plant as the rate of flow increases in future years.

7.4.1. *First stage screening.* The raw sewage entering the works passes through "comminuters" which are fixed, vertical cylindrical screens incorporating continuously revolving cutting knives. This type of screen produces a maximum particle size of approximately 9 mm.

7.4.2. *Grit extraction.* Due to the sandy nature of the subsoil in many of the areas within the region it is anticipated that the grit content of the sewage will be somewhat above average.

The velocity of discharge will, of course, vary with the static head which, in turn, is dependent upon a combination of tide level (range 9 m) and the reducing level of the sewage in the balancing tank and outfall pipe during discharge (range 33 m). As a result of the foregoing the anticipated velocity range is 4.15 m/s maximum (top sewage level in tank combined with lowest tide level) to 1.52 m/s minimum (lowest level of sewage in outfall pipe combined with highest tide level). The combination of high velocities and grit content would tend to cause abrasion of the lining of the submarine pipeline and at the diffuser nozzles. It was, therefore, decided to instal equipment capable of grit extraction of particle sizes above 0.02 mm.

The grit extractors are situated immediately downstream of "comminuters". Grit is automatically washed and stored in containers for transportation to a disposal tip.

7.4.3. *Fine screening.* The fine screens consist of static, semi-circular stainless steel perforated plate with axis horizontal. The screens are constantly swept by revolving arms tipped with nylon brushes. Three such screens form a single screening unit. Screened sewage passes from one screen to another, each successive screen having smaller diameter perforations (9 mm; 5 mm; 3 mm). The screened sewage passes directly to the balancing tank.

Sufficient hydraulic head is allowed between the screens and the balancing tanks for possible future inclusion of a sedimentation process.

7.4.4. *Disintegration of solids.* Screened solids are conveyed to a solids sump and are subjected to very fine disintegration, all as described in Section 5.1.

The disintegrators are based on the principle of domestic kitchen sink waste disposal units. They consist of a high speed cutting impeller bearing up against very closely spaced, fixed cutting knives. Several such units can be placed in series when the gaps between the cutting knives are reduced on each successive unit.

7.4.5. *Grease.* In coastal resorts the gastronomic inclination of the holiday populus is inclined towards fried foods and consequently a much greater proportion of grease is contained within the sewage from such an area. It is, therefore, essential that velocities, both in the main bore and through the diffuser ports, are sufficiently high to prevent the build-up of grease.

The frequency of discharge from the balancing tanks depends upon the incoming flow. This can vary considerably from season to season and during any period of 24 h, depending upon peak flow pattern. As the periods between discharge become greater the tendency for grease build-up on the tank walls increases. In consequence the balancing tanks are compartmented into four units so that at times of very low flow the retention period between discharges is reduced. This system also helps to prevent the possibility of the sewage becoming septic.

Grease removal equipment is provided in the form of high pressure jets ejecting hot or cold water, with or without detergent. It is usual to de-grease the tanks at the beginning and end of each holiday season and usually once every 6 weeks during the holiday season.

7.4.6. *Provision for future additional treatment.* Provision has been made for future sedimentation tanks to deal with the ultimate anticipated rate of flow. In the unlikely event of this additional treatment being required comminution, grit extraction and screening would be retained, the screened sewage being passed to the sedimentation tanks. Screened solids would also be disintegrated (probably to a lesser degree) and the disintegrated solids passed to the sedimentation tanks. Sludge would be drawn off and pumped to a selected site approximately 2.5 km from the balancing tanks. The sludge would be processed either by heat treatment or vacuum filtration into a handleable state and would be disposed of, either in a regional incineration plant or buried in refuse tips.

7.5. Emergency power supplies

In the event of loss of mains power due either to technical failure or industrial action three emergency systems of power supply have been provided. In the case of technical failure a secondary supply has been arranged from the national grid.

In the case of total technical failure of both the main and secondary supply or as a result of industrial action, diesel standby generators would automatically take over the power load.

In the case of a localized accident (say, fire or explosion in the main control room) a set of rechargeable accumulator batteries housed in the control valve chamber would automatically open the valves which would remain open until mains power and diesel standby generator facilities had been reinstated. In this case sewage would bypass all treatment processes and flow continuously through the submarine pipeline by gravity — the balancing tanks becoming non-functional.

7.6. Hydrographic investigations

Extensive hydrographic and oceanographic investigations were necessary in order to design and establish the diffuser location. Detailed description of these investigations and the application of the field data to design are beyond the scope of this paper and, therefore, these investigations can only be described in briefest outline. With reference to Fig. 1 extensive seismic and seabed investigations were carried out between Freshwater East and Tenby. Possible routes for the submarine pipeline were found only at Tenby and the route selected originating near Manorbier. The seabed at Tenby consisted of sand to some considerable depth, but would have involved a submarine pipeline 5000 m long. In addition Tenby is somewhat removed from the "centre of gravity" of the reticulation system. After careful study of the geology of the area it was anticipated that a channel should exist through the rock near Manorbier. After intensive seismic investigation and bottom probing this channel was found to exist with a width of 60 m at its widest point.

The usual observations of temperature, salinity, dissolved oxygen etc., at all depths and over extensive areas were recorded over full tide cycles at different seasons of the year.

The direction and speed of currents was observed at all depths within a 2 km radius of the proposed diffuser location.

Extensive current tracking was carried out over full tide cycles and during various conditions of weather, wind and sea state. The author has always felt that as sewage has a specific gravity of approximately 1.0 it is far better to observe current patterns and tracks utilizing a colour dye mixture of fresh and sea water rather than pieces of wood, half empty beer bottles, drift cards, etc. Various coloured dyes were mixed with fresh water and ejected on to the sea from a vessel anchored over the diffuser system. A different colour dye patch was released every half hour over full tidal cycles. High speed launches and helicopters observed the track of the dyes. The helicopters were equipped with crop-spraying gear in order to re-plenish the dye patches which very quickly dissipated — even under calm sea conditions. However, for sub-surface current patterns floats with deep drogues were used with the minimum of projection above the sea surface.

Generally it was found that currents are predominantly parallel with the coast on both ebb and flood tides and that there is quite severe variation of current speed and direction with depth at the diffuser location selected.

There is no period when the velocity of seawater in the diffuser area is less than 0.25 m/s. The average surface velocity is 0.75 m/s and the maximum velocity recorded was 1.78 m/s.

7.7. Reticulation scheme (Fig. 1)

The ultimate designed maximum rate of flow is 3×10^6 m^3/day. This represents approximately $6 \times$ dry weather flow (DWF). The bulk of the flow is generated from Saundersfoot and Tenby in the East and Pembroke Dock in the West. During the height of the holiday season there is approximately twice as much flow from Saundersfoot and Tenby as there is from Pembroke and Pembroke Dock. The present maxi-mum rate of flow of the whole area is approximately 0.9×10^6 m^3/day.

There are no major rivers in the area, the only water courses being very small streams. These streams have been utilized for the discharge of effluent from a multiplicity of small sewage disposal works servicing small inland villages. However, the majority of these works are overloaded, placing a heavy pollution load on the few small streams within the area. Saundersfoot and Tenby have sedimentation tanks installed, but discharge the sludge, together with sewage, through short outfalls (just beyond the low water mark) at certain states of the tide. Pembroke and Pembroke Dock discharge crude sewage into Milford Haven over the ebb tide. This is achieved by storing the sewage in a tank sewer with motorized penstocks con-trolled by a lunar clock.

A further extensive area to the North and East of Saundersfoot may be taken into the scheme and allow-ance has been made in the design for this eventuality.

7.8. New and abandoned works

In addition to the balancing tanks and pre-treatment works the new regional scheme will entail the establishment of twenty six new pumping stations. .

However, when the regional scheme is fully operational it will be possible to completely abandon twelve sewage disposal works, nine pumping stations and six major crude sewage discharges.

7.9. Cost comparisons of alternatives examined

The various alternatives examined are briefly outlined below. The costs have been up-dated to February 1975. Particular note should be taken of the various total capacities.

7.9.1. *Disposal by one submarine pipeline and diffuser.* This is the regional scheme outlined in this paper.

Total capacity 3×10^6 m³/day.

Cost £12,621,000.

7.9.2. *Disposal by one submarine pipeline and diffuser.* This scheme has a capacity in accordance with the flow equivalent of the population and industrial expansion forecast by the local planning authority before incorporation of the additional capacity for unforeseen development.

Total capacity 2.1×10^6 m³/day.

Cost £10,277,340.

7.9.3. *A separate sewage works to each village and town.* This is a hypothetical situation as there are no water courses to provide sufficient dilution to the effluents. It is included for comparison purposes only.

Total capacity 1.888×10^6 m³/day.

Cost £13,281,250.

7.9.4. *Three separate sewage disposal works.* Again this is a somewhat hypothetical case in that it assumes it would be possible to group the whole area into three units and that the resulting secondary effluent could be diluted by means other than submarine pipelines and diffusers.

Combined total capacity 2.1×10^6 m³/day.

Cost £14,843,750.

7.9.5. *One submarine pipeline and diffuser discharging settled sewage.* This assumes the same regional scheme as outlined in this paper and subject of 7.9.1 above, but including settlement of 50% of ultimate flow capacity and treatment of sludge.

Total capacity of submarine pipeline 3×10^6 m³/day, but with only 50% capacity provided for settlement and sludge treatment.

Cost £15,078,100.

7.9.6. *Full biological secondary treatment with shortened submarine pipeline.* This assumes that a suitable site could be found near the proposed balancing tank site and that only 50% purification capacity would be installed initially.

Total capacity 3×10^6 m³/day.

Cost £16,328,100.

8.   CONCLUSIONS

In the U.K. disposal of raw sewage containing disintegrated ("liquefied") solids through extended submarine pipelines and diffuser systems has proven to be a successful means of achieving the "Californian" standard of bathing waters.

As a safeguard it is recommended that provision is made for additional treatment in the unlikely event of this being required in the future.

The characteristics and dynamics of nearshore oceanographic conditions around the U.K. coast afford continuous replenishment of un-polluted seawater — a situation which does not exist in some areas throughout the world.

Economic appraisal of various degrees of pre-treatment up to secondary treatment stage is advocated as being an essential phase in the design considerations of every project.

Improvements in constructional techniques have established the economic viability of extended submarine pipelines enabling diffusers to be located in very deep water at considerable distances from the shore. Gravity discharge (where feasible) further contributes to the economic viability of utilizing extended submarine pipelines. This factor alone may well afford the means of economic and safe disposal of wastes in the marine environment in areas with very little tidal range.

The sea is probably the greatest resource of nature and as such must be respected, conserved and properly harnessed for the use and benefit of mankind.

REFERENCE

1.   Snook, W.G.G., Design and Construction of submarine outfalls and diffusers. *Third International Congress, Marine Municipal and Industrial Wastewater Disposal, Sorrento (South Italy)*. 1975.

INDUSTRIAL SOURCE WATER POLLUTION CONTROL

William J. Lacy
Principal Engineering Science Advisor
Environmental Protection Agency
Washington, D.C.

Summary – Manufacturing operational reliability must consider the pollution control systems reliability of performance, particularly as this may impact the manufacturing process. The effects of an open or closed-cycle water pollution system on manufacturing operations should be essentially the same if effluents standards are strictly enforced. The impacts of water pollution control on plant operations are not projected to be significant. Rather slight increases of 1–5% in manufacturing costs, plant space, manpower and energy needs are foreseen. Environmental standards concerning discharges of wastewaters are expected to accelerate the pressure on industry to reduce both the pollutional discharge loads and the magnitude of effluent volumes in order to minimize impacts on the environment. Industrial water quality requirements for reuse are less demanding, as a general rule, than for municipal supplies. Accordingly, direct industrial water reuse (i.e. closed-cycle industrial water systems) should be technically and economically achievable earlier than comparable municipal water reuse systems. Wastewater reuse is resource conservation and method of pollution control. Additional R & D activity in this area is the key to accelerating the implementation of extensive wastewater reuse systems and eventually the totally closed cycle. The latter, which will result in no effluent discharge, would comply with any water quality standards in any country now, or in the future.

1. INTRODUCTION

The environmental control laws in all industrialized nations are placing a new responsibility on industry. Namely, to clean up, and to do so within a finite time frame.

In spite of the rhetoric on the feasibility of cleaning up industrial wastewater discharges, the process of cleaning up has started. Industry has the capacity to clean up. Industry will most likely respond to the intent of the legislation. The question now remaining is how best to proceed to do the job, what effects, consequences, and/or impacts the clean up will have on industrial productivity – if any at all.

In many cases there will be an increase in the complexity of manufacturing operations via the effects of increased requirements for energy, manpower, space, and regulatory requirements, all of which would reflect in higher manufacturing costs.

On the other hand, opportunities exist in the form of a better public image, improved product quality and/or yield, beneficial process changes, and resource use optimization, all of which may reflect in lower manufacturing costs or increased profitability.

Major program goals of the EPA's industrial program include research into industrial water reuse and product (or by-product) recovery methods. We are striving to demonstrate the practicability of "closed cycle operations by industry".

The goals were formulated on the basis of the alternatives available to an industry for pollution control, as shown in Fig. 1.

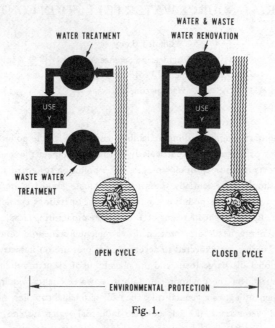

Fig. 1.

In viewing Fig. 1, it must be remembered that waste disposal operations normally result in a net cost to the industry producing the waste. However, by-product recovery and utilization techniques can reduce the net cost of treatment and may prove to be less expensive than the alternative method of disposal. In some cases a profit may be expected by the implementation of waste resources recovery as a pollution control method. Recovery of by-products from wastewater residues is in the scope of the present program.

Recycled water may also be a valuable resource as a result of intake water supply shortages, increasing water supply and water treatment costs, and mounting municipal sewerage charges. The recovery of usable water and thermal energy are key methods of reducing overall waste treatment costs and should also be considered.

The keys to industrial closed cycles are therefore wastewater treatment for recovery and reuse, and by-product recovery of waste materials for beneficial purposes.

Industries are becoming increasingly more aware of the need for water reuse and by-product recovery as a method for pollution control. This fact has arisen not only becuase pollution affects the environment, but also because pollution affects the general public, who are the customers for industrial products. In addition, industries also depend upon our nation's rivers and streams for suitable water for their manufacturing processes. Finally, the emergence of the fact that closed cycle operations may not only be economically competitive with treatment for discharge, but may be less costly.

2.   APPROACHES TO WATER POLLUTION CONTROL

First lets look at a few basic approaches to water pollution control and water management in manufacturing operations. The traditional approach has been what may be classed as an open cycle water system (see Fig. 1a). This is analogous to the once-through system much discussed. It calls for a volume of intake of water approximately equal to the volume of discharge of wastewater. More often than not the intake is treated to meet manufacturing water use quality needs. In many cases in the past some treatment prior to discharge has been made. At present about 40% of water utilities invested capital is for intake water treatment and the rest is for waste water treatment [1, 2]. Treated intake water is directly associated with manufacturing operations which are reliant on this supply at an acceptable quality. As for waste water treatment, until now there has been little reliance of manufacturing operational continuity on this end of the business. In summary, intake water quality is directly keyed into plant operations, whereas waste water effluent quality has not been.

Another approach is the closed-cycle system where, starting with an inventory of water, water is managed in continuous recycle with the intake requirement being equal to the net consumptive use of a plant. Figure 1b illustrates this approach. In actual practice today we have in existence a combination of both approaches, with the long term approach favoring the closed-cycle water management system. This system can comply with the toughest of any effluent regulation which may be imposed now or in the future. It does so merely by not having a discharge which is subject to monitoring and or regulatory compliance to a set of effluent quality criteria requirements. In spite of the resistance some industries may have on the acceptability of the latter system as a means of meeting pollution standards, the facts are that industry has been moving in the direction of closed-cycle operations over the last two decades. Table 1 is a history

Table 1. Industrial water reuse-recycle trends.

Industry	Year	Gross water use, BGY	Water Intake, BGY	Reuse ratio
All U.S.	1954	27040	11570	1.81
	1959	26350	12130	2.16
Industries	1964	29850	14010	2.13
	1968	35700	15470	2.30
	1973	46900	15010	3.1
Chemicals	1954	4290	2690	1.59
and allied	1959	5280	3240	1.61
products	1964	7670	3900	1.96
	1968	9460	4510	2.09
	1973	11100	4200	5.30
Petroleum	1954	4150	1250	3.32
and coal	1959	5780	1320	4.37
products	1964	6160	1400	4.40
	1968	7220	1370	5.27
	1973	8200	1300	6.30

of this progress over the last several Department of Commerce statistics on Water Use in Manufacturing. Particularly note the large increase of water reuse by the Petroleum Industry in a period of 12 years as depicted by a 90% increase in the reuse ratio (water used/water intake) of water in the industry.

Industry is diversified in type, size and output. It utilizes many processes sometimes to produce the same product. Therefore water quality requirements for different industries of for various processes within a plant, and for the same process in different plants, vary widely.

Existing treatment technology can treat water of virtually any quality to any desired characteristics required by an industrial process. Generally the cost of treating water for a specific purpose is acceptable to industry, because it represents only a small part of the production and marketing costs.

Even though water quality is stringent for many industrial processes at a point of use, normally industrial process water quality demands are less stringent than those for public, recreational, agriculture, etc., uses. Therefore, it would be not only prudent but good business to investigate water quality requirements for closed-cycle operations within the plant.

Regardless of which approach evolves the use of either approach may require meticulous attention of an industrial plant to the water quality of their discharges, or of the reused water if they choose not to discharge. In the case of a discharge, as with the open-cycle approach, standards of effluent quality, based on units of production, will have to be met and are anticipated to become increasingly stringent. Now for the first time American industry may have to shut a plant down if its water effluents are not within a prescribed limit of acceptability as required by a regulatory agency. What does this mean and what must industry do to plan for averting adverse impacts on production operations?, which we all know can be a costly affair. Unscheduled plant shut downs are always a production man's nightmare.

Now what does all this have to do with pollution control? Well, if you are required to produce a waste effluent within a specific set of specifications (effluent standards) then you are in effect incorporating your water pollution control activities as an integral process module of your production operation. If a plant is not going to be operating to meet its allowed limits for discharge you may be forced to shut down your plant until you can meet the local regulatory requirements.

This integration of pollution control operations to manufacturing operations is now one further consideration for which you must plan in your plant production reliability evaluations. In addition, capacitance in water-management systems will need to be provided to allow temporary relief in situations of upsets in pollution control devices and/or manufacturing systems.

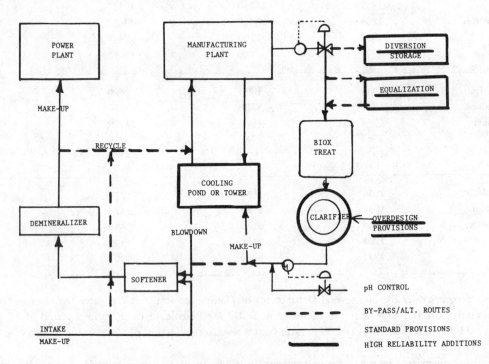

Fig. 2.

Let us take a specific hypothetical example. Assume an industrial complex is to operate a closed-water cycle and the basic water conditioning and treatment operations consist of biological oxidation, alkalinity and hardness control, and distillation. As shown schematically in Fig. 2, if we consider each water conditioning or treatment operation as a separate operation, what should be done to increase the operating reliability of these operations to reduce or eliminate the potential of the water utilities from impacting the production operations? The latter are highly dependent on water as a service or process raw material.

There are various design and operational tools. For example, we should consider providing for:

(1) equalization — to provide for storage, mixing and dampening of water quality and flow fluctuations;

(2) diversion storage — to contain off-standard quality waters and/or discharges from operating upsets, and to rework water;

(3) inventory storage — to assure an available supply of quality water while upsets, repairs, or temporary shutdowns of a water utilities operation becomes necessary;

(4) appropriate by-passes — to temporarily allow the water utilities to operate while diverting, storing, or reworking water. In extreme cases to provide for a relief valve;

(5) process stability — each selected water conditioning and or treatment process selected should have over design features in critical areas and be characteristically insensitive to minor fluctuations in feed quality while having a high range of volume through put capacity;

(6) process control — provisions to monitor water quality and volume flow so as to initiate corrective actions when necessary.

As you will note from Fig. 2 — the water utility operations have all the features just described. With these provisions incorporated sufficient "capacitance" is inherent in the water utility system to prevent forced shut down resulting due to poor water quality resulting from production malfunctions etc., and this would allow for less probability of interfering with manufacturing operations.

Industrial waste treatment source control often tends to be practised in terms of the residual characteristics of separate manufacturing processes. In some industries, segregation, rather than collection, of industrial waste streams may be the most effective method of minimizing net residuals management costs. Under such conditions, each residual stream tends to receive only that treatment which is appropriate for its volume and constituents, and uncontaminated waste waters can be segregated and discharged directly or recycled. Because residual streams may be segregated and treated according to waste characteristics, some processes become integral parts of the manufacturing operation rather than waste treatment.

3.   COSTS OF TREATMENT

The costs of controlling industrial residual discharges may be altered at the tail-pipe end of a production process. However, effective cost reduction can also be achieved by selecting and/or modifying the production processes *per se*, by changing the raw materials used and/or by changing the end products that are produced. Further alterations in control costs may be achieved by recycling or recovery and reuse of certain residuals generated in the production process.

Since some of these changes may significantly affect the costs of industrial pollution abatement, estimates of the effects of these changes must be made. The impact on treatment costs of the following alternatives should therefore be analyzed.

1. Alternatives in technology (kraft *vis-a-vis* sulfite process in pulp production).

2. Alternatives in raw materials and fuels (SO_x content alternatives in coal).

3. Alternatives in end-products produced (different finishing processes in steel products).

4. Materials recovery (recovery of mercury used in production of caustic soda).

5. By-product recovery (recovery of blood and related by-products in slaughter houses).

Many industries have developed production technologies, which produce fewer residuals. For example, the older mercury cell production process of caustic soda results in significant generation of mercury wastes; the newer diaphragm technology for caustic soda production does away with mercury residuals entirely.

A more complex example of the impact of alternative technologies on industrial pollution abatement costs can be seen in steel processing. Steel can be produced by either the prevailing ingot casting technology or by continuous casting. The use of continuous casting technology has increased rapidly since its introduction in the early 1960s and by 1985 is expected to surpass the ingot casting process. When estimating pollution control costs for the steel industry, it is important to realize that continuous casting generates less lubricating oils and produces superior metal surfaces requiring less surface treatment such as scraping, sand blasting or treating with acids and other corrosive chemicals. Consequently, continuous casting results in fewer acid wastes, surface scale particles and less process waste water, all of which reduce the pollution abatement costs associated with given levels of steel production.

Two additional examples of significant impact on pollution abatement costs resulting from alternate technologies can be seen in the pulp and paper manufacturing industry. In the case of wood pulp production, the use of the sulfate or kraft process of producing wood pulp, which increasingly dominates the sulfite and other methods, reduces the more common waterborne residuals (and therefore results in lower treatment costs) albeit at a cost of generating other residuals. The kraft process brings a major reduction of waterborne dissolved solids (DS), which are difficult and costly to treat. Moreover, it results in large reductions in the generation of solid wastes, sulfur dioxide and the wastewater load. However, it increases particulates, and creates odorous sulfides and chlorine gas.

Several examples illustrate the possibilities for such reduction. In the canned and frozen foods industry, a shift from water conveying to dry conveying suction systems, reduces residual water flows and BOD and DS; similar results, with considerably reduced residual water loads, can be obtained from dry caustic or cryogenic peeling, and blanching with hot air instead of steam or hot water.

In the production of plastic materials and polyvinyl chloride (PVC) resins, switching to the bulk method or producing PVC resin can considerably reduce waste water − BOD, suspended solids (SS), and DS. Since the use of each of these (or similar) production subprocesses will result in a reduction in water pollution control costs, estimates of pollution control costs must include such alternatives for production subprocesses, and, as in the case of changing technologies, projected future abatement costs must also reflect the future use of such alternatives.

4. POLLUTION CONTROL VS ENERGY DEMAND

Although environmental concern has caused and will cause small increases in the U.S. energy demand and has restricted our energy supply to some extent, other factors have been more significant. Not the least of these has been the public's continually escalating demand for energy.

Based on an analysis of the situation it has been reported that the energy required to operate (1) sulfur dioxide (SO_2) emissions from power plants; (2) municipal wastewater treatment plants; and (3) solid waste collection and disposal systems; for pollution control would be only 1.04% of today's total national demand [3]. This low energy consumption rate assumes all the U.S. power plants have SO_2 scrubbers, all municipal waste water is treated to the tertiary level, and that all municipal solid waste is properly collected and landfilled. This power consumption would approximate 747 trillion BTU's per year — less than twice the amount of energy used in 1968 to air condition our homes [4].

The energy demand associated with waste water treatment depends on the degree of treatment and the unit processes involved. The major use of energy is for electricity to operate equipment such as pumps, scrapers, compressors, chlorinators, etc. An estimate of the 1968 consumption of electrical energy for various municipal waste treatment processes contained values from 0.018 kWh/day/person for minor treatment to 0.226 for tertiary treatment. The average value was 0.057. For the study population served, the total electrical energy consumption was about 7.5 × 10^6 kWh/day or about 29 trillion BTU's per year. This represents about 0.04% of the total U.S. demand for energy or about 70% of the annual electricity consumed by electric blankets in 1968 (see Table 2). If today's total population were served by tertiary treatment, the electrical energy consumption for municipal water pollution control would approximate 182 trillion BTU's per year or 0.25% of today's energy demand.

Table 2. Electrical consumption of selected small appliances, 1969.

Appliance	Annual kWh per item	Number of items (millions)	Total annual consumption (billion kWh)	(trillion BTU)
Bed coverings	147	27.0	3.97	42.6
Blenders/clocks/fans	75	146.0	4.41	47.2
Broilers	100	14.0	1.40	15.0
Coffee makers, automatic	106	50.0	5.30	56.7
Dehumidifiers	377	3.8	1.43	15.3
Food disposers/hair dryers	44	36.0	0.73	7.8
Frypan/hot plates/irons	420	105.0	15.70	167.9
Heaters (portable)	176	17.0	2.99	32.0
Humidifiers	163	4.0	0.65	7.0
Knives/mixers/shavers	39	86.0	1.17	12.5
Radios	86	57.0	4.90	52.4
Toasters/toothbrushes	44	69.0	2.19	23.5
Vacuum cleaners	46	53.0	2.44	26.1
TOTAL			47.28	506.0

Hirst [5] estimated that the total operational energy demand for primary and secondary treatment for all municipal plus industrial waste water plants in the United States is about 290 trillion BTU's per year or 0.4% of total energy use. If all waste water were to be treated to a tertiary level, the total operational energy demand would be about 0.8% of the total energy demand.

It should be concluded that maintenance of environmental quality does have an energy cost, both on the supply and demand side, it is relatively small, it is well worth the price. Since it is small in relationship

Table 3. Residue management check list

---

1. Can you avoid the generating of residues?

    a. By improved housekeeping.
    b. In-plant changes.
    c. Process modification (dry vs wet caustic potato peeling, etc.).
    d. Nonaqueous process systems.

2. Do you really need to use so many chemicals in your process?

    a. Unnecessary acid-base neutralizations.
    b. Free local alkalinity.
    c. Salt producing water and wastewater treatment processes.
    d. Other chemical substitutes.

3. Are you sure you will have an inorganic (salt) blowdown?

    a. Preliminary material balances can be deceiving.
    b. Buildup within a close-cycle limited to salts added or corrosion products.
    c. What about losses (purged from system) due to: leaks/cooling tower drift/ sludge blowdowns/loss to *product.*

4. Have you looked at the market to sell or give them away?

    a. Road salt — local or on site needs.
    b. Brine feedstock for Cl_2/NaOH production.
    c. Micronutrient for fertilizers.
    d. Soil conditioner or construction material fillers.

5. Can you reuse it as a raw material — after reconstitution?

    a. For Cl_2/NaOH production.
    b. As salt for salting-out process.
    c. For an exchange regeneration.
    d. As fuel supplement for steam production.

6. Can your site contain it?

    a. By permanent impoundment — concentrated storage.
    b. By incorporation into plant construction materials — asphalt, concrete, etc.
    c. By drying it with waste flue gases for dry storage.

7. Have you considered paying a specialized waste service institution to handle for you? Such services are or can be made available.

8. Will the local regulatory agency permit controlled disposal?

    a. By sanitary landfill.
    b. Salt cavern deposits.
    c. Salt disposal into salt environments (ocean, etc.).
    d. Seasonal discharge of detoxified salts to high water streams for dilution.
    e. Deep well introduction.

---

to the energy needs of manufacturing operations it is not visualized that the energy needs for water pollution control will affect the operational reliabilities of plants which install first rate water pollution control systems.

5.   IMPACTS OF POLLUTION CONTROL ON MANUFACTURING INDUSTRIES

During the process of promulgating the EPA effluent guidelines and standards, assessments were made on the impacts of costs, space, and energy needs.

In general, the projected impacts on the cost of manufacturing operations to meet the 1977 water pollution control requirements have been estimated to be less than 3% of the product sales costs. Similarly energy needs, although reflected in the costs, are also in the range of less than 3% of the total energy requirements for manufacturing. The space requirements have also been nominal, however there are many cases for older plants located on crowded real estate where the additional space need may not be forthcoming. The EPA studies of the industries proposed for regulation do not indicate that effects on product quality or yield will occur. If such effects do occur, they could manifest themselves in a manner which reflects greater yield of raw materials into saleable products as the pollutants may be prevented from entering wastewater by in-plant controls, process changes, and/or by reuse of pollutants as raw materials otherwise required. It should be noted that these techniques should result in material and energy savings to help off-set their cost of implementation. They are conservational in nature.

In view of these projected impacts it is contemplated that industries will proceed to meet the new water pollution control laws except where definite hardships are evident. In those cases the local regulatory agencies may have to judge the circumstances on an individual case-by-case basis.

6.   WASTEWATER TREATMENT RESIDUALS PROBLEM

Before an industry becomes overly concerned about a potential waste problem, it should resort to assessing its impact (economically vs environmentally) using a check-list of alternative actions such as shown in Table 3.

In spite of the importance of residues from the treatment of industrial wastewaters their relationship to the overall industrial water pollution control cost picture should be kept in perspective. In general, for most industries the major effort, and costs undoubtedly are in the handling and management of the industrial water.

7.   CONCLUSIONS

Manufacturing operational reliability must consider the pollution control systems reliability of performance, particularly as this may impact on the manufacturing process. The effects of an open or closed-cycle water pollution system on manufacturing operations should be essentially the same if effluents standards are strictly enforced.

The impacts of water pollution control on plant operations are not projected to be significant. Rather slight increases of 1–5% in manufacturing costs, plant space, manpower and energy needs are foreseen.

Environmental standards concerning discharges of wastewaters are expected to accelerate the pressure on industry to reduce both the pollutional discharge loads and the magnitude of effluent volumes in order to minimize impacts on the environment.

Industrial water quality requirements for reuse are less demanding, as a general rule, than for municipal supplies. Accordingly, direct industrial water reuse (i.e. closed-cycle industrial water systems) should be technically and economically achievable earlier than comparable municipal water reuse systems.

Wastewater reuse is, resource conservation and a method of pollution control.

Additional R & D activity in this area is the key to accelerating the implementation of extensive wastewater reuse systems and eventually the totally closed-cycle. The latter, which will result in no effluent discharge, would comply with any water quality standards in any country now, or in the future.

8.   SUMMATION

Industrial process design must provide manufacturing facilities that generate minimum waste materials. The designer is challenged in providing these facilities for all types of operating conditions. Continued interest is directed toward recovery, reuse, and recycle of wastewaters. With the increase in activity on the part of all nations and local regulatory agencies, as well as the general populace in this area, it is vital that the chemical and engineering professions keep meeting these new challenges.

The continued growth of industry all over the world and its resulting impact upon available natural resources — water, air and land — together with ever more stringent environmental regulations on effluents will provide the incentive to search for new technology to combat pollution.

REFERENCES

1.   Kollar, K.L. and Brewer, R., Achieving pollution abatement. *Construc. Rev.* July (1973).
2.   1974 Census of Manufacturing, Water Reuse in Manufacturing April 1975. U.S. Dept. of Commerce.
3.   *Environmental Protection Agency News*, National Environmental Research Center, Cincinnatti, Ohio, January 11, 1974.

4.   *Ibid.*

5.   Hirst, E., The energy cost of pollution control. *Environment* 14 (8) 37-44, 1973.

DESIGN OF MARINE WASTE DISPOSAL SYSTEMS FOR INDUSTRIAL WASTES

Harvey F. Ludwig

Southeast Asia Technology Co. (SEATEC), Bangkok, Thailand

Summary — The most important factor in designing a marine disposal system for industrial wastes is to gain access to the unconfined deeper offshore waters, beyond the zone of nearshore water to be protected, where the dilution, dispersion, and absorption capacities are relatively high compared to discharge into shallow confined waters. For discharge into confined waters, the treatment requirements will be virtually the same as for discharge to inland fresh waters, hence secondary or even tertiary treatment will often be required. For discharge to unconfined waters only a few parameters are applicable, such as floatables, color, and toxicity, hence the treatment requirements will be greatly simplified. Monitoring of the affected receiving water is important to "prove out" the validity of the design, especially because of the present confusion from the regulatory point of view as to the extent to which marine waters may properly be utilized for waste disposal. Examples are cited illustrating these principles.

1. INTRODUCTION

How to design a proper treatment and disposal system for discharge of industrial wastes to the marine environment is a good question to ask these days, both because (1) there is little concensus on the extent the ocean waters of the world should properly be available for waste disposal, not even for "simple" sanitary sewage, and (2) only in the past few years has there been a real in-depth effort (by the U.S. EPA) to investigate, understand, collate, and codify information on treatment of the various types of industrial wastes by categories and sub-categories, so that for a particular kind of factory the engineer can have specific guidelines as to which parameters are meaningful and should be used as the basis for design. In other words, both subjects — how to use the ocean for receiving polluting materials, and what are the specific pollutants in a given industrial waste that significantly affect environment — are under continuing study and debate and it will be some years before any real consensus can be expected. In the meantime, of course, the engineer must proceed with design of industrial waste disposal systems, using his best judgement. It is the purpose of this paper to assess the current situation and to develop the best possible guidelines for the engineer's current use.

2. PARAMETERS OF SIGNIFICANCE IN MARINE WASTE DISPOSAL

Over the past few years — since about the time of the first of these conferences at Trieste — a sizeable controversy has been going on in the U.S.A. due to the proposal by EPA that all wastes should receive a minimum of secondary treatment before being discharged to *any* receiving water whether this be a small stream or the deep open ocean. Enforcement of this concept would cost huge sums of money for many of the coastal cities of the U.S.A., for example, Los Angeles and San Diego in California, where primary treatment together with long submarine outfalls have been giving good service, where even extensive scientific studies have not demonstrated any serious environmental impairment. Accordingly the U.S. Government has conducted extensive hearings and considerable research and investigation is underway to try to get enough facts so some rational policy decisions can be made.

The extensive U.S. Government hearings, conducted by the U.S. Senate and published in 1974 in the report, "The Impact of Secondary Treatment on Wastes Discharged into the Ocean" [1], is significant in bringing together for the first time in a single meeting practically all available pertinent information relating to marine waste disposal. While the hearings were by no means conclusive from the regulatory point of view, the total scientific information available did indicate a number of very basic concepts including (1) the design of waste disposal systems to open ocean waters represents a vastly different set of conditions than discharge to confined inland waters and estuaries, and hence should be approached not by reference to experience with confined waters but as a new technology based on marine science *per se*; (2) most of the traditional parameters for dealing with waste disposal to confined waters have little if any significance, including BOD, COD, nutrients, suspended solids, pH, acidity, etc.; instead the meaningful parameters are those relating to the known values of ocean waters, for example, the aesthetic value of beaches and recreational areas, wherein floatables becomes a significant parameter, and the biological productivity of ocean waters, especially the shallow coastal and estuarine areas which are so important for fisheries, wherein toxicity accumulation can be very important; (3) whereas the discharge of heavy metals and trace elements into shallow biologically reproductive areas can be harmful, their discharge into deep ocean waters likely may be harmless — there is a buildup of heavy metals in the benthos in the vicinity of the point of discharge, but their concentrations are still within the range of these elements in natural marine sediments around the world; nevertheless, careful attention needs to be given to this subject especially through monitoring and continuing study of possible adverse effects on marine ecology; (4) whereas nutrients cause serious problems of eutrophication in confined waters, eutrophication is no problem at all in open waters and such nutrients actually enhance biological productivity; (5) whereas organic matter (measured as BOD, COD, etc.) is a major problem in depleting DO in confined waters, DO is no problem in open waters and the organic material enhances biological productivity; (6) secondary treatment is actually less desirable for discharge of organic wastes than primary treatment because the organic content in primary effluents enters the natural food chains of marine ecology without disruptive effects whereas secondary effluents are likely to result in algal blooms which may be undesirable in shallow or recreational zones; (7) thermal discharges, while significant in confined waters, have insignificant effect on temperatures in the ocean except in the vicinity of the point of discharge but even this effect is not necessarily harmful; also, if thermal discharges are derived originally from deeper cooler ocean waters they can enhance marine productivity by serving to bring nutrients in the deep waters up to the surface; and (8) toxic organic materials, such as chlorinated organic substances including pesticides and plasticizers typified by DDT and PCB, likely are harmful to marine ecology, hence much more attention should be given to eliminating such materials from waste discharges especially through control at the sources.

All of the above scientific guidelines are useful to the engineer in the design of industrial waste marine disposal systems, to the extent that the design may be governed by the scientific approach. In summary, these guidelines indicate it is very important, if at all possible, to locate the point of discharge in unconfined rather than in confined ocean waters, and having done this, particular attention must be paid to those parameters which affect (1) aesthetics (floatables, color, etc.) [3], (2) to toxicity which may be harmful to marine ecology, especially chlorinated hydrocarbons and to some extent the heavy metals and trace elements, and (3) where pathogenic factors may be involved, to public health criteria especially coliforms.

3. INDUSTRIAL WASTE TREATMENT PARAMETERS

Over the past several years the U.S. EPA's program of research and development has sponsored a large number of investigations of treatment needs and methods for particular industries, which eventually will cover virtually all types of industries in the U.S.A. The results, which are being presented in a series of

"Point Source" reports, represent the first comprehensive and systematic approach to this difficult subject. Each of these reports, for a particular type or category of industry, describes the processing methods, the sources of the wastes, their amounts and characteristics, the significant parameters from the point of view of protecting receiving waters, and the applicable treatment technologies. These publications represent a set of "bibles" on the subject of industrial waste treatment, incorporating all known pertinent information. They are an important contribution to environmental engineering.

Table 1.

SIGNIFICANT PARAMETERS FOR TREATING WASTES FROM VARIOUS
INDUSTRIES FOR DISCHARGE TO CONFINED WATERS
(After Point Source Publications of US/EPA)

INDUSTRY	pH	ACIDITY/ALKALINITY	TSS	TDS	BOD	COD	TOC	TEMPERATURE	COLOR	GREASE/OIL	PHOSPHORUS	NITROGEN	SURFACTANTS	CHROMIUM	LEAD	COPPER	MERCURY	ARSENIC	MANGANESE	ZINC	POTASSIUM	ALUMINUM	HEAVY METALS	CYANIDE	PHENOLS	FLUORIDE	SULFATE	COLIFORMS (FECAL)	CHLORINATED HYDROCARBONS
Inorganic Chemicals	●	●			●	●								●	●		●							●					
Organic Chemicals	●	●	●	●	●	●	●			●	●	●	●											●	●				
Aluminum — Bauxite Refining	●	●	●	●																						●			
Aluminum — Primary Smelting	●		●																							●			
Aluminum — Secondary Smelting	●		●		●							●				●						●				●			
Feedlots			●	●	●	●					●	●															●		
Fruit/Vegetable Canning	●		●	●	●																						●		
Plywood	●	●	●	●	●	●		●		●	●	●		●		●		●		●						●			
Iron Smelting	●		●													●			●					●	●				
Cement	●	●	●	●			●													●							●		
Asbestos Building Materials	●	●	●	●	●	●	●				●	●									●	●							
Cane Sugar Refining	●	●	●	●	●	●					●	●																	
Fiberglass Insulation	●		●	●	●	●		●	●	●																●			
Meat Processing	●	●	●	●	●	●	●		●	●	●	●																	
Grain Processing	●	●	●	●	●	●	●		●		●	●																	●
Soap and Detergents	●	●	●		●	●				●			●																
Synthetic Resins	●		●		●	●						●		●						●									
Steam Power								●																					
Pulp and Paper	●		●		●	●			●																				
Brewery			●	●	●																								
Copper Wire	●		●	●										●		●				●									

Table 1 is a compilation, made from a random selection of point source reports, which shows the significant parameters to be considered in the design of a waste treatment system for particular industries, based on the assumption that the treated effluent is to be discharged to confined inland waters such as a river or lake or confined marine waters such as a shallow bay or estuary. Some twenty nine parameters are listed, along with twenty one different types of wastes.

Examination of Table 1 shows the most important parameters, applicable to most of the industries, include pH, acidity/alkalinity, total suspended solids, total dissolved solids, BOD, COD, and phosphorus and nitrogen. With the possible exception of suspended solids, none of these most important parameters has any particular meaning nor application for discharges to open coastal waters. Of all the twenty nine parameters, only a few, namely floatables (grease and oil), color, coliforms, chlorinated hydrocarbons, and possibly the toxic metals, need be considered. However, as stated above, for discharge to confined marine waters, especially biologically reproductive zones, all of the parameters checked in Table 1 will need to be considered. For such cases the point source publications are the best references.

4. DESIGN OF INDUSTRIAL WASTE TREATMENT SYSTEMS

As indicated earlier, the first step in the design, whenever possible, is to gain access to the deeper ocean waters, either through discharge to a municipal or other nearby system which already utilizes a deep-water outfall, or through construction of a submarine outfall for serving the particular industry. The second step is to provide a treatment plant to effect the removals needed to suit the particular zone of discharge. This may range from very little or even no treatment to a complex processing system where the place of discharge is in a sensitive receiving water zone.

Generally it will be far cheaper to discharge into an existing municipal or other nearby system should one exist, not only because of the economies of scale but because of the dilution factor generally available. Often the volume of industrial waste is small compared to the total volume of sanitary flow in a municipal system, hence much dilution is available which in itself may eliminate need for special treatment by the industry. Also, the municipal treatment system may already incorporate facilities for removal of floatables, certain heavy metals, etc., so that the needs for treatment by the factory prior to discharge to the municipal system will be much simpler than for direct discharge.

Assuming a new submarine outfall must be built, the design procedures will follow those for sanitary outfalls except that the controlling parameter will likely be, not coliforms, but to reach the deeper zone at a point achieving sufficient dilution so there will be no discernable adverse effects in the nearshore zone to be protected. An important step is to delineate the zone to be protected, and similarly to delineate the zone of mixing and dilution to be used beyond the zone of protection. Usually an oceanographic survey, together with sampling and analyses to establish the existing background on environment and ecology, will be essential to proceeding with design.

Selecting the treatment method will be governed by the applicable parameters as previously discussed. Assuming the waste includes no significant floatables nor toxicants such as chlorinated hydrocarbons, e.g. the waste discharges from cane sugar mills and from many pulp and paper mills where organic loading is the problem, no treatment may be needed, i.e. the ocean-waters will serve to do the work of biological stabilization.

Assuming floatables are significant in the wastes, e.g. from a palm oil processing plant or from a meat processing plant, it will be almost certain that treatment to remove floatables will be needed, e.g. by gravity separation (with or without chemical flocculation), by pressure floatation (with or without chemical flocculation), or other means. Due to the extreme variations in the behavior of different wastes with respect to removal of floatables, only by actual testing of representative samples of the waste can the design criteria be established for the best treatment processes. Sometimes laboratory "test-tube" scale testing will suffice but usually larger bench-scale testing is necessary for reliable results. A simple, convenient, and inexpensive methodology for trying out the applicability of floatation, developed by the Envirotech Corporation, is included with this paper as Appendix I.

In the event the waste contains significant heavy metals, such as lead or mercury, or other toxic inorganic chemicals such as chromates and cyanides, the local regulatory agencies will likely require some level of treatment to meet specified limits. Often the specified limits have been set arbitrarily, without regard to receiving water capabilities; however if they exist they must be considered in the design to the extent such removals are really needed to protect environment, or may be needed, or may be enforced regardless of need. A recommended approach is first to make the technical evaluation to determine the treatment removal needs (both known and potential) based on evaluation of the receiving water environment, then compare these with the stated regulatory requirements (both known and potential), then develop a phased design which may be implemented in steps beginning with the "absolute musts".

A good design must accommodate both sets of constraints, the scientific and the regulatory, and thus far in World pollution history the two sets of conditions rarely agree. While the regulatory criteria are often too severe because of their abitrary nature (especially the desire of "good" bureaucrats for uniformity for purposes of easing administration), sometimes they are far too lax and often ignore real hazards such as toxicity. Hence simply following the regulations, while legally admissible, will usually not produce the desired result, namely competent protection of the environment in a manner which will both protect the industry legally and actually eliminate damages to environment so the industry's record will stand the test of time. As mentioned earlier, the present massive study of industrial wastes by U.S. EPA represents the first comprehensive technical study of the problem, hence it will likely be many years before industrial waste regulations as practised by governments around the world will become scientifically based.

Another factor which may influence the design of the treatment system for an industrial waste containing toxicity is the possible need for bioassay testing. Discharge of the waste effluent into deeper waters usually will suffice, but in some cases it may be necessary to "prove" the validity of the design by conducting bioassays using selected local fish specimens. The most practical approach is to employ standard acute toxicity testing, to determine the dilution at which a given percentage of the specimens survive over a given period (usually the TL_{90} value), then to achieve a dilution in the design to reduce the concentration (after initial mixing) to a specified fraction of the test value (usually 25 times). While this procedure is strictly empirical its validity has been demonstated in the extensive toxicity studies made of San Francisco Bay. Sometimes such bioassay testing is necessary for convincing the regulatory authorities that the design is specific and competent for the local situation.

For many wastes involving heavy metals chemical precipitation, to achieve gross removal, may be an adequate first step. Additional removals will require use of more complex processing methods including applications of ion exchange, carbon adsorption, and special proprietary methods. Similarly, if chlorinated hydrocarbons are significant, the best control will be source control, to keep the material from entering the waste. Otherwise sophisticated and expensive tertiary treatment, including carbon adsorption, may have to be used. In all cases where toxicity is significant, whether from inorganic or organic sources, attention should be given to the in-plant processing methods to reduce the loadings through source control or modifications in plant processing. In caustic soda plants using mercury cells, for example, a number of surveys by Akzo Chemie of the Netherlands [4] have shown that in-plant control not only is cheaper than removal of mercury from wastes (usually accomplished with special exchange resins) but in addition the salvaged mercury represents a sizeable additional saving.

Where the factor of aesthetics in the marine receiving water zone is important, as at a beach or skin-diving area, then special attention must be given to removal of floatables and to any other parameter, such as color or turbidity, which may affect scenic and transparency values. Wastes from textiles, for example, may be important for their color contribution which may need to be reduced by appropriate treatment such as chemical precipitation, carbon adsorption, and other means. Again, textile wastes vary greatly from plant to plant and only by pilot testing can the best solution be developed.

Turbidity and/or color has been a factor in the discharge of industrial wastes affecting marine aesthetic values in the vicinity of Monterey, California. In one case, at a bay near Moss Landing, a beautiful scenic/ fishing area, an aluminium plant (producing aluminium from seawater) discharged a whitish-colored mineral residue into the bay with pronounced effects on the beauty of the harbor. The solution was either to remove this material from the discharge or to relocate the point of discharge into the open ocean where this material could be absorbed without any adverse effects.

A second problem with turbidity/aesthetics occurred along the famed 17-mile drive of California, possibly the World's No. 1 ecology-scenic attraction, where a company mining beach sand disposes of white clay mined with the sand by separating out the clay and discharging a clay slurry offshore through a submarine outfall. In this case the design procedure involved first, evaluation of the offshore receiving waters and currents to characterize the existing ecology, then design of an outfall and diffusion system which would achieve sufficient dilution not to create an offshore white-colored zone, and with the outfall sufficiently far out to prevent return of the clay materials to the very sensitive tidal zone which is a virtual showplace of marine organisms. A continuing program for monitoring of the receiving water has been carried out, which has proven that the disposal system has been performing well.

5. DISPOSAL OF RESIDUALS

Industrial waste treatment will result of course in production of concentrated residuals which in themselves constitute a difficult disposal problem. A National Conference held in February 1975 at Washington DC, was devoted to this particular problem. Review of the presentations at this conference indicates, where the receiving waters involved are of a confined nature and hence cannot absorb such wastes, that the cost of disposing of the residual materials will be a sizeable portion of the total cost of waste disposal. Usually the answer will be in landfilling under controlled conditions, or in re-cycling, or in reuse for fuel or as the basis for a new by-product, etc.

For disposal to the coastal marine environment the problem may be vastly simpler in that the large removals of organics achieved in conventional treatment, which lead to large amounts of residuals, will likely not be necessary.

6. SUMMARY AND CONCLUSIONS

(a) The best choice for disposal of industrial wastes into marine waters, when possible, is to discharge into a large nearby system, such as the municipal system, which already utilizes a prolonged submarine outfall. Treatment by the industry will usually be greatly simplified resulting in much lower costs. When the industry must employ its own system, consideration should be given to a joint regional system shared by several industries.

(b) The most important factor in the design is to reach a discharge point in deeper water, beyond zones which are sensitive either because of aesthetic use, biological productivity, or other reasons. Once this is achieved the treatment needs are simple compared to discharge to confined waters. For discharge to open coastal waters the usual set of parameters, so important for conventional treatment design, has little applicability. The key parameters will be those such as floatables, color, and toxicity, which relate specifically to effects on specific marine environmental values.

(c) The existing regulatory requirements for discharging industrial wastes into marine waters have often been arbitrarily prepared, hence may have little scientific basis. The professional approach to design is to determine treatment/disposal needs for both sets of constraints, the scientific and the regulatory, then develop a phased design so that the system can be progressively expanded from a first stage representing the best initial investment. The end product should not only protect the industry legally but "wear well" protecting environment.

(d) In many cases, regulatory approval of the proposed design is best obtained by providing for continuing monitoring of the affected marine environment following plant start-up, with the commitment that additional treatment will be furnished should the monitoring program show any such need. The overall treatment and disposal plan submitted for regulatory approval should include preliminary design information for such additional treatment, and site layouts showing that adequate space is available, to present a convincing case that follow-up treatment facilities will be provided if shown to be needed. In most instances, where the first stage investment has been soundly conceived, the monitoring program will prove its validity and assure all concerned that environmental values are being protected.

REFERENCES

1. The impact of secondary treatment on wastes discharged into the ocean. U.S. Senate Committee on Public Works Hearing, 18 March 1974 (Serial No. 93-H36).

2. Coastal water research project, 1974 annual report, Southern California Coastal Water Research Project, El Segundo, California.

3. Ludwig, H.F., Aesthetics — a key parameter in marine waste disposal. Proceedings of Sanremo Conference on Marine Waste Disposal, IAWPR, 1973.

4. Akzo Imac TMR process for removal of mercury from waste water. Akzo Zout Chemie Nederland, Hengelo, Holland.

H.F. Ludwig

APPENDIX I

FIELD FLOATION TESTING PROCEDURE

(As developed by Mr. Robert Emmett of the Envirotech Corporation)

The test unit, shown in the accompanying drawing (Fig. 1), comprises a steel pipe of approximately of 0.5 l volume, with appropriate valves and fittings so that it can be filled with a suitable solution, pressurized, shaken violently for about 30—60 s to ensure complete solution of the air, and then the effluent released slowly through the stop-cock valve into a suitable container.

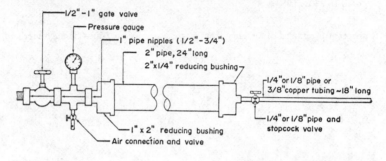

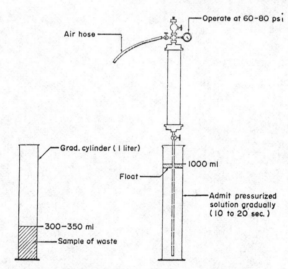

Fig. 1. Improvised dissolved air flotation test apparatus.

Two modes of operation may be used, either pressurization of the total feed, or pressurization of a bleed stream or side stream, which is essentially the solution collected from the floatator after the solids have separated out. For a trial test, water can be used and pressurized alone before being introduced into the waste stream. To do this, add about 2/3 l. of water to the pressure vessel, and place about 1/3 l. of the waste effluent into a 1-l. cylinder. After pressurizing the water, this can be released gradually over a period of about 10 s into the graduate cylinder as shown in Fig. 1 and the results observed. If the oily materials separate alone, free of other floatables, this would be highly desirable. The test should be repeated using the effluent as the pressurization liquid, to see if this has any adverse effect on the floatation of the oil, due to the high shear as the liquid passes through the nozzle when it is released from the pressure vessel.

The use of demulsifying agents at this point can be considered to assist in achieving separation, or even flocculation of some sort if it were desired to flocculate the entire mass and trap the oil with the solids. The mixture of oil and floated solids might then be filtered on a pressure filter, and the oil removed in this manner.

DEVELOPMENT OF OILY WATER SEPARATING SYSTEMS FOR SHIPBOARD USE

Robert S. Lucas

Naval Engineering Division, USCG Headquarters

Washington, DC, U.S.A.

Summary — This paper discusses the solution to shipboard oily waste disposal problems, as applied to U.S. Coast Guard ships and boats. The first part discusses the requirements, both international and for the United States. Secondly, the development of three oily-water separator systems is traced; conclusions reached include the opinion that filter-coalescer systems can be successfully applied aboard all but heavy-fueled ships.

1. MANDATE TO ELIMINATE POLLUTION OF THE SEA BY OIL

By its Resolution A. 176 (VI) of 21 October 1969, the Assembly of the Inter-Governmental Maritime Consultative Organization (IMCO) decided to convene in 1973 an International Conference on Marine Pollution. This Conference was held in London from 8 October to 2 November 1973. As a result of its deliberations, the following instruments were adopted:

> International Convention for the Prevention of Pollution from Ships, with its Protocols, Annexes and Appendices; and Protocol Relating to Intervention on the High Seas in Cases of Marine Pollution by Substances Other than Oil.

Other Resolutions were adopted which include (1) urging implementation of the 1969 Amendments to the International Convention for the Prevention of Pollution of the Sea by Oil, 1954; (2) recommending that Governments and other interested bodies undertake concerted efforts to reduce the discharge of oil from ships into the sea with a view to the complete elimination of intentional pollution not later than the end of the present decade; (3) recommending the development of efficient oil content monitoring arrangements; and (4) several others, to the total of twenty six, that include items concerned with both oily and non-oily pollution.

1.1. Regulations concerning oil pollution

Regulations adopted by the International Conference on Marine Pollution cover many areas, but include the following applicable to the subject matter at hand: (1) provides for surveys of every oil tanker of 150 tons gross tonnage and above, and every other ship of 400 tons gross tonnage and above, to ensure that the equipment and associated pump and piping systems, including oil discharge monitoring and control systems, oily-water separating equipment and oil filtering equipment fully comply with IMCO requirements, and are in good working order; (2) provides for issuance of an International Oil Pollution Prevention Certificate (1973) to ships that comply with the requirements; (3) specifies requirements for control of operational pollution; (4) requires oil discharge monitoring and control system and oily-water separating equipment on certain ships; (5) requires retention of oil on board in some cases, i.e. slop tanks fitted with oil/water interface detectors, and tanks for sludge or oil residues and (6) specifies that the oily-water separating equipment must produce an effluent that has an oil content of not more than

100 ppm, and that the oil filtering system, when receiving the effluent from the oily-water separating equipment, must produce an effluent the oil content of which does not exceed 15 ppm.

1.2. Requirements for control of operational pollution

The key requirements for control of discharge of oil are the following (for non-oil tankers, of 400 tons gross tonnage and above): Any discharge into the sea of oil or oily mixtures is prohibited except when all of the following conditions are satisfied: (1) the ship is not within a special area; (2) the ship is more than 12 nautical miles from the nearest land; (3) the ship is proceeding enroute; (4) the oil content of the effluent is less than 100 ppm; (5) the ship has in operation an oil discharge monitoring and control system, oily-water separating equipment, or other approved installation. The above does not apply if the oil content of the oily mixture without dilution does not exceed 15 ppm.

2. LEGAL REQUIREMENTS FOR U.S. SHIPS

In the United States, in Public Law 92-500, the "Federal Water Pollution Control Act Amendments of 1972", Congress declares "that it is the policy of the United States that there should be no discharges of oil or hazardous substances into or upon the navigable waters of the United States, adjoining shorelines, or into or upon the waters of the contiguous zone." The law further prohibits discharge of oil "in harmful quantities", which has been determined in regulation (40 CFR 110.3) to include discharges which (1) violate applicable water quality standards, or (2) cause a film or sheen upon or discoloration of the surface of the water or adjoining shorelines or cause a sludge or emulsion to be deposited beneath the surface of the water or upon adjoining shorelines.

2.1. United States' goal

The stated U.S. goal is to prohibit the discharge of oily wastes into all waters of the world by no later than the end of the decade. The time is approaching when the discharge of oily bilge and ballast water even on the high seas may no longer be legally permissible.

2.2. Status of the problem [1]

The problem of oily water disposal has plagued engineers and captains of ships for years. Some of the efforts taken by shipboard engineers have included reducing oil and water leaks and distributing engine-room bilge water to other bilges or to an empty fuel tank when in port for later disposal at sea.

The Coast Guard has a vigorous pollution prevention and enforcement program. Routine anti-pollution flights are conducted by both fixed-wing aircraft and helicopters. New technology is being used as equipments are developed. For instance, Coast Guard aircraft are now being equipped with remote sensing devices in the infra-red and ultra-violet ranges to detect oil slicks. Civil penalties may be anticipated. Criminal offenses with more severe penalties may also result if the Coast Guard is not notified when an oil discharge occurs. Costs of clean-up of even comparatively small oil spills can be substantial.

2.2.1. *Transfers of waste ashore.* Very few shore facilities are available to dispose of large quantities of oily water. Fewer yet are convenient to use and none are inexpensive. Even if shore disposal facilities were available and convenient, the bilge water accumulated at sea or in remote ports still poses a disposal problem. Coast Guard ships, for instance, may spend long periods of time in coastal waters and are unable to pump bilges. Even though slop-oil tanks are required on certain ships, this approach really only transfers the problem ashore, and in any case presents space problems on many ships and boats.

2.3. Fuel contamination a secondary consideration

The case of ballasted ships is complicated by the twin factors of oil pollution and fuel contamination. They are related since many ships are required to fill their empty fuel tanks with sea water to maintain the ship's intact and damaged stability characteristics. Many ships avoid ballasting since the residual salt water remaining after deballasting contaminates the new fuel load. The diesel fuel de-contamination equipment now in use cannot always adequately remove this water. The discharge of ballast water must be done beyond 50 miles offshore in order to avoid contravening the rules of international conventions. That last 50 miles of the voyage has the highest potential for collision yet the ship is in its poorest stability condition to receive such damage. This same situation is generally considered as the prime cause of the sinking of the *Andrea Doria* after colliding with the *Stockholm*. The ideal situation would be to retain ballast water until moored and then discharge it. Secondly, it would also be desirable to be able to remove any residual ballast water from the fuel tanks after the vessel has been refueled. Thirdly, it is desirable to be able to discharge the bilge water when moored to prevent flooding of the vessel's machinery. All of which, of course, must be done without discharging oil into adjacent waters and should be done with one machine for space, weight and cost considerations.

3. DEVELOPMENT OF OILY WATER SEPARATOR EQUIPMENT

Many principles of operation have been used in the attempts to remove oil from water. Some separators have been tried on ships. Most of these were large and employed gravity settling or absorptive methods. Until recently, none had been successful in eliminating the discharge of visible oil.

3.1. Coast Guard involvement [2]

The Coast Guard search for an oily water separator and oil-in-water monitor began in September 1970 and was originally intended to serve as a solution to the stability problems of some of our ballasted cutters. The program was ultimately expanded to include disposition of bilge wastes on all cutters, down to and including 65-ft tugs. A background investigation and source search located nine oily water separators and four oil-in-water monitors with potential for shipboard use. Shoreside screening tests were conducted as the first phase of a three phase program. The shoreside tests were conducted to avoid the fruitless shipboard installation of equipment not capable of adequate performance.

3.1.1. *Initial testing program.* The shoreside testing spanned a 12 month period and was conducted by Coast Guard Naval Engineers at the U.S. Army Mobility Equipment Research and Development Center, Ft. Belvoir, Virginia. The first test phase identified equipment utilizing the filtration/coalescence principal as having the best potential for more detailed testing. Monitors utilizing the principal of infra-red light absorption were similarly identified as having the most potential for detecting non-soluble oil in water. Two filter/coalescer separators were given extensive testing during the second phase of the program. Both of the separators would produce an oil-free water effluent when a low-emulsifying test pump was used. Cumulative test data indicated that the poorest water effluent quality could be expected when processing cold, fresh water that had been contaminated with used, diesel engine lubricating oil and processed by a centrifugal pump. One system produced an acceptable effluent under these very difficult conditions. Using a low emulsifying pump and reducing the flow rate through the system, further improvements were obtained. This equipment was thus selected for a shipboard prototype system.

3.1.2. *First shipboard installation.* In October 1972, a prototype separator system was installed in the

Coast Guard Cutter *Alert*. A monitor device was provided to continually analyze the effluent water for oil and to automatically recirculate unsuitable water products. The oil product from the system was to be returned to the ship's fuel tank or a waste oil tank, depending upon purity. The selected separator consists of three pressure vessels in series. The first pressure vessel protects the following filter/coalescer stages by filtering out dirt and removing entrained air. The dirt-free oil-in-water emulsion is then forced into the first coalescing stage (second pressure vessel). This stage contains the filter/coalescer cartridges which are the heart of the system. These cartridges function to break the emulsion by forcing the minute oil drops to merge (coalesce) together. The resulting oil drops rise to the top of the pressure vessel until sufficient oil is collected to cover a capacitance probe. Upon sensing oil, the capacitance probe will simultaneously open the oil discharge valve and close the water outlet valve. The valves will remain in this position until sufficient oil is discharged to uncover the capacitance probe. The probe will then cause each valve to return to its normal position. The third pressure vessel (second coalescer stage) is a duplicate of the second vessel and functions as a polishing stage.

3.1.3. *Control system details.* The selected monitor consists of a pump, sensing station and associated amplifying/switching circuitry. The pump extracts a continuous sample of the system's effluent water and emulsifies it. The sample then passes through the sensing station and returns to the effluent stream. The sensing station has two clear ports to allow infra-red light to pass through the sample stream. A photo-cell detects any reduction in the transmitted light caused by the presence of oil. The resulting signal is amplified and displayed as a linear indication of oil content on a meter and recorder. Adjustable alarm points are provided for process control. The *Alert's* system uses this alarm signal to close the separator water outlet valve. With this valve closed, the supply (bilge or ballast) pump increases system pressure. A pressure sensitive regulating valve was provided to relieve this pressure to permit recirculation of unsuitable water effluent to the pump suction. The system will remain in the recirculation mode as long as the oil content of the water effluent exceeds the alarm set point on the monitor. Recirculation through the separator diminishes the oil content of the water. The monitor will return the water discharge to overboard automatically when it becomes satisfactory. The system is totally automatic after start-up.

3.1.4. *Operating results.* After installation of the system on the *Alert*, both bilge and ballast mixtures were processed. It was determined empirically that the threshold for producing a visible sheen of oil on the water adjacent to the ship with this system was no less than 30 mg oil/l (mg/l) of water. Since the threshold will vary under many operating and detection conditions, the alarm point was conservatively set to recirculate all water effluent exceeding 20 mg/l of oil. This alarm point has since been adjusted downward to 15 mg/l. A low-emulsifying pump was installed for use when processing the more difficult bilge water (as compared to ballast water). This pump is a double-diaphragm, air-driven type. A three-fold reduction of oil in the effluent water was recorded when using this pump in lieu of the centrifugal bilge-ballast pump. Further improvement in effluent quality was accomplished by reducing the flow-rate of the low-emulsifying pump. The reverse occurs however, when reducing the flow-rate from the installed centrifugal pump. All emulsions were broken satisfactorily from both bilge and ballast water with the exception of a surfactant-stabilized emulsion caused by the introduction of detergent soap. This mixture was ultimately carried to sea for disposal. Aside from test work, the system has never discharged sufficient oil to create a visible sheen on the water. Discharged water quality has been below 10 mg/l and usually below 5 mg/l. Recirculation time is less than 5% of operating time. All product oil from the separator system to date has been returned to the ship's fuel tanks and consumed without apparent difficulty. On at least one occasion the system was used successfully to decontaminate diesel fuel. In addition, residual ballast water that usually exists in fuel tanks is removed after refueling and settling has been accomplished.

4. SECOND GENERATION EQUIPMENT DEVELOPMENTS [3]

The *Alert's* system accomplished all of the basic goals of the program and more. The solution to the problem of oily waste disposal from larger ships, i.e. those that could be fitted with the large 100-GPM systems, was solved. However, the Coast Guard has approximately 210 cutters and boats 65 ft or over in length that do not require such large systems. It was obvious that smaller and cheaper systems had to be developed. The limited data available indicated that a 10-GPM system could handle the bilge water produced by our non-ballasting units from 200 ft down to about 130 ft in length, and that even smaller units were desirable for cutters under that size.

4.1. Development of a 10 GPM system

Working closely with industry, a 10-GPM unit was procured and installed in a 114 ft river buoy tender, the CGC *Foxglove*, during September, 1973. This system is identical in concept with the 100-GPM unit, differing only in size of the pressure vessels, number of filter/coalescer cartridges and control system. The manufacturer of the monitor redesigned his equipment to produce a smaller, lighter unit but utilized the same internals as previously. The controls for this system were similar to the 100-GPM system except that proportioning probes and valves were used. This system was found to be very successful in this application, but the original, simpler capacitance probe scheme has been utilized in all production models.

4.2. Development of a 5 GPM system

Last but not least, a small, simple, inexpensive separator system had to be developed for Coast Guard cutters and boats that did not require even a 10-GPM system. Minimum cost was an overriding goal because of the large number of these systems required. Our experience to this point indicated that the monitor was an inseparable part of a flow-through system, since it prevents the discharge of oil from the separator when processing difficult emulsions. However, its cost is relatively high, being about one-third that of the 100-GPM system and over half the cost of a 10-GPM system. Our experience also had shown that, with but little exposure or training a man could visually detect oil in water in a test tube at a concentration well below that which would result in a visible sheen upon the surface of the water. The design problem became, then, how to prevent the discharge of a separator effluent that contained enough oil to cause a visible sheen without using an oil-in-water monitor.

4.2.1. *Small system details.* A system was designed in-house that is simple, cheap and manually operated. The concept of the system is to pump the oily bilge water to a small holding tank, then recirculate this emulsion through the separator until the contents of the tank test (visually) oil free. The holding tank is then discharged through the separator and overboard. Valves and controls are arranged so that it is impossible to pump the oily bilge water directly overboard. The separated oil is transferred to a dirty oil tank.

5. PROBLEM AREAS

It is known that surface-active agents (surfactants) have a deleterious effect on any oil and water separator. This apparently affects a filter/coalescer by coating the filter fibers and changing the fiber's lipophylic/hydrophobic (oil-loving/water-hating) balance which starts the coalescing action. There is some evidence that surfactants also prevent the agglomeration of the smaller coalesced oil droplets by reducing the interfacial tension. Under a given set of conditions, the rate of separation is primarily governed by the oil droplet size (per Stokes law). The smaller sized droplets have insufficient buoyant force to separate from continuous phase (water) within the separator. These smaller oil drops are carried out with the water phase, recontaminating the effluent water. Some surfactants that are common aboard ship that may reduce the effectiveness of the separator are: bilge cleaners (Gamlet C-clean etc.); liquid and powder detergents; hand soap; protein fire fighting foam; and aqueous film-forming foam (AFFF).

5.1. Surfactants and recirculation

In general, compounds that increase the pH of water may be considered surfactants and should be prohibited from entering the bilge areas being serviced by the separator system. This is not to say that total failure of the system will always occur. More likely a greater period of recirculation would be encountered. The length of this period would be proportional to the "toxicity" of the surfactant. In the event that the oil content of the water could not be reduced below the alarm set point, the ship would have to proceed beyond the 12-mile limit where discharges containing up to 100 mg/l of oil are currently allowed. For ships in port, discharges up to 100 mg/l are sometimes permitted to be introduced into municipal sewer systems.

5.2. Restrictions on system use

One limitation of these systems is their inability to handle heavy fuel oil except in very low concentrations. This is not a problem on diesel-propelled ships, or for bilge water only on any ship, but does prevent their use in the deballasting of heavy-fueled ships.

6. CONCLUSIONS

The systems developed by the U.S. Coast Guard are capable of meeting the U.S. national as well as international requirements for oily-water separators, filters, and monitors. The systems are comparatively small, light in weight, and relatively inexpensive. Similar equipment is obtainable from several U.S. manufacturers at competitive prices; all production units for the Coast Guard were obtained by the competitive bid process. There are certain drawbacks to the filter-coalescer systems, mainly their inability to handle heavy oils, but they represent the best state-of-the-art for shipboard use at present.

REFERENCES

1. Norton, L.B., A marine system for eliminating oil pollution, presented at the New York Metropolitan Chapter, American Society of Naval Engineers, New York, N.Y., 12 July 1973.

2. Test and evaluation of oil pollution abatement devices for shipboard use, phases I—II—III, Naval Engineering Division, USCG Headquarters, Washington, D.C. (1972-1973).

3. Lucas, R.S., Separation and monitoring of oily bilge water, presented at the Offshore Technology Conference, Houston, Texas, as OTC paper 2200, 5-8 May, 1975.

THE CONTROL OF WASTEWATER DISCHARGES TO MARINE WATERS
THE APPLICATION OF LEGAL STANDARDS AND CRITERIA
TO DISCHARGES OR TO MARINE WATERS

S.H. Jenkins

Bostock Hill & Rigby Ltd.

288 Windsor Street, Birmingham B7 4DW, England

The object of this paper is to discuss the question of marine pollution standards or criteria as applied to discharges of sewage or other wastewater or sewage sludge into estuaries, coastal waters or the open sea. In this context standards are defined as parameters of pollution which may be expressed numerically or described in words for the purpose of setting legally permitted limits to the content of polluting matter in a discharge or in the receiving water. Criteria, though not legally enforceable requirements of water quality, may be expressed in identical terms. Both standards and criteria are instruments of policy. The clearer the objectives of policy can be defined the more accurately can standards or criteria be laid down. However it does not follow that because water quality standards or criteria have not been promulgated or published, a country does not have a policy to control marine pollution.

A general definition of "criteria" adopted by the Group of Experts on the Scientific Aspects of Marine Pollution (GESAMP) is "the required scientific information on which a decision or judgement may be based concerning the suitability of the environment to support a desired use, recognizing that the health of man is paramount and that the latter can be affected either directly or indirectly". For the marine environment, criteria "should include consideration of all aquatic compartments rather than water alone".

Aspects of marine pollution that are the concern of international organizations, such as the disposal of radioactives wastes, the transport of oil or chemicals in ships or the exploration or exploitation of the sea bed, will not be considered in this paper.

Reference is made in the final section of this paper to various legal and administrative procedures that may be used to control water pollution. These procedures may include measures to protect inland as well as coastal waters. Differences in the methods of control arise from differences in systems of government and differences in emphasis on the objectives of control measures. One may identify at least four methods of control exercised by governments which may have a bearing on the question of marine water quality standards. These are (1) pollution control; the construction of works to abate pollution and financial control of public works is the responsibility of one government department, (2) an agency of government sets the required goals public works must achieve and grants loans for works designed to meet these goals, (3) an agency of government approves public works such as a sea outfall in order to ensure that all the necessary legal requirements are met, (4) an agency or committee is given powers which may be limited to seawater quality control so as to protect fishery or other interests.

The question of standards or criteria for marine discharges will be considered from the point of view of a scientific and technical adviser who is required to state the limiting concentration of pollutants which may be permitted in a discharge or in a receiving marine water to safeguard the use of the water for certain purposes.

PRESENT POSITION REGARDING STANDARDS FOR MARINE DISCHARGES

Legal standards for marine water quality have already been promulgated or have been in use in some states or countries and action to enforce them has been reported. Such standards are under consideration or have been recommended in other countries. Views have been expressed that the objectives of these standards are attainable without their use.

Although each country adopts the measures it considers most suitable to deal with its own problems, the author believes that an objective enquiry into the subject of marine quality standards is desirable in order to discuss their reliability as parameters of pollution and the factors which may influence them. In general terms answers to the following four questions will be sought:

1. Are marine water quality standards having the force of legal requirements necessary or desirable to control marine pollution?

2. Are the same objectives equally well attainable by the use of criteria which do not have the force of law?

3. What methods may be used to predict how such standards may be attained and what is the accuracy of prediction?

4. What administrative measures are necessary to ensure that the standards are met; what surveillance and sampling techniques and methods of examination should be used and does the pollution caused by bathers themselves have to be taken into account?

PARAMETERS OF POLLUTION USED IN STANDARDS OR CRITERIA

Bacterial or microbiological parameters

Of the various parameters that are in use or have been proposed for marine waters the most common is a bacterial measure of water quality. Its use is supported either on the grounds that bathing will be safe in water that meets the required bacterial standard or such a standard ensures that the water will satisfy amenity requirements; thus, it becomes less necessary to depend upon verbal definitions of water quality requirements. Such a standard according to McKee (1960) is a measure established by authority. Although it "does not necessarily mean that the standard is fair, equitable or based on sound, scientific knowledge" one presumes that an authority would have a good reason to select a particular standard. Measures that were not regarded as fair, equitable or based on sound, scientific knowledge might be acceptable for short periods or in an emergency but it is doubtful whether efforts would be made to obtain standards for sea water that did not represent reasonable requirements for particular circumstances, attainable at reasonable cost.

PUBLIC HEALTH SIGNIFICANCE OF POLLUTED SEA WATER,
MEASURED BY ITS BACTERIAL CONTENT

The most frequently used numerical indicator is the coliform bacteria count as defined in Standard Methods (1965, 1971). In this publication it is recognized that the coliform bacteria enumerated may include bacteria of non-intestinal origin and attention is drawn to the method of differentiating the intestinal from the non-intestinal for purposes of evaluating pollution of sewage origin in raw water supplies, streams, and for special purposes.

But this differentiation is not recommended for raw water supplies destined for public use, and it is rarely used in marine water quality standards or criteria proposed in the U.S. (see Table 1) or in Israel (Shelef) or

Table 1. Some standards or criteria

BALTIC (Svansson)

A Convention of Baltic States agreed to ban:—

PCB's

DDT

Other pesticides and some heavy metals. No coliform standard.

MARYLAND, U.S. Shellfish beds (Carter)

Median coliform count in the overlying waters must not exceed 70 per 100 ml.

The 90 percentile must not exceed 230 per 100 ml. Carter (1975) recommends limits for specific organisms instead of a coliform standard.

CALIFORNIA, U.S. Surface bathing waters

In 20% of the samples taken the coliform count must not exceed 1000 per 100 ml.

FLORIDA, U.S.

The Median Coliform Count (MPN) of not less than 10 samples must be less than 1000 per 100 ml.

ONTARIO, Canada

The suggested median coliform limit (MPN) is 1000 per 100 ml with a limit of 2400 per 100 ml for individual samples.

NEW YORK CITY, U.S. Bathing Waters

Class A Group 1. With less than 1000 coliforms per 100 ml: epidemiology satisfactory.

Class A Group 2. With the average coliforms greater than 1000 per 100 ml, but less than 2400: epidemiology satisfactory.

Class B. With the average coliforms greater than 2400 per 100 ml and 50% of the samples with more than 2400, the beach is not recommended.

Class C. Epidemiology poor. The beach should be closed.

Table 1 continued.

ITALY
Proposed National legal limits
Circular 105 — Ministry of Health 2 July 1973

Beach water

Faecal coli, not to exceed 100 per 100 ml

Nutrients in effluents discharged to sea

Ammonia as NH_4	5 mg/l
Nitrite as NO_2	2 mg/l
Nitrate as NO_3	50 mg/l
Phosphate as PO_4	20 mg/l

Legal Limits for Venice
prescribed by decree of the
President of the Republic

Effluent analysis	To lagoon Venice	To public sewer
BOD mg/l	35 – 50	500
NH_4	2 – 5	50
NO_3	20 – 50	–
PO_4	1.5 – 5	50
CN	0.2 – 0.5	5
Cd	0.02	0.1
Cr^{3+}	1	2
Cr^{6+}	0.1 – 0.2	0.2
Hg	0.005	0.01
Ni	2	4
Pb	0.1	0.2
Cu	0.05	0.1
Zn	0.5 – 1	1

LOS ANGELES, U.S.
Bargman (1975) Selection of standards for effluents diffused into the ocean

	Limits for 50% of time	Permitted for 10% of time
Total Grease	10	15
Floating Grease	1	2
Cadmium	0.02	0.03
Chromium	0.005	0.01
Copper	0.2	0.3
Mercury	0.001	0.002
Nickel	0.1	0.2
Zinc	0.3	0.5
Cyanide CN	0.1	0.2
Phenols	0.5	1
Chlorine	1	2
Ammonia N	40	60
Chlorinated carbon compounds	0.002	0.002
Toxicity concentration	1.5	2

The effluent, after dilution so as to give a concentration of 4% of effluent must not cause more than 50% mortality in a standard toxicity test of 96 hours duration.

ISRAEL – Criteria
Sea Water Quality Report to Water Commission Office Tel Aviv 1972, Shelef (1975)

Classes of Sea Water

Class	Description or use	Properties
A	Aesthetic value	50 m from coast water to be free from visible waste material
B	Coastline of natural reserve	As A but definition of waste more specific
C	Commercial fishing	pH / Temperature / Dissolved oxygen toxicants controlled
D	Industrial use	As C with ammonia limit added
E	For bathing and direct contact	As above floatables limited. Coliforms median limit 1000/100 ml. Not more than 2400/100 ml in 20% of samples.

Table 1 continued.

ITALY (continued)

Chlorinated hydrocarbons	0.05 – 1	0.2
Coliform organisms per 100 ml	20 000	—
Faecal coli per 100 ml	12 000	—
TLm^{50} per 100 ml (bathing water 6 hr)	100	—

Metals (total)

Conc. of each metal = <3
Limit for each metal

YUGOSLAVIA
Legal Standards (Stirn) 1975

Receiving Water Quality Limits	Class I Fish Culture	Class II Bathing recreation
Visible oil	0	0
Hazardous compounds	0	0
Coliform MPN per 100 ml	10	2000*
Phenols mg/l	0.001	0.001
Arsenic	0.01	0.01
Copper	0.01	0.01
Nickel	0.001	0.001
Cadmium	0	0
Lead	0.05	0.05
Chromium	0.1	0.5
Zinc	0.01	0.1

* too generous a standard according to Stirn

SLOVENIA
Legal Standards (Stirn) 1975

Coliform limit (MPN) for Class II water shown for Yugoslavia is 500 per 100 ml

CROATIA
Legal Standards (Stirn) 1975

Coliform limit (MPN) for Class II water shown for Yugoslavia is 500 to 1000 per 100 ml.

U.K.
Criteria Suggested by Key (1970)

Criteria suggested on the basis of coastal investigations carried out in the U.K.

90 percentile limit 1000 per 100 ml

in a number of other countries (Ludwig), or Yugoslavia (Stirn). In Italy, however, such differentiation is essential for in the proposed national standards for marine waters the bacterial limits are expressed as the number of faecal coliform bacteria per unit volume. In the U.K. both *E. coli* and coliform counts have been used in coastal water studies. Smith (1975), who prefers a faecal coliform (*E. coli*) count to a coliform count, assumes that the ratio of *E. coli* to coliform count is 1 to 2.5 and supports his preference on grounds of the success in the use of *E. coli* in improving the hygienic quality of food and water. It should be emphasized that public health authorities in the U.S. have for many years successfully used the coliform count as a means of determining the sanitary quality of the water at bathing beaches. According to Ocean Dumping (1970) "many beaches have been closed to swimming because of the high coliform count of the water".

In a book recently published, (Gameson 1975), Shuval (1975) states arguments in the following four numbered paragraphs in favour of a bacteriological standard, stating that:—

1. Although there is as yet no clear cut association between bathing in polluted seawater and enteric disease, the consumption of shellfish grown in polluted waters has caused typhoid fever and infectious hepatitis. Furthermore, drinking water was protected by a strict coliform standard long before all the scientific facts to justify its use were established.

2. In order to arrive at a coliform standard that might be related to a health risk at a hypothetical bathing beach Shuval (1975) makes the assumption that sewage contains 10^8 coliforms/100 ml, that these are reduced to 10^6/100 ml by dilution on discharge and by a factor of 100 due to natural die-away in travelling from the point of discharge to the beach. Therefore on this basis 10,000/100 ml would be present at the beach. The virus count of the sewage is taken to be 1000/100 ml, containing 1% of a virulent virus, i.e. 10 such viruses per 100 ml, each capable of causing infection.

    Dilution reduces the virulent virus to 1 per 1000 ml. Assuming 10 ml sea water is ingested per bather per day, one bather in 100 may ingest an infectious dose of virulent virus. If one in ten such bathers may become ill on this account 1 bather in 1000 may become ill when the coliform count is 10 000/100 ml. By the same reasoning the risk of illness would be reduced to 1 in 10 000 bathers when the coliform count is 1000/100 ml.

3. According to Shuval, although an epidemiological basis for a rational bacterial standard is lacking, there is sufficient knowledge of the coliform content of sewage, the extent to which the bacteria disappear or are reduced in number as a result of primary treatment, dilution, dispersion, sedimentation, die-away rate in sea water to suggest that a standard or goal of 1000 coliforms/100 ml is attainable. The assumptions made by Shuval in arriving at a standard will be discussed later by the author.

4. Since pathogens survive in sea water and some, especially viruses, are infective at very low doses, it is reasonable to take preventive action to protect bathers by reducing the concentration of organisms of known faecal origin to the lowest practicable level, without waiting for more positive epidemiological evidence to support the numerical standard.

Some of these views receive wide support. McGregor (1975) for example, after referring to the contamination of shellfish in temperature zones with *E. coli* when taken from beds with overlying sewage-polluted water and the possibility of additional health risks in warm climates, concludes that no excuse justifies delay in reducing pollution or permitting it to continue at its present level and he advocates the use of a bacterial measure of sewage contamination. The connection between polluted water and the contamination of shellfish by faecal bacteria and sometimes by pathogenic organisms is well established. Therefore for economic and public health reasons, and to give some protection from shellfish harvested illegally, apart from those under public health authority surveillance, McGregor believes it is necessary to control the bacterial quality of all waters overlying shellfish beds. Covill (1975) regards it as reasonable to require bacteriological standards and monitor microorganisms in marine waters as chemical standards. Wakefield (1975) has expressed a commonly held point of view that bacterial standards need not be related to public health, since they are a measure of aesthetic quality. Israel appears to be the first Mediterranean country to propose classifying marine waters according to water quality, recognizing that there are gradations of contamination by different components of sewage and industrial effluents. Sea water of aesthetic value is placed in the lowest quality category, in which no bacterial limits are stated; a bacterial standard is reserved for sea water likely to come into contact with human beings. For health reasons this water has to meet the highest of five quality requirements (Shelef, 1975).

The use of bacterial standards or criteria has been growing in recent years. Ludwig (1975) referred to their employment in Southeast Asia; Stirn gives the promulgated coliform standards for discharges to the Adriatic (1975). Proposed Italian faecal coliform standards and those adopted in certain provinces have begun to influence the design and location of marine outfalls. To an increasing extent coliform criteria at bathing beaches are being regarded as the most important parameter in sewer outfall design even in parts of the world where such criteria are not recognized (Camp). In Mediterranean countries, where the public health aspect of pollution is considered by some to be of paramount importance because of the tourist industry and the small tidal range, increasing importance is being attached to bacterial standards in order to protect bathing beaches.

The arguments against relying upon microbial standards for protecting bathing beaches have been summarized by Moore (1975). These arguments apply to the use of standards having legal status which may be enforceable and subject to legal sanction for infringement of the conditions. Reference will be made later to the alternative of using criteria, which may be interpreted in the light of local circumstances without infringement of the law.

The objections that have been made against the use of bacteriological standard, are as follows:

(a) The classification of bathing waters in terms of permitted median and 90 percentile levels "is simplistic to the point of being spurious" (Moore 1975).

(b) No evidence links the risk of communicable disease with sewage contaminated sea water; to this extent microbial standards are irrelevant.

(c) Years of study has led to the conclusion that there is no logical justification for rigid bacteriological standards, for by selecting sampling positions and times of sampling the quality of the beach water could be graded upwards or downwards; the coliform counts are log normally distributed from day to day and within a period of a day as well as at different points on a beach. The ratio of the 90 percentile coliform count to the median count was found to be 9 to 1. Such highly variable coliform contamination on any one beach cannot be realistically accommodated within a rigid bacteriological standard.

(d) Little evidence of a health risk associated with bathing in polluted seawater has been found, nor has any pointer to a higher incidence of minor illness among sea bathers been disclosed. After noting that the full report of the U.S. Public Health Service investigations of 1948–50 and the nature of the illnesses had not been published (The results had, however, been reported by Stevenson (1953) and, after giving consideration to subsequent work, Moore (1975) concluded that it is difficult to show any cause and effect relationship between sea water bathing and illness.

(e) Concerning the possible risk of virus infection from sewage polluted sea water Moore refers to the absence of evidence that enterovirus infections can be waterborne and also to the fact that such viruses would only be present in polluted waters when infections due to these viruses were already circulating in the population, so that they would be monitored in sea water only as part of a wider epidemiological study of the causative agents. Bacterial indicators of sewage contamination would therefore indicate the possible presence of enteroviruses only when these infections were already present in the community.

(f) If control over sewerage schemes and the siting of outfalls is satisfactory, the microbiological condition of bathing waters need cause no concern. Mackay (1975), criticizing bacteriological standards as applied to estuaries and the cost that might be involved in meeting them under certain conditions, concluded that beaches that satisfy aesthetic requirements are suitable for bathing. The conclusion reached by Agg (1975) is that statistically meaningful standards that could be justified are not yet available although a well defined bacteriological criterion could be included as one of several specific criteria of use in the design of new sea outfalls.

STANDARDS AND CRITERIA – PROPOSED OR SUGGESTED

It is of interest and of some practical importance to compare standards or criteria that are in use or that have been proposed, in order to consider the factors that influence the various parameters used or how to attain them.

A selection of actual or suggested standards or criteria is given in Table 1.

A positive and helpful statement on the question of criteria for coastal waters is contained in a recent report of a working group of the World Health Organization (1975) after reviewing the position concerning methods for the protection of bathing beaches and coastal waters in several countries, it was concluded that it was desirable to set broad upper limits for the number of faecal indicator organisms, (E. coli being one of the most sensitive indicators) and that the limits should be expressed in broad terms as orders of magnitude in preference to rigidly stated specific numbers. Highly satisfactory bathing waters would have fewer than 100 E. coli/100 ml; to be acceptable, waters should have counts consistently below 1000 E. coli/100 ml. Emphasis was placed on the aesthetic and physico-chemical properties of bathing beaches and sea water, even though the risk of injury to bathers from pollutants which made the water unsightly or unpleasant was slight. Among the subjects which this group recommended deserved further study were the evaluation of health risk arising from polluted sea water, the effects of sea disposal of wastes, the establishment of codes of practice for sea disposal of wastes, and research on standards and the evaluation of toxicity measurement techniques.

The need to implement some of the recommendations of this working group, especially an investigation of the public health aspects of bathing in polluted sea water and the significance of bacteriological sea water quality standards is emphasized by the recent publication of proposals in a Council Directive (1975) according to which it would be recommended that Member States of the Commission of the European Communities (Belgium, Denmark, France, Germany (FRG), Ireland, Italy, Luxemburg, Netherlands, United Kingdom) lay down values for parameters to which authorized bathing waters should conform within 8 years after the directive became effective. In view of the fact that for the first time a group of nations might be recommended to adopt guidelines and mandatory minimum values for bacteriological and physico-chemical parameters of pollution it is desirable that they should be set out in full. They are given in Table 2.

The information given in the preceding tables shows that the most commonly used parameter of sewage pollution in sea water is the coliform or the E. coli count. It has a bearing on the subject of standards to know whether techniques are available that permit the design of outfalls to provide the conditions needed to meet the required standards.

Table 2. Commission of the European Communities proposed water quality objectives for sea bathing waters at authorized or tolerated bathing waters

Parameter	Units & expression of results per	Method of examination	Minimum frequency of sampling if population is <10 000	>1000	Water temp. °C	Guideline values which may be made more stringent (G)	Mandatory values which may be made more stringent but not exceeded (I)
Total coliform	Total coliform per 100 ml	Multiple tube: confirm positive results	B	A		2000	10 000
Faecal coliform	Faecal coliform per 100 ml	Multiple tube	B	A	>20	500	2 000
Fae cal streptococci	Faecal streptococci per 100 ml	1. Latsky method MPN 2. Membrane filtration & culturing	D	C		100	
Salmonella	Salmonella per litre	Membrane filtration, inoculation & enrichment					0 (2)
Viruses	PFU per 10 litres	Concentration & confirmation					0 (2)
pH	pH	Calibration at pH 7 & 9	B	A			6 – 9
Colour		Visual or photometric	B	A			No visible change
Mineral oil	mg/l	Visual inspection	B	A			No visible film
		By extraction & weighing	B	A			<0.3
Surface active agents	Na lauryl sulphate mg/l	Visual inspection	B	A			No lasting foam
		Spectrophotometric	B	A			<0.3

Table 2 continued.

Parameter	Units	Determinand	Standard	Method	B	A	Notes
Phenol indices	mg/l	C_6H_5OH	No phenol odour	Spectrophotometric	B	A	
Tarry residual			No visible tarry residue on foreshore	Visual inspection	B	A	
Floatables			None	Visual inspection	B	A	
Pollution Index			As, Cd, Cr^{6+}, Pb, Hg, CN^- must not exceed seawater values 1 km from coast				
Pesticides	mg/l			Extraction & chromatography			(2)
As, Cd, Cr, Pb, Hg	mg/l	As metal		Atomic absorption			(2)
Cyanide	mg/l	CN		Absorption spectroscopy			(2)
Transparency	m			Secchi disc			(3)
Dissolved oxygen	% sat.			Winkler			(3)
Nitrate	mg/l			Spectroscopy			(3)
Phosphate	mg/l			Spectroscopy			(3)

N o t e s :

Sampling frequency: A = weekly; B = every 2 weeks; C = monthly; D = 6 monthly.

(2) = Positive results should be checked by a competent authority.

(3) = Positive results should be checked by a competent authority if a tendency towards eutrophication is shown.

G = 90% of samples must have values below those given in column G. Remaining 10% must not be more than 50% above the values stated, except pH and dissolved oxygen.

I = 95% of samples must have values below those shown. Remaining 5% must not be more than 50% above the values shown except for microbial values, pH and dissolved oxygen.

An outstanding example of success in meeting a coliform standard is at Los Angeles where a sewer conveys 1.64 million m³/d of partly and fully treated sewage 8 km out to sea, dispersing it under 61 m of water, while an 11 km sewer deals separately with the raw and digested sludge. The examination of samples taken over a period of nearly 30 years shows that the mean of the annual mean values of the coliform count at beach stations fell from 2100/100 ml in 1947 to 25/100 ml in 1973, and that for over 10 years the counts have been well within the State limit requirement of 250/100 ml (Bargman, 1975).

The variability of the results of coliform determinations at a single point is also a factor of importance. Counts of coliform bacteria reported by Gameson (1975) show that at a given beach station, coliform bacteria are distributed lognormally approximately. The variability in the coliform counts at a single beach station at each of five different sites is shown by curves in Fig. 1. The distribution of the results at Site C, where the median count for 1197 samples was 85/100 ml was lognormal, just as it was near the old short outfall at Site D where the median count was 15,000/100 ml.

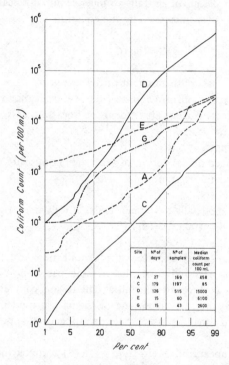

Fig. 1. Logarithmic probability plot of coliform showing the variation in coliform count
on different days at one station at five different sites. The inset table shows the number
of days, samples and the median count. From Gamson (1975).

If it is necessary to develop a mathematical model from which predictions concerning the behaviour of bacterial indicators in sewage discharged from a sea outfall can be made, considerable costs may be involved in obtaining adequate data for the model by means of field tests. Oakley and Staples (1975) cited, within recent years, costs of £82,000 for a physical model which could reproduce the conditions arising in the Tyneside area of England; £31,000 for a physical mathematical model for the adjacent Teeside; £40,000 for a mathematical model relating to conditions in the Edinburgh area; £45,000 for a mathematical model for South Hampshire which required 9 months investigation, and £35,000 for a 17 month investigation in Hong Kong which involved intuitive and graphical methods of interpretation using metering records.

If the costs quoted above were updated they would be increased substantially. However it appears that the cost of such investigations would be of the order of 5—7% of the total cost of an outfall. Even so, the bacteriological data obtained may be difficult to interpret; an Australian study costing $A1,000,000 gave inconclusive information, Oakley (1975).

According to Paolletti (1975) the first 400 m of a marine outfall including 40 m of diffuser, was built at Maiori, Italy. Because the faecal coliform count of water at the shore was found to be 330/100 ml, in excess of the permitted limit of 100/100 ml the outfall was extended by 300 m which was in accordance with the length of outfall previously predicted by the design engineer. In a different study, of a preliminary nature, for an outfall at Horse Sand, South Hampshire, England, Camp (1975) recommended primary treatment of the sewage in order to reduce the *E. coli* count from 2×10^6/100 ml to 2×10^5/100 ml. The outfall and diffuser system was designed so that under given conditions of wind blowing from two directions the estimated *E. coli* count at the coast would not exceed proposed levels.

A more general approach to the design of outfalls to meet required faecal coliform limits is the use by Aubert and Breitmayer (1975) of a formula given by the equation:

$$\log C = \log (C_0/b) - \frac{By}{1.1} t$$

where C_0	=	faecal coliform count of the sewage
C	=	faecal coliform count of the sewage at time t hours after discharge
B	=	a coefficient dependent on the flow and which is 1 for a discharge of 1 m^3/s and
b	=	depth of outlet, metres
y	=	bactericidal action of sea water (a measure of die-away) which is 1 for the Mediterranean.

Applying this formula thirty new outfalls have been built; in only one case has the faecal coliform count exceeded the required limit of 1000/100 ml 50 m offshore.

It is clear, however, that there are many situations throughout the world where complex tidal conditions make generalizations of the kind referred to in the previous paragraph difficult to apply. For example van Dam (1975), after 12 years work on the outfall at The Hague, concluded that the observed coliform distribution data are difficult to interpret from the mathematical model used since movement of coliforms towards the coast occurs in a manner which does not appear to happen with dye tracers.

Thus, with the exception of Aubert and Breitmayer's (1975) formula most outfalls designed for large flows in which a coliform limit at given points must not be exceeded (or acceptable ocean and beach conditions in the absence of such limits) begin with a preliminary study of the dispersion of conservative substances under a variety of conditions, usually with the object of formulating a mathematical model from which, by a process of extrapolation, the results can be predicted for conditions which at present do not exist or which it has not been possible to reproduce in carrying out tests.

In order to meet a given coliform limit at a particular location Shuval (1975), as already mentioned on p. 242, has suggested a method of approach which deserves to be examined in detail in order to determine how the predicted results are likely to vary if the known variations in the magnitude of some of the factors are taken into account. Considerations of such factors should help to clarify the issues involved in defining standards, assess their significance and determine whether arguments expressed in favour of one standard or another, or no standards at all, represent fundamental differences in objectives or merely different approaches to attain the same objectives. Shuval suggested that in a hypothetical situation the count of 1000/100 ml would be attainable by partial treatment to remove floatable solids and die-away. In these calculations the coliform density was taken as 10^8/100 ml, a figure also used by Josa (1975) after finding that the average coliform count of sewage was 67 million/100 ml or 5×10^8 per person and the water consumption 500 l/d. Camp (1975) took the *E. coli* count of sewage to be 10^6/ 100 ml.

A significant fact in calculations involving coliform numbers is the seasonal variation in the count. Gameson (1975) showed that the coliform count in sewage rose steadily from 5×10^6/100 ml in winter to 50×10^6 in summer. Shuval (1975) referring to work by Kehr and Butterfield (1943) stated that the "coliform excretion per person remains constant in the population". Observations carried out at two sewage works over a 4 year period — one with a contributory population of over 40,000, the other more than one million — showed that in both cases the *E. coli* count remained fairly steady, whereas the coliform count varied considerably from low values in winter to 10—20 times these values in summer (private communication). The sewage from the population of over one million contains industrial effluent, partly derived from metallurgical industries and greater in volume than the domestic sewage. This diluting effect probably accounts for the lower *E. coli* counts and the smaller summer peaks. These figures indicate that calculations using the bacterial counts would give different results, depending upon whether the coliform or the *E. coli* counts were used. These results are shown in Figs. 2 and 3.

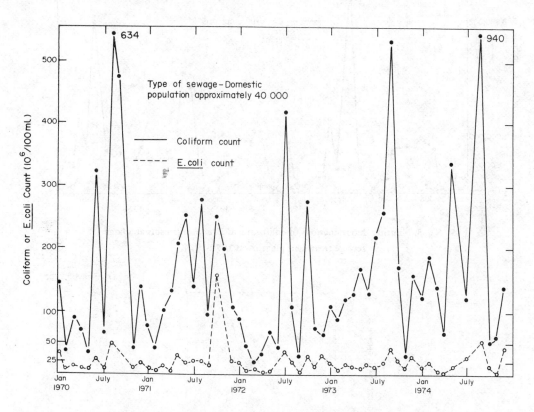

Fig. 2. Monthly determinations of coliform and *E. coli* in sewage 1970-1974.

The total numbers of coliform bacteria and *E. coli* reaching a sewage works having a contributory population of approximately 40,000 were determined at hourly invervals. The results are plotted in Fig. 4. They show peak values between 10.00 and 12.00 h, were in agreement with the impurity load determined by chemical tests, with a second, evening peak at 21.00 h. The results also showed a close relationship between the coliform and the *E. coli* counts (private communication).

In calculating the probable coliform density of sea water at an outfall it is important to note that there may be considerable diurnal variation in the flow and impurity content of sewage. This variation is related to the size of the drainage area, the sewerage system and local circumstances such as the proportion of visitors in the total population at a holiday resort. Fig. 5 typifies the flow of sewage at works of different sizes.

The maximum rate of flow coincides with the maximum strength resulting in the rate of delivery of impurity that at peak flow may be several times that of the average. Below (Fig. 6) are shown typical variations in the load of impurity in the sewage received at a sewage works receiving sewage from a population of about 66,000 inhabitants through a sewer 8 km long. This sewer was designed for an ultimate population of 350,000.

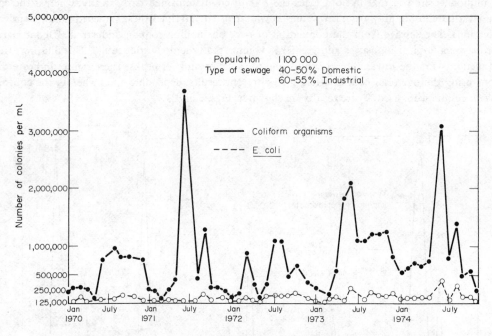

Fig. 3. Monthly determinations of coliform and *E. coli* organisms at a large
sewage treatment plant from 1970-1974.

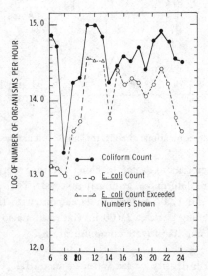

Fig. 4. Number of coliforms and *E. coli* reaching a sewage works, every hour,
population approximately 40,000.

Continuous passage of the sewage through a tank of several hours capacity does not balance the strength of the sewage. This is shown by Figs. 7 and 8 in which sewage was passed at a constant rate of flow through tanks providing (a) 7.5 h and (b) 2.5 h retention respectively.

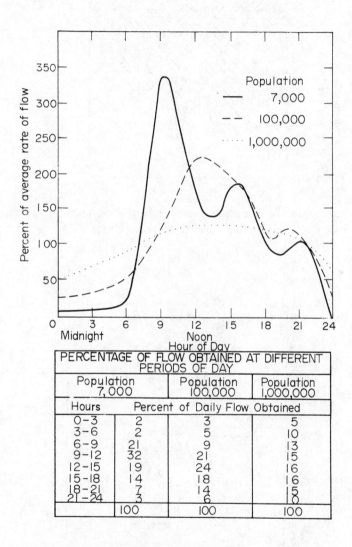

PERCENTAGE OF FLOW OBTAINED AT DIFFERENT PERIODS OF DAY		
Population 7,000	Population 100,000	Population 1,000,000
Hours	Percent of Daily Flow Obtained	
0–3	2	3	5
3–6	2	5	10
6–9	21	9	13
9–12	32	21	15
12–15	19	24	16
15–18	14	18	16
18–21	7	14	15
21–24	3	6	10
	100	100	100

Fig. 5. Indication of variation in rate of flow of sewage with population.

Very little mixing occurs in a sewer. pH records taken over several years showed that intermittent discharges of acid and alkali several kilometres from the point of measurement produced wide fluctuations in the pH of the sewage, and variations ranging from pH 2 to 11 were observed within a minute or two. In a test on a sewer 8 km long, salt added momentarily as a strong solution at the head of the sewer was not detected at the sewage works because it arrived in between samples taken at 15 min intervals (private records).

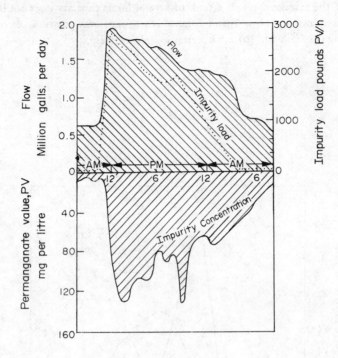

Fig. 6. Variation in flow, strength and load of domestic sewage over a 24 h period.
From Whitehead and O'Shaughnessy (1935).

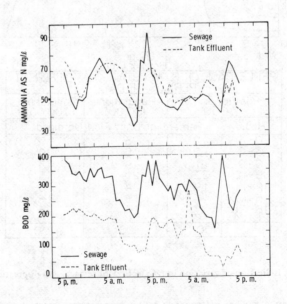

Fig. 7. 48-h continuous settlement test with 4550 m^3 sewage/day flowing through a rectangular tank
at a constant rate giving 7.5 h retention.

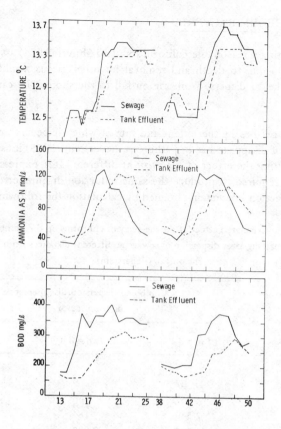

Fig. 8. 50-h continuous settlement test with 13 650 m³ sewage/day flowing through
a rectangular tank at a constant rate giving 2.5 h retention.

REDUCTION OF ORGANISMS ON DILUTION

Studies on the dilution obtained on discharge of sewage through outfalls usually give results that agree
with the calculated dilution. For instance Paoletti (1975) using the natural silicate content and the faecal
coli count of sewage as a tracer to estimate the dilution obtained above the diffuser length of a marine
outfall obtained results that approximated to the calculated dilution using silica but using the faecal coliform data the dilution appeared to be unacceptably greater than the expected dilution.

REDUCTION OF BACTERIA BY DISPERSION

In a comparison of the relative importance of dispersion and die-away in reducing the content of a nonconservative component of sewage such as coliform organisms, Feuerstein (1975) has made a calculation
based on an average horizontal current speed of 9 cm/s for a discharge flow rate of 0.33 m³/s from an
outfall diffuser length of 100 m and an initial horizontal diffusivity of 0.01 cm/s. Without die-away the
coliforms would be reduced in numbers by factors of 3.85, 8.06, 13.3 and 26.2 at distances of 1, 2, 3 and
5 km from the diffuser due to dilution and dispersion from the point of discharge. The importance of
die-away is dealt with in the following section.

DIE-AWAY RATE OF COLIFORM BACTERIA

In his approach to formulating a reasonable coliform standard Shuval (1975) assumed that a reduction in numbers by a factor of 100 due to death and removal by other means would occur in the 2 h it was estimated sewage would take to disperse from the outfall to the shore. The causes of this die-away are many.

The importance of the proces lies in the extent and rate at which it occurs under different conditions and its relation to the rate of disappearance of pathogens. Feuerstein (1975) illustrates the overall importance of die-away by calculating the effect of die-away at different rates, expressing the result in terms of the dilution that would be required to produce the same reduction in numbers. Making the assumptions referred to in the previous section Feuerstein obtained by calculation the following results:

Table 3. Calculated effect of the die-away of coliforms on the numbers
remaining after dispersion of sewage at different distances from
the outfall (Feuerstein).

Distance from outfall	Combined dispersion/disappearance dilution factor		
	$T_{90} = 4$ h	$T_{90} = 8$ h	$T_{90} = 00$
km			
1	22	9	4
2	267	46	8
3	2 540	184	13
5	164 000	2 080	26

It is evident that the die-away rate used in a formula for the purpose of estimating the numbers of coliform bacteria at different distances from an outfall is likely to be an important factor.

PUBLIC HEALTH SIGNIFICANCE OF COLIFORM COUNT IN SEA WATER

Views that have been expressed on the significance of coliform counts for monitoring seawater pollution have varied from their rejection to dependence on their use to ensure safe conditions for swimmers. Between these extremes the view is held that by observing a coliform standard aesthetic requirements are also satisfied. The most frequently occurring type of coliform organism in sewage may be differentiated from other organisms which are included in the coliform count. For example the method described in the U.S. Standard Methods (e.g. 1965) separates coliforms of faecal origin from those derived from non-faecal sources. The method described in *The Bacteriological Examination of Water Supplies* (1956, 1969) differentiates between *E. coli* and other organisms which may be present in sewage.

In establishing bacterial standards or criteria for sea water the question should be asked whether the indicator organism enumerated should be one whose normal habitat is the human or animal intestine, in which case the obvious choice is *E. coli*, or should the organisms be any that have formed part of the sewage, in which case the whole group of coliform organisms would be used. The smaller seasonal variation that was found to occur in the numbers of *E. coli* in sewage would favour their use as the more constant tracer, in addition to their undisputed reliability as an indicator of faecal pollution, but the coliform count would be a more delicate indicator of sewage (as opposed to faecal) pollution and their greater numbers in summer (in a temperate zone) might be used to provide an additional public health safeguard.

Various methods exist in different countries for the bacteriological examination of waters. The importance of obtaining continuity of results as an essential basis for experience in monitoring water quality ensures that once satisfactory techniques have been established there must be compelling reasons to change them. However, in the case of bacteriological standards or criteria that are applied to sea water there are now adequate reasons for considering the use of uniform, internationally agreed techniques and agreed criteria which may be suitable for particular regions. It would facilitate cooperation in areas such as the Mediterranean, where climatic conditions are similar, if continuing improvement of techniques led to the adoption of international methods for the determination of all the important parameters of pollution.

The importance of this matter is illustrated by the fact that an assumption is sometimes made that a relationship exists between the coliform count and the *E. coli* count. The ratio has been taken as 1000 − 400/100 ml. An actual example given below shows that the ratio may vary over an extremely wide range and be far from constant even in the same geographical area. This example shows the results of a recent study in Southampton Water.

Table 4. Southampton Water, ratio of coliform to *E. coli* counts, McGregor (1975)

Coliforms per per 100 ml	*E. coli* (Type I) per 100 ml	Ratio coliform/ *E. coli*
50	8	6
2250	1700	1.5
170	17	10
9000	5500	1.8
900	40	22
3500	3500	1
554	4	138
1100	800	1.4
1800	90	20
5500	3500	1.5
16,000	5500	3
1100	40	28
2500	1300	2
1300	350	4
3500	1100	3
2250	500	4
3500	2500	1.4
16,000	3500	4
9000	1700	5
5500	2500	2
2500	1300	2
250	20	12
1600	1600	1
350	8	44
2500	250	10

Ratios
Range   1 − 138
Within range of   1 − 5 (60% of values).

It will be seen that the use of a factor to convert coliform counts to *E. coli* counts, or vice versa, might introduce difficulties, particularly since the greatest ratios are in the range of results for coliform counts that have been proposed as limiting values for sea water. The ratio of coliform to *E. coli* counts calculated from the results of Petrilli *et al.* (1975) varies from 2 to 15 with 80% of the results falling within the range of ratios of 2—5. This same difficulty would apply in using a ratio to estimate the number of faecal streptococci. Petrilli *et al.* (1975) found that the ratio of *E. coli* to faecal streptococci varied from 0.5 to 7 with 70% within the range of 1—3. Using the method of Taylor and Burman (1964) Gameson (private communication) has reported that the ratio of *E. coli* to coliforms is approximately unity.

HEALTH RISKS

The emphasis that has been placed upon bacteriological standards for sea water specifically for protecting bathers from enteric or other infections has made the subject controversial. Although Moore (1975) states that no good evidence exists to show a quantitative relationship between coliform counts and health risks, a growing concensus of opinion favours the use of bacterial limits without waiting for supporting evidence. From their findings that tests for enterovirus gave positive results when *E. coli* counts exceeded 1000/100 ml and the coliform count 5000 — 10 000/100 ml, Petrilli *et al.* (1975) concluded that although any evaluation of health hazard was not yet possible they favoured steps such as the use of a bacterial standard which could help to reduce viral infection and supported the view of sanitary authorities "in dreading mussel cultures not situated in quite pure waters".

SAMPLING TECHNIQUES

If bacterial standards or criteria are to be used for sea water, agreed procedures should be laid down to determine the location of sampling stations, the number of samples to be taken and the conditions under which the samples are taken so that, for instance, sediments are not re-suspended and included in the sample examined. Little published information appears to be available concerning the variation in coliform count determined at very frequent intervals around one sampling point, other than that of Gameson's (1975); at five different sites the coliform counts were found in a large number of samples taken at a single station and the results are reproduced in Fig. 1. These show that the distribution is approximately lognormal. Such data should be regarded as essential in considering the question of bacterial standards.

If the results of examination of legal standards for sea water quality at bathing beaches were to be used as evidence in a court of law it might be necessary to lay down the method of taking samples, to make provision for any potential defendant to be present in order to verify sampling in accordance with legal requirements, and to permit a potential defendant to have a portion of the sample for his own examination. It might be necessary to define regulations in sufficient detail to minimize disagreement. These difficulties might be sufficient to deter a public health authority from insisting upon standards, if the numerical reliability of the results had to be proved. However, it should be possible to frame regulations which stated the range within which results would be regarded as satisfactory, the proportion of results falling outside this range that would be permitted and the maximum value that such non-complying results might reach. Defining a standard along such lines would appear to protect the quality of sea water where it was necessary by law to state the permitted limits for pollutants.

The difficulties that arise in defining a bacterial standard as a legal requirement do not apply to bacterial criteria that may be included in a code of practice. Such criteria are usually modified in the light of new information or adjusted to suit local circumstances.

POLLUTANTS OTHER THAN MICROORGANISMS

Over a period of many years a number of officially recognized organizations have studied and made recommendations on the control of marine pollution resulting from the discharge of strong biodegradable organic wastes that lower the oxygen concentration of sea water or cause ecological changes. These bodies have given particular attention to non-degradable compounds that are injurious to marine biota and human beings, heavy metals that directly or indirectly threaten the well-being of some communities or that may produce long term undesirable consequences. Therefore it is not necessary to refer to any aspect of these discharges except that of their method of control.

A few examples of legal limits and criteria for sea water and waste water discharged to sea by pipeline have been given in Table 1. If these limits are compared with the following annual average results of analysis for metals determined in the domestic sewage from two works it is apparent that most of the heavy metals in some industrial discharges would have to be removed before the factory effluent could be accepted into sewers discharging through ocean outfalls.

Table 5. Average annual results for metals in sewage at sewage works A, population 7000 and B, population 30 000

		Works A			Works B	
	1962	Year 1966	1967	1961	Year 1962	1963
		mg/l			mg/l	
Cr		0.021	0.025			
Cu	0.5	0.23	0.24	0.3	1.2*	0.1
Ni	0	0.03	0.04	0.5	0.08	0.05
Zn	0.9	0.29	0.31	0.8	0.8	0.1
Cd		0.003	0.004			
Fe	3.4	1.3	1.3	3.8	3.4	2.7

* 0.42 excluding high figure.

Control of the discharge of non-microbial pollutants to marine waters may be necessary for the following reasons: (a) aesthetic, i.e. to prevent oil discharges, grease slicks or floating matter, or odour nuisance; (b) short term toxic effects such as direct or indirect interference with fishing or fisheries, causing tainting of fish or the accumulation of undesirable substances in fish; or (c) long term effects which may cause irreversible changes in the marine fauna and flora or threaten the existence of species of wild life. Substances responsible for conditions in category (a) could be controlled by defining the appearance of the receiving water at, or at a fixed distance from, the outfall. This method would be open to a pollution control agency that did not have jurisdiction over the admission of substances to a public sewer or the disposal of its contents. Authorities which are responsible for making discharges of sewage through outfalls usually have more or less complete control over what is admitted to the sewer and the treatment of the sewage.

The importance of aesthetic considerations, especially the avoidance of slicks and detectable odour has been overshadowed by the emphasis that has been placed on the bacteriological quality of sea water. Agg (1975) states that slicks are visible with a dilution of 1 in 100 and, in certain conditions, of 1 in 250. In the absence of adequate dilution some form of pretreatment may be essential to remove slick forming substances. The odour of sewage is said to be detectable in extremely high dilution and may require a dilution of up to 1 in 1000 to prevent it. Traces of methylamine, propylamine and butylamine, liberated into the atmosphere from sewage sludge containing milk waste solids were detectable 1½ km from the source; a thioketone, produced by the interaction of traces of solvent and sulphide in the sewage, is stated to be detectable by smell at a concentration of 1 in 100 million parts of air; the odour from a sludge

processing plant, objectionable at a distance of 3 km from the source, has compelled the local authority
to close down the plant. Traces of butyric acid, derived from the treatment of carbohydrate wastes,
brewery effluents and sewage sludge have caused odour problems several kilometres away from the
source (private sources of information). To sum up, it may be said that the appearance of the sea is just
as important to some as the quality of the water and to such persons any evidence of the discharge of
sewage to sea such as that revealed by the "boil" at an outfall is objectionable. In the author's view the
public and their lay representatives are much more concerned with the aesthetic quality of sea water and
bathing beaches than with the bacterial quality of the water, possibly because they take the latter for
granted.

In order to control substances within category (b) the level of concentration at which specified potential
pollutants may have toxic effects must be known. Much information on the toxicity of many substances
to several species of test animals, including shellfish is available [e.g. Portmann (undated)] and several
test procedures are in use for this purpose. The conditions of carrying out these tests may be standardized
with sufficient precision to permit the results being used for legal purposes. In a legal judgement in the
United Kingdom it has been ruled that fish tests may be used for determining the quality of a receiving
water and an official test defines how the toxicity of an effluent may be carried out. Since these tests give
information concerning the concentration k of a particular substance which causes 50% mortality (or
specified symptoms in a specified time) it becomes possible to define a legal limiting concentration of the
pollutant which may be any fraction of k.

In establishing limits or criteria based on tests such as those just described, three important factors must
be taken into account. These are (1) the ionic state of a substance used in a standardized laboratory test
may be substantially different from that of the same substance in sea water so that the effect of the
substance under a wide range of conditions may be required before limits can be laid down. (2) Freshwater studies have shown that the total toxicity of a liquid containing several pollutants may be obtained
by summating the percentage toxicity in which each of the toxicants is present, the percentage toxicity
being the ratio between the concentration of the pollutant in the liquid and that required to cause 50%
mortality to fish under test conditions, expressed as a percentage. A combination of limits placed on the
discharge of individual substances and satisfaction of the requirements of a test applied to all the constituents, whether specified or not, would make it practicable to define the conditions of discharge
sufficient for most purposes. (3) Many effluents may contain substances that are difficult to identify and
which may be different from the raw materials used. In the manufacture of halogenated herbicides it was
found that only 10–20% of the total organic carbon content of the effluent could be accounted for in
the form of the ten or eleven products made (private communication). Therefore, for control purposes
it is important to carry out toxicity tests using industrial effluents or sewage containing such effluents.

Substances that may have long-term harmful effects may present special difficulties in control because of
insufficient information concerning their safe concentration or their effects in the presence of other
potential toxicants. In controlling such substances it may be necessary to make use of methods such as
that described by Aubert (1973), by inference from *in situ* ecological studies (Stirn *et al.* 1975) or by
inference from the chemical analysis of marine organisms. Such methods may be adequate to deal with
existing discharges. To make provision for the control of future discharges it is considered essential that
manufacturing and other interests are represented on national or international bodies created for the
purpose of advising on marine pollution so that the long term views of marine ecologists and pollution
control agencies can be taken into account in the planning of industrial activities.

SEA DUMPING OF SLUDGE AND LIQUID INDUSTRIAL WASTES

Discharges of industrial wastes made through sea outfalls may be subject to the same conditions of control
as those that apply to sewage, except that the possible presence of pathogenic organisms in industrial
effluents is rarely a matter for concern. It is not considered necessary in this paper to discuss the control

of these discharges since this subject will be dealt with at this symposium in a paper by Professor L. Mendia.

The discharge of liquid industrial wastes and sewage sludges into marine waters by dumping from vessels has been carried out for many years. These operations are controlled in some parts of the world by international agreements, the terms of which are implemented and enforced by their embodiment in the national laws of the participating countries which give each country power to take action for infringements of the law against individuals and companies within its jurisdiction. The general principle followed in such legislation is to prohibit dumping within a defined area, provide a list of substances which may be dumped, list those which require a permit if they are present in excess of a given limit, require information concerning the volume and composition of the wastes to be dumped and specify the area in which dumping may take place. Thus, dumping is permitted only with the approval of some recognized national authority which is able to control the operation so that harmful substances are excluded or accepted only at safe levels. These conditions may act as a deterrent to dumping. Thus, a feasibility study carried out for sea dumping a strong organic waste of agricultural origin from a factory near an estuary showed that sea dumping was much more costly than pretreatment on site followed by disposal to a municipal sewer and payment for the cost of treating the waste (private information).

For the protection of the Mediterranean it appears that incorporation of the provision of the Oslo Convention of 1972 for preventing pollution by dumping in the northeast Atlantic and those of the Baltic Convention for the control of pollution in the Baltic Sea would be desirable, with the inclusion of suitable protocols to safeguard against dumping and accidental spillages (author's personal conclusion of lecture by Dr. J. Ros, March 22 1975 at Conference on the Pollution of the Mediterranean, La Manga Spain). It is to be expected that the attitude of nations and individuals in the region of these two seas would be much more critical if the sea dumping of material compared with the same operation carried out on the Atlantic seaboard.

The dumping of sewage sludge differs from the disposal of most industrial wastes in that it is only economically feasible if the operation is carried out on a large scale.

Other differences are that almost all the impurity is present in suspension, most of it is organic and the greater part of that is biodegradable, leaving harmless residues; the sludge is likely to contain heavy metals, possibly chlorinated hydrocarbons and numerous organic substances that have been derived from industrial effluents admitted to the sewers and adsorbed by the sludge; in addition sewage sludge is rich in coliform bacteria and may contain pathogenic bacteria, viruses and the cysts or eggs of parasites.

Following recommendations for setting standards for the dumping of wastes in U.S. coastal waters and on the issue of permits which were made in a report to the U.S. President in 1970 (Ocean Dumping, 1970) drawing attention to the serious economic consequences of ocean pollution and the fragmentary legislation that existed to control it, wide powers were given in The Ocean Dumping Act of 1972 that controls the dumping of wastes in U.S. ocean waters of the territorial sea and the contiguous zone, besides the control of all U.S. vessels dumping in all ocean waters. The objectives of this Act and a parallel Water Act of 1972 are to control all U.S. ocean disposal and disposal in U.S. marine waters and to eliminate all harmful discharges to the sea by 1985 (Lacy and Rey, 1975).

One of the intentions of the 1972 Oslo Convention (already referred to) signed by twelve nations, is to regulate by national legislation the dumping of sludges of known volume and composition into assigned waters under licence. In due course, it is possible to forsee similar concerted action on the part of the nations having a seaboard in the Mediterranean which could include sludge dumping and it is probable that comparable regional agreement elsewhere will eventually occur in response to a common need.

Therefore, it is timely to discuss the subject of sewage sludge dumping objectively in order to consider the possibility of control of the operation by reference to the quality or composition of the sludge, the quality of the seawater or other factors.

Examination of the conditions resulting from sludge dumping in different parts of the world show greatly varying effects. Serious ecological consequences and economic losses have resulted from many years of dumping of sludge from sewage works in the New York City area, several kilometres off the coast, in New York Bight. Clearly, the absorptive capacity of this area to render sludge harmless without ecological damage has for long been exceeded and the call for regulation of the process is overdue.

Sludge dumping is not such a widespread practice and the dumping sites are not so numerous as to make it impracticable to study each site with sufficient care to take into account all the important factors concerned in making the operation a satisfactory one. The need for such a study in order to exercise control over sludge dumping is that some municipalities in densely populated conurbations favour treatment of liquid wastes and discharge to local rivers with off-site disposal of sludges.

Control of sludge dumping by permit under a voluntary scheme or under statutory powers is now practised in several countries and the operation carried out without ill effects. The sewage sludge from London from 1887–1915 (28 years) was dumped in the Barrow Deep, 15–20 km offshore in 13–18 m of water and in the adjacent area of Black Deep from 1915–1967 (52 years). The Barrow Deep has again been in use since 1967. The quantities of sludge dumped by licence under Section 2 of the Dumping at Sea Act, 1974 are as follows:

		Tonnes per day	
Wet Sludge:	Raw	1 258	
	Digested	10 989	= Dry weight of 270 tonnes
	Total	12 247	

The analysis of a major fraction of this sludge was as follows:

	mg/kg wet wt.
Zinc	2390 – 5860
Copper	505 – 2500
Chromium	200 – 1065
Nickel	150 – 350
Cadmium	30 – 70

The organic muds in the Black Deep have diverse populations which are not indicative of gross pollution and the bottom water is aerobic. No completely anaerobic sediments are found in the Barrow Deep and the benthos is normal.

The sewage sludge from greater Glasgow has been dumped in the outer Firth of Clyde in 76 m of water since 1904.

The annual quantity of 1,100,000 tons contains solids derived from industry which contribute PCBs and dieldrin. Apart from a small area near the dumping ground in which bottom sediments have become anaerobic and gained in heavy metals content, little adverse change has occurred, without affecting commercial fisheries within the area. At several points in the estuary, however, mussels have been reported to contain DDT, dieldrin and PCBs, Porter (1973).

Similarly, the results of a comprehensive investigation by Caspers (1975) into the effect of dumping 250,000 t of sewage sludge per annum from Hamburg into 20 m of water in the German Bight off the North Sea have shown that no ecologically harmful effects could be detected and that the great development of certain species encouraged by the sludge may be regarded as a beneficial enrichment of the sea.

The practical difficulties in controlling pollution from sludge dumping, by any method other than a permit for one particular type of sludge from a known source dumped into one defined area, are well illustrated by one of the most comprehensive investigations on the subject ever undertaken (Working Party 1972). At present 550,000 wet t of sludge per annum, or approximately 40,000 t of dry matter, are dumped in Liverpool Bay which is a rich fishing area off the north-west coast of England. The study was carried out in order to be able to predict the effect of a sixfold increase in dumping, due to a proposed centralized sludge collection and disposal scheme. The Bay also receives crude sewage discharge by pipeline from many sources. By examining the rate of oxidation of organic matter it was concluded that the level of dissolved oxygen in the immediate area of the dumping ground could be reduced from its present value of 97% of saturation to 75% if the equivalent of 250,000 t of dry matter is dumped. This sixfold increase would result in the sea water suspended solids content rising in the dumping area to about 1000 mg/l after dumping and falling to below 50 mg/l before the next dump under the worst conditions. The sludge solids are dispersed over an area of 8×30 km. A sixfold increase in dumping could increase the sludge solids in the surface layers of the ocean bed to 0.5% in the dumping area but much less outside this zone. As to the effects on algae, fish and shellfish it was concluded that sludge would only have any effect at concentrations of suspended matter higher than those found in the sea water after release of the sludge, that the flesh of fish is unaffected by dumping and that the heavy metals content of molluscs taken from stations near the disposal area were within the range found in other areas. With an increase in sludge dumping the possibility of an increase in persistent substances in food species was recognized and in that event monitoring such substances in foods would be necessary.

Reference to the presence of persistent chemicals in sewage sludge raises the question as to whether such substances should be controlled by means of a sea water limit or by restricting the content of persistent chemicals in the discharge. Several persistent organic and inorganic substances are volatile and are present in rain water. Some, such as the organochlorine pesticides or PCBs which have widespread use in industry, are present in sewage. A sea water quality standard applied to sludge dumping would thus suffer from the drawback of having to take into account pollutants that might be introduced from sources other than sewage sludge. An advantage of exercising control over the composition of the sludge is that most of the non-degradable objectionable substances admitted to sewers become concentrated in the sludge and are therefore easier to detect there. Portmann (1975) gives a figure of 0.1–1.0 mg/l of organochlorine pesticides in sludge which indicates a concentration factor of 40–400. Another advantage of limiting the content of substances in sludge is that it becomes easier to control the discharge of unacceptable chemicals at source and monitor the effects of control by examining the sewage sludge. A third advantage is that analysis of the sludge provides information on the total potential hazard because sludge solids may settle rapidly or disperse quickly after release in sea water.

ADMINISTRATIVE ASPECTS OF CONTROLLING MARINE POLLUTION

In the preceding sections of this paper the important technical and scientific factors that have to be taken into account in controlling marine pollution have been considered. These factors help to determine what steps should be taken to control the pollution. Although in matters of pollution control governments and agencies responsible for national and local policies may have to respond in varying degree to the demands of special interests, pressure groups and public opinion informed by the watchdogs of community interests, it is important that marine pollution prevention should not be seen as an isolated branch of pollution control, practised for the purpose of protecting bathing beaches, marine resources or navigation. Instead it should be regarded as one aspect of the national control of all surface waters so that steps can be taken

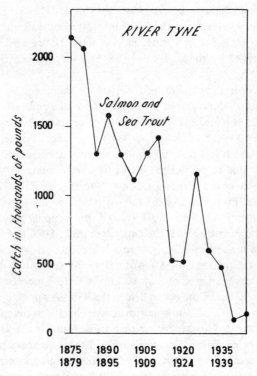

Fig. 9. Annual catch by nets of migrating fish, averaged over 5 year periods,
in the River Tyne. From Jenkins (1965).

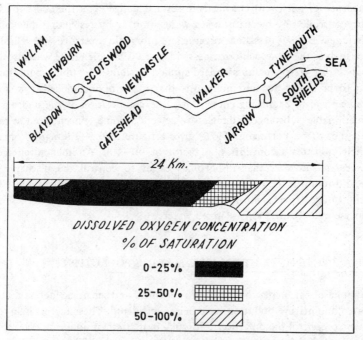

Fig. 10. Dissolved oxygen in the tidal reaches of the River Tyne estuary,
August 1938. From Jenkins (1965).

in the orderly development of the industry, commerce and agriculture of a country consistent with enhancement of living standards and the enjoyment of national amenities.

Such considerations lead to the question as to the steps that can be taken by a nation to control marine pollution from (a) Rivers, (b) Sea outfalls, (c) Sea dumping, into its own territorial waters or in the open or high seas beyond territorial waters.

Control of pollution of rivers

A considerable proportion of the pollution of estuaries and the sea is derived from polluted rivers. The first step in controlling marine pollution is to legislate against the pollution of rivers and estuaries by requiring local authorities to treat their sewage so as to comply with effluent quality conditions. These conditions should be such as to compel industry to exercise some control over industrial effluents discharged to public sewers, and if necessary to pretreat its effluents. From the point of view of the industrialist and that of the pollution control authority the policy of treating mixtures of sewage and industrial waste is to be encouraged, because of the benefits of large scale operation and because by treatment or legal sanction the amounts of potentially harmful substances in industrial effluents can be reduced. This method of control allows additional restrictions on the discharge of such substances to be made when these are proved to be necessary.

An essential step in the control of pollution of rivers and estuaries is that the control agency should have defined legal powers in the planning of housing and industrial development so that at a sufficiently early stage any department of national or local government responsible for authorizing expenditure on house or factory construction is compelled to obtain the support of the pollution control agency. This support would depend upon whether adequate sewers or sewage treatment facilities existed or whether development should be discouraged because the resultant effluents might cause unacceptable conditions. As a result of such early collaboration between planners and pollution control agencies, regional schemes might be encouraged leading to greater efficiency in treatment and more satisfactory treatment of industrial effluents.

Action being taken to clean up estuaries in some parts of the world is a reversal of past mistaken policies that permitted the discharge of sewage and industrial waste into tidal estuaries in the belief that the large volumes of sea water available for dilution would prevent any ill effects. There is evidence that much pollution of coastal waters occurs in the vicinity of heavily polluted estuaries. The removal of such pollution would improve local amenities and could produce economic and ecological benefits by restoring fisheries and fish breeding grounds and encouraging the rich fauna and flora in the intertidal zones in many estuaries. Figure 9 illustrates the need that exists to control pollution in estuaries. It shows the decline of the fishing industry in the River Tyne over nearly a century due to pollution by sewage and industrial waste, some of it toxic. Figure 10, which represents the dissolved oxygen conditions over 40 years ago shows that the pollution of this estuary is of long standing.

In Italy a good example of pollution caused by an estuarine discharge is given by Petrilli (1975) who states that it will be necessary to pipe the dry weather flow of the estuary out to sea in order to reap the benefits of a 700 m outfall that conveys town sewage away from the local beach. The serious economic consequences of pollution in destroying marine resources can be seen in the estuary of the River Kalyan, India where industrial effluent discharged into the tidal part of a river has almost destroyed a once thriving fishing industry in a long estuary (private communication).

Control over outfalls

Most countries provide legal and administrative powers to control the size and location of pipes through which sewage or industrial effluent is discharged into their territorial waters or to the open sea. The legal powers may be divided between different sections of local and national government. Thus, at an early stage it may be necessary to obtain permission from a local planning authority to lay a sewer in land over which the authority has control. A separate pollution control authority may have to give consent to construct such a pipeline within its area and it may also have jurisdiction over the route the pipeline follows in its territorial waters. If these legal powers exist and are used a government department or state agency responsible for marine pollution control may have all the powers necessary to control (a) the volume and composition of the contents of a pipe discharging industrial wastewater or domestic sewage and (b) the point at which it leaves the coast and the whole of its route. In the case of a public sewer, any government department authorized to grant financial assistance for the construction of the sewer may withold such support until the local authority has taken steps to implement national or regional policies or undertaken to carry out conditions required by the pollution control agency in fulfilment of its policy.

In the case of pipelines that extend beyond the territorial zone, legal powers could still be exercised by the requirement of a pollution control agency that certain conditions of discharge would be observed before the sewer could be laid through that section of the area within the control of the agency.

By means of national legislation enacted for the purpose of protecting fisheries and marine ecology, coastal amenities, navigation or public health, together with those other powers of a pollution control agency or a local authority already referred to it should be possible to ensure that sewage contains industrial effluents that have been adequately controlled, that before discharge the total flow of sewage receives whatever preliminary treatment is necessary and that the point of discharge and the immediate dilution given on discharge are all in accordance with the control authority's requirements.

The statements in the preceding paragraphs show that if legally enforceable standards or criteria for receiving waters are to be used to determine whether a discharge to such waters complies with the standards or criteria, then any agency endowed with legal powers that has shared in determining the point of discharge and the conditions of discharge must accept some responsibility for the results. In order to guarantee that these results are satisfactory there is always the possibility that schemes that are unnecessarily cautious from engineering and financial points of view could be proposed. The same difficulty might arise if standards or criteria for receiving waters were set by a government department which had no involvement in the financial implications of meeting its standards; however, where responsibilities of this kind are divided between different departments the machinery usually exists for resolving such differences at a higher level.

SUMMARY AND CONCLUSIONS

1. (a)   From a review of the published evidence on the methods of controlling marine pollution caused by the discharge of sewage and industrial wastewater through pipelines it is concluded that although the pollution may be controlled by the promulgation of legal standards applicable to the wastes discharged or the receiving waters, this is not the only method of control.

   (b)   An irrefutable reaon in some countries for using legally defined standards to control polluting discharges to any water — surface, underground or marine — is that the only way control can be legally exercised is by defining the limiting concentration of impurity permitted in a discharge or receiving water.

   (c)   A second reason is that in some countries one department of government without responsibility for the construction, siting or costs or treatment of the contents of a sewer may be required to set standards in order to determine what is safe water for bathing or shellfish culture. Its judgement of what is safe is unlikely to be influenced by some of the considerations that would have to be taken into account by a department having wider or overall responsibilities that included finance, the siting and construction of the outfall in addition to desirable standards or crtieria the outfall was intended to meet.

2. It has been shown to be feasible to make discharges of sewage to sea through outfalls under conditions that give results acceptable to marine biologists and users of the water for all recreational purposes without any obligation to comply with numerical standards. However, essential design criteria have to be met in order to obtain such results.

3. The design of sea outfalls may necessitate detailed preliminary study of existing conditions and on-site investigations of all the factors necessary to select an outfall location that will meet the essential design criteria — initial dilution, dispersion, freedom from slicks, odour and absence of numbers of coliform bacteria indicative of excessive concentrations of sewage, freedom from faecal matter, floating substances and chemical pollutants.

The enormous numbers of coliform bacteria in sewage make them natural tracers for studying the processes of dilution and dispersion. These organisms are initially associated with the suspended solids portion of the sewage so that changes in the state of aggregation of the solids, or settlement, could affect the numbers of bacteria in a given volume of sewage after discharge. In addition the bacteria die off in sea water at a rate which is not constant. These facts are unimportant if outfalls are located so far out to sea that the possibility of any coliform bacteria surviving near the coast line is remote. But if the outfall is designed with the intention of meeting design criteria without too expensive a margin of safety it becomes important to know the accuracy with which the required design criteria can be predicted. Investigations to provide the answer to the question of accuracy of prediction are in progress and consequently any tentative conclusions that may be drawn are likely to require modification in the future.

It is certain, however, that at a number of sites where detailed investigations have been in progress for several years the numbers of coliform bacteria in sea water near the shore at any one site varies enormously. This fact stands out so clearly that any standard or criterion of quality based on the numbers of coliform bacteria in bathing water should be made to accord with the fact that the numbers may be distributed lognormally. This fact should also be borne in mind when interpreting the results of examining water which is expected to meet a given numerical standard.

In the absence of clear cut evidence that there is a relationship between the coliform count of sea water at bathing stations and enteric disease it is reasonable to regard the coliform count as a sensitive numerical indicator of sewage pollution, the implication being that the greater the coliform count the greater the proportion of sewage present. This rational practical approach might encourage a search for other constituents of sewage that could serve as tracers. Silica, for example, which exists in sewage at concentrations much greater than in seawater has already been successfully used for tracing the path of sewage dispersed in sea water and in determining the initial dilution of sewage on discharge from an outfall.

4. To ensure that standards are met requires surveillance and monitoring at critical parts of the shore or area in the vicinity of the discharge, as well as sampling at specified distances at a given frequency and arrangements to confirm the results obtained by some authorized organization. The more the method and frequency of sampling and examination can be codified and put on to a routine basis the quicker it will be to establish norms and identify unsatisfactory conditions. In this connection it is interesting to note that a proposal has recently been submitted by the Commission of the European Communities to its Council concerning a directive that would bind its member states to adopt micro-

biological and chemical limits for sea bathing waters and to lay down conditions of frequency and method of sampling and examination. The guidelines and criteria for recreational sea water quality reported by a working group brought together by the Regional Office for Europe of the World Health Organization in 1974 have an important bearing on the microbiological aspects of sea water quality criteria.

5. A brief reference should be made to steps that have been taken when sea water has failed to meet bacterial standards set for the purpose of protecting bathers or shellfish culture. The action taken has been to prohibit bathing from certain parts of the beach or the harvesting of shellfish from the sea-bed underlying polluted waters. It is usual by means of public health regulations to prohibit the exposure for sale of shellfish that contain more than a prescribed limit of *E. coli* per unit weight of fish flesh.

6. It is suggested that marine pollution should be regarded as one aspect of the pollution of all surface and underground waters. Such an approach has the advantage of focusing attention on measures for ensuring adequate treatment of sewage and industrial effluent discharging to rivers and estuaries. The same powers can be used to ensure that industrial effluents are properly controlled before acceptance into sewers that outfall to the sea.

7. By using legal powers to control the construction of houses and factories by granting planning permission only where adequate sewerage and sewage treatment facilities exist, by the exercise of similar powers over the route followed by sewers laid through territorial waters and beyond these waters, and by granting or witholding financial aid for public works, pollution control authorities, planning authorities and other government departments are able to determine the conditions under which marine discharges may be made. The exercise of such powers places on the authorities involved some responsibility for the results obtained and it is difficult to see how legal standards could be used in these circumstances.

8. The question of sludge dumping at sea is examined. This process can be carried out satisfactorily, it is concluded, within the capacity of some parts of the ocean and the sea bed to absorb the sludge provided legal control is exercised by each nation over what is dumped, where it is dumped and in what quantity. Control is also required over the composition of the sludge so as to avoid any harmful ecological effect.

The ability of an area to absorb sludge is capable of being determined by *in situ* and laboratory investigation. Dumping sludge in the sea at a rate greater than the ability of the water and the sea bed to absorb it is likely to produce serious economic and ecological effects. Because of the differences that exist in different seas and climates it is difficult to relate experience in sludge dumping from one part of the world to another.

Acknowledgements — The author is grateful to his present and former colleagues for permission to refer to private sources of information.

Mr. H. Fish, Director of Scientific Services, Thames Water Authority kindly supplied information concerning sewage sludge statistics and analyses relating to sludge dumped in Barrow Deep and Black Deep.

REFERENCES

Agg, A.R. (1975) Criteria for marine waste disposal in Great Britain. *Marine Pollution and Marine Waste Disposal*, Ed. E. de F. Frangipane and E.A. Pearson, Pergamon Press, Oxford.

Aubert, J. and Breittmayer, J.P. (1975) The Mediterranean. *Marine Pollution and Marine Waste Disposal*, Ed. E. de F. Frangipane and E.A. Pearson, Pergamon Press, Oxford.

Aubert, M. and Donnier, B. (1973) Evaluation of ecological consequences of marine pollution, pp. 337-345. *Progress in Water Technology*, Vol. 3, *Water Quality; Management and Pollution Control Problems*, Ed. S.H. Jenkins, Pergamon Press, Oxford.

The Bacteriological Examination of Water Supplies, Report on Public Health and Medical Subjects No. 71, H.M.S.O. London, 3rd Ed. 1957, p. 29 *et seq.*
4th Ed. 1969, p. 25.

Bargman, R.D. (1975) Experience with marine waste disposal systems in the City of Los Angeles, *Marine Pollution and Marine Waste Disposal*, Ed. E. de F. Frangipane and E.A. Pearson, Pergamon Press, Oxford.

Camp, I.C. (1975) *Pollution Criteria for Estuaries*, p. 7.23. Ed. P.R. Helliwell and J. Bassanyi, Pentech Press, London.

Carter, H.H. (1975), Prediction of far-field exclusion areas and effects, *Discharge of Sewage from Sea Outfalls*, Ed. A.L.H. Gameson, Pergamon Press, Oxford.

Caspers, H. (1975), Dumping of sewage sludge in the North Sea. Paper given at the F.I.A.C. Seminar on Pollution in the Mediterranean Sea, La Manga del Mar Minor, Spain, March 20-22, 1975.

Commission of the European Communities (1975), Proposal for a council directive relating to pollution of sea water and fresh water for bathing (quality objectives), submitted to the Council by the Commission, Brussels, 3rd February, 1975.

Convention for the Prevention of Marine Pollution by Dumping from Ships and Aircraft, Oslo, 15 Feb. 1972.

Covill, R.W. (1975), Bacteriological, biological and chemical parameters employed in the Forth Estuary. *Pollution Criteria for Estuaries*, pp. 9.1-9.62, Ed. P.R. Helliwell and J. Bassanyi, Pentech Press, London.

Discharge of Sewage from Sea Outfalls (1975) Ed. A.L.H. Gameson, Pergamon Press, Oxford.

Feuerstein, D.L. (1975) Functional design of outfall and treatment systems, *Marine Pollution and Marine Waste Disposal*, Ed. E. de F. Frangipane and E.A. Pearson, Pergamon Press, Oxford.

Gameson, A.L.H. (1975) Experiences on the coast of Great Britain, *Marine Pollution and Marine Waste Disposal*, Ed. E. de F. Frangipane and E.A. Pearson, Pergamon Press, Oxford.

Helliwell, P.R. (1975) *Pollution Criteria for Estuaries*. London.

Italian Proposed Limits for Waters (1973) circular 105, Italian Ministry of Health, 2 July, 1973.

Joint Group of Experts on the Scientific Aspects of Marine Pollution (G.E.S.A.M.P.).

Kehr, R.W. and Butterfield, C.T. (1943) Notes on the relationship between coliform and enteric pathogens, *Publ. Hltb Rep.* 58, 589-607.

Key, A. (1970) Water Pollution Control in coastal areas; where do we go from here? *Proc. on Symposium on Water Pollution Control in Coastal Waters, Bournemouth, 1970*, p. 111 Inst. Water Poll. Cont.

Lacy, W.J. and Rey, G. (1975) Technology of marine disposal of industry wastes, *Marine Pollution and Marine Waste Disposal*, Ed. E. de F. Frangipane and E.A. Pearson, Pergamon Press, Oxford.

Ludwig, H.F. (1975) Criteria for Marine waste disposal, *Marine Pollution and Marine Waste Disposal*, Ed. E. de F. Rangipane and E.A. Pearson, Pergamon Press, Oxford.

Mackay, D.W. (1975) Techniques for pollution control in estuarial waters, *Criteria for Estuaries*, pp. 11.1-11.12. Ed. P.R. Helliwell and J. Bossanyi, Pentech Press, London.

McGregor, A. (1975) Medical aspects of estuarine pollution, *Pollution Criteria for Estuaries*, pp. 3.1-3.18. Ed. P.R. Helliwell and J. Bossanyi, Pentech Press, London.

McKee, J.E. (1960) The need for water quality criteria, *Proc. Conf. Physical Aspects of Water Quality*, Ed. H.A. Faber and J. Bryson, U.S. Publ. Health Serv., Washington D.C. 244 pp.

Moore, B. (1975) The case against microbial standards for bathing beaches, *Discharge of Sewage from Sea Outfalls*, Ed. A.L.H. Gameson, Pergamon Press, Oxford.

Oakley, H.R. and Staples, K.D. (1975) Interpretation of field data, *Discharge of Sewage from Sea Outfalls*, Ed. A.L.H. Gameson, Pergamon Press, Oxford.

Ocean Dumping (1970) A National Policy Report to the President prepared by the Council on Environmental Quality, Oct. 1970, Washington, D.C., p. 65.

Paolletti, A. (1975) Studies of marine waste disposal for the city of Maiori utilizing a continuous monitoring system. *Marine Pollution and Marine Waste Disposal*, Ed. E. de F. Frangipane and E.A. Pearson, Pergamon Press, Oxofrd.

Petrilli, F.L., De Flora, S. and Lemori, L. (1975) Pollution of coastal waters in Italy. Bacteriological and virological research, *Marine Pollution and Marine Waste Disposal*, Ed. E. de F. Frangipane and E.A. Pearson, Pergamon Press, Oxford.

Porter, E. (1973) *Pollution in four Industrialised Estuaries. Four Case Studies Undertaken for the Royal Commission on Environmental Pollution*, H.M.S.O. London.

Portmann, J.E. The toxicity of 120 substances to marine organisms. Report of the Ministry of Agriculture, Fisheries and Food Laboratory, Burnham-on-Crouch.

Portmann, J.E. (1975) Persistent organic residues, *Discharge of Sewage from Sea Outfalls*, Ed. A.L.H. Gameson, Pergamon Press, Oxford.

Shelef, G. (1975) Çriteria for marine waste disposal in Israel, *Marine Pollution and Marine Waste Disposal*, Ed. E. de F. Frangipane and E.A. Pearson, Pergamon Press, Oxford.

Shuval, H.I. (1975) The case for microbial standards for bathing beaches, *Discharge of Sewage from Sea Outfalls*, Ed. A.L.H. Gameson, Pergamon Press, Oxford.

Smith, P.R. (1975) Pollution Cirteria for Estuaries, p. 11.17. Ed. P.R. Helliwell and J. Bossanyi, Pentech Press, London.

Standard Methods for the Examination of Water and Wastewater, American Public Health Association, Inc. 12th Ed. 1965, 13th Ed. 1971.

Stevenson, A.H. (1953) Studies of bathing water quality and health. *Am. J. Pub. Hlth*, **43**, 529.

Stirn, J. (1975) Criteria for marine waste disposal in Yugoslavia, *Marine Pollution and Marine Waste Disposal*, Ed. E. de F. Frangipane and E.A. Pearson, Pergamon Press, Oxford.

Stirn, J., Avcin, A., Kerzan, I., Marcotte, B.M., Meith-Avcin, N., Vriser, B. and Vukovic, S. (1972) Selected biological methods for assessment of marine pollution, *Marine Pollution and Marine Waste Disposal*, Ed. E. de F. Frangipane and E.A. Pearson, Pergamon Press, Oxford.

Svansson, A. (1975) Some problems in the Baltic, *Discharge of Sewage from Sea Outfalls*, Ed. A.L.H. Gameson, Pergamon Press, Oxford.

Taylor, E.W. and Burman, N.P. (1964) The application of membrane filtration techniques to the bacteriological examination of water, *J. Appl. Bact.* **27**, 294.

van Dam, G.C. (1975) The Hague outfall, *Discharge of Sewage from Sea Outfalls*, Ed. A.L.H. Gameson, Pergamon Press, Oxford.

Wakefield, J.A. (1975) *Pollution Criteria for Estuaries*, Ed. P.R. Helliwell and J. Bossanyi, p.9.59, Pentech Press, London.

Working Party (1972) Out of sight out of mind, *Report of a Working Party on Sludge Disposal in Liverpool Bay*, Vol. 2 appendices. Department of the Environment, H.M.S.O., London.

World Health Organization (1975) Guides and criteria for recreational quality of beaches and coastal waters. Report of a Working Group, Regional Office for Europe, Copenhagen, 1975.

CRITERIA FOR INDUSTRIAL WASTEWATER DISPOSAL IN THE MARINE ENVIRONMENT

Luigi Mendia
"Centro Studi e Ricerche di Ingegneria Sanitaria",
Faculty of Engineering, University of Naples, Naples, Italy

1. INTRODUCTION

The type and importance of the detrimental effects caused by wastewater disposal on the receiving waters depend, as known, on the categories of pollutants characterizing the discharges as well as on the geographic and natural characteristics of the receiving waters.

It is therefore obvious that a number of criteria adopted, e.g. for the ultimate disposal of sewage, cannot be applied to all municipal discharges and *a fortiori* to industrial effluents.

Similarly, the use of the sea as the ultimate receiving body of discharges makes some conditions either unfavourable or favourable to their disposal (depending on the type of effluents), but anyhow different from those found in fresh-water environments.

1.1. Effects of marine pollution

The negative effects caused by the disposal into the sea of industrial effluents may be summarized as follows:

    (a) hazards to public health

    (b) damages to natural resources

    (c) loss of recreational uses and amenities

    (d) hindrance to maritime activities

    (e) obstacles to the use of sea waters in the external environment

The above mentioned damages may be only seldom connected with a particular class of pollutant; more often, instead, the damage is a consequence of a series of joint-causes or factors of pollution, e.g. oxygen deficiency, temperature, euthrophication, synergic and/or inhibitory actions.

1.2. Classification of discharges

A general classification of industrial effluents on the basis of the damages they may cause should be as follows:

    hazardous to public health   (pathogenic micro-organisms, parasitic forms, irritants)

directly toxic or noxious to natural resources and indirectly noxious to public health	(chromium VI, mercury, cadmium, cyanides, phenols, pesticides, radionuclides)
indirectly noxious to natural resources	(suspended materials, thermal energy, acids, bases)
causing oxygen deficiency	(reducing agents, biodegradable compounds, oils, synthetic detergents, nutrients)
altering the aesthetics	(gross solids, floatables, foams suspended materials, colour)
possibly interfering with maritime activities	(gross materials, plastics, hydrocarbons, acids, nutrients)

As may be seen, pollutants are quite different not only with regard to their effects, but also to their physical and chemical behavour; therefore, once again it is quite obvious how waste disposal mostly requires a specific regulation depending on a given typology of cases.

1.3. Power of the marine environment to solve pollution phenomena

With regard to the capacity of the marine environment to solve pollution phenomena, we should like to mention some peculiar aspects:

    (a) extent of the diluent/dilute ratio;

    (b) phenomena of dynamic oceanography;

    (c) degree of salinity;

    (d) complexity of the pelagic life.

The former two mostly play a favourable role in the solution of pollution phenomena. However, an inadequate evaluation of the degree of reliability of such factors is the basis of some incorrect disposal solutions. As a matter of fact, not all pollutants are treatable by dilution and dispersion phenomena without any damage to the marine environment, although a number of striking pollution phenomena may thus be solved in a shorter time and within a narrower space. This is one of the reasons why some discharges should not be disposed to the sea without preliminary treatment, as in the case of toxic substances with an accumulating action.

With regard to the positive influence of meteomarine phenomena, any generalization made on the basis of the experiences acquired under different geographic conditions, may be misleading. In our opinion, many useless controversies which have developed during these last few years, especially in the Mediterranean countries, on the criteria for marine wastewater disposal, have arisen from the indiscriminate use of experiences that have been acquired, e.g. with disposal systems in deep and open seas with wide tidal cycles. It should be obvious that systems and techniques successfully adopted along the Pacific and Atlantic coasts, or in the North Sea, cannot be adopted as such in the Mediterranean, which is characterized by a slow water exchange and by tides of the order of a few tens of centimetres. Furthermore, in our opinion, on critically evaluating such solutions, one should not forget to take into account the socio-economic and temporary conditions in which choices have been made. In particular we refer to the different uses of the coastlines, mentioned above, with respect to those of seas like the Mediterranean.

By recalling the two other aspects, i.e. salinity and the complexity of pelagic life, we should remember that they hinder the solution of marine pollution phenomena by industrial effluents.

Due to chemical and biochemical reasons, as is well known, salinity delays biodegradation of organic substances. However, the most delicate aspect consists in the variety and complexity of the marine ecosystems compared with the fresh water ones. The more varied the levels of the food chains in which accumulation phenomena take place, and the particular importance that marine resources have as foodstuffs for population, the greater are the possibilities for industrial effluents to become a menace to public health — by direct and indirect action — certainly the risks are greater than those derived from sewage disposal. Not by chance the movement for the protection of natural environments in general and of the sea in particular, which lately forced public attention, after the appearance of the Minamata disease was caused by pollution from mercury. After this alarm, others followed concerning cadmium, DDT, PCB, various pesticides, etc.

Pollution phenomena, which were once evaluated on a traditional sanitary (microbiological) basis have of recent years been evaluated according to more modern and comprehensive criteria of ecological and toxicological type. Hence, this new trend, i.e. the detection of the consequences and of the detrimental effects brought about by pollution, caused as a consequence, the evolution of measurement techniques and of methods of approach to the problem of abatement of pollutants, especially with regard to industrial effluents.

2. STRATEGY AND GUIDELINES OF THE PROGRAMMES OF POLLUTION ABATEMENT

It is clear that the guidelines for this problem, beyond the permanent problems concerned with the technical methods for purification (which are numerous) are conditioned by obvious economic considerations connected with the costs of technical measures.

Therefore, the definition of the norms is the most critical and, I dare say, the most informed moment of environmental protection. Production and even siting choices depend upon it. Since one wishes to use as well as to protect one's environment, the strategy of marine pollution abatement must be a matter of management of coasts; in its turn, such management must be framed within the territorial order forming the hinterland. Furthermore, on considering the amount of pollution caused by rivers, it is evident that environmental protection of the sea must begin upstream, that is it is necessary to tackle the problem of pollution of the water bodies flowing toward the sea, and even before that, i.e. to control emmission of industrial effluents into the drainage system.

In turn, the definition of the whole coastal order depends on environmental policy guidelines formulated at a regional and national level. Any technical action outside such guidelines runs to risk of being unfruitful, unprofitable or even troublesome.

Tables 1, 2, 3, 4 show schemes of the related methods.

The importance of the planning phase, which, as already mentioned, is certainly the most "delicate", apart from geographical differences, varies depending on the socio-economic situations in which action must develop: areas of recent development, areas with defined and restricted use, areas zoned for multiple use. For example, the Mediterranean is characterized by the whole spectrum of situations. Several Mediterranean towns exist in which a number of problems must be simultaneously solved, such as:

settlement of the basic sanitary infrastructures;

development of industrialization;

preservation of natural activities;

abnormal growth of the urban network (megalopolite).

In order to develop adequate norms for situations of this type some compromises may be necessary and the best solution found for each situation. This is possible only in the light of knowledge of local conditions.

As mentioned above, when norms are formulated the development of control and technical measures to implement policy becomes possible. (In particular, among the precautionary measures that may be taken, the adoption of alternative less polluting producing technologies may be considered).

Table 1. Development trends of the strategy for the environmental protection of sea coasts

---

Localization choices along the shore

Definition of the strategy for pollution abatement

Survey of natural conditions, of pollution sources and of the environmental quality

Planning of preventive measures; studies on the evolution of environmental quality

Planning and realization of technical measures

Staff training, research activity

Table 2. Survey and monitoring of the environmental conditions, of pollution sources and consequences

Survey of environmental conditions	Oceanographic, meteorological, physical, chemical, microbiological, ecological
Characterization of pollution sources (earth: direct, indirect; pelagic)	Cartography, Definition of pollutant loads
Evaluation of damages and hazards	Public health, natural resources amenities, activities, maritime external uses of sea water

Evolution patterns of the state of environmental quality (on the basis of the present settlements).

Table 3. Planning of preventive measures

Definition of uses of coastal strips

Range of proposals for settlements and uses of the coast

Preliminary evaluation of the consequent hypothetical pollution situations (evolution patterns of the environmental quality)

Choice of environmental quality criteria for different uses

Technical norms for discharges and standards for effluents

Definitive choice of settlements and of the uses of the coast

Definitive evaluation of pollution situations provisional patterns)

Preliminary proposals for technical solutions

Studies on the feasibility of solutions

Table 4. Planning for technical measures

Funds allocation

Definition of priorities

Management trends (unions, pre-treatments, recovery of material)

Definition of technical choices

Process Research (eventual)

Planning of works

Construction (phased)

Start-up and short-term evaluation of the efficiency of adopted solutions (choices of possible alternatives)

Management of plants

Monitoring of effluents

Systematic and overall control of the quality of receiving waters

Long-term evaluation of the efficiency of adopted solutions (possible adjustments of plants)

3. CRITERIA

Three main types of regulation trends may be considered, that is: *conformity criteria, absolute criteria, mixed criteria.*

3.1. Conformity criteria

According to what has been stated above, the succession of the actions in the case of coastal environment, should be:

1. definition of the quality criteria of coasts;

2. provisional patterns of the evolution of the quality of the environmental situation of the coast:

    polluting loads (rivers and drainage inclusive)
    site of outfalls
    oceanographic characteristics

3. partition of polluting loads among users (riparian);

4. adoption of effluent standards by individual polluters;

5. calibration of interventions;

6. control of observance of quality standards assessment of schemes or proposals.

In the case of municipal sewage disposal, the problems to be solved are essentially technical, and obviously involve both the economy and public administration. Hence, difficulties of a different type arise: programming, scientific.

In one sense, the most "delicate" problem is the partition of polluting loads among users.

At present, the extent of difficulty mostly varies depending on the political-economic system, both regional and national. The role of compromise cannot be hypothesized. The regulation trend is to define the quality criteria for receiving waters depending on the optimal uses, to evaluate the natural capacity of the environment to metabolize possible and/or real pollution, to define the polluting loads that, in view of and in observance with the two quoted presuppositions, may be discharged into the environment considered. Hence regulation tends to conform to (and hence varies with) the environment to be protected. When, as regards their negative effects on the coastal environment, the pollutants involved are not "equivalent" to those of sewage, difficulties arise in defining the quality states of the receiving waters in relation to the constituents of industrial discharges. It is essential to make provision for continuous monitoring. The organization required for data collection and interpretation, though costly, is also necessary for the control of municipal discharges.

Regulation of this type is correct in theory and may be adopted in virgin areas or where overall planning of both territory and coasts is possible, and where an adequate scientific and technical organization is available. In this case, typologies, priorities, and the extent of settlements are defined before the appearance of pollution and therefore even before starting a policy of wastewater purification, the economic burden involved in purification may be predicted.

3.2. Absolute criteria

Far greater difficulties arise when regulations concern waste disposal control in damaged environments from more or less known activities which are anyhow varied and of different kinds, in areas where growth has not followed a reasonable guideline of economic and social development but was dictated by profit and commercial expediency and sometimes by exploitation of the environment.

As an example, we quote some Mediterranean towns, such as Barcelona, Marseille, Genoa, Leghorn, Naples, Palermo, Venice, Trieste, Athens, Istanbul, Tel Aviv.

Among them, the Gulf of Naples constitutes the most thrilling example. When disposal regulations must be applied to situations of this type, it makes no sense, as it is practically impossible, to follow the logic mentioned above.

Certainly, from preliminary considerations, the acceptable limits of polluting loads suggested are practically as severe as those appropriate for the most rigorous conventional standards for effluents.

To quote Naples again, even simple surveys indicate how the displacement of existing settlements would affect pollution and suggest alternative technologies to those already adopted. Therefore, the use of *absolute criteria* seems to be the only remaining possibility of environmental protection, before the damage becomes irreversible (as is already the case in some coastal areas) in order to plan remedial works that cannot be deferred. It is of little importance to probe too deeply into the study of the natural features beyond certain limits, and to follow every detailed parameter of coastal pollution; already, in some cases, there is a sufficiently accurate cartography of outfalls indicating the type and values of the polluting loads.

But is is always convenient to establish quality criteria of the marine environment, just to start planning uses of the water. In practice, this has been the procedure for pollution abatement in the Gulf of Naples: in fact, nothing else could be done in the situation as it now exists in this area.

3.3. Mixed criteria

When criteria must be formulated concerning different types of ultimate receiving waters, such as lakes, rivers, sea, different standards may be adopted for the three types of receiving bodies; on the one hand it must be borne in mind that lakes are more sensitive to pollution in general, on the other hand the marine environment has a certain capacity to withstand particular types of pollutants, such as those present in sewage or in sludges. As to some parameters, e.g. pH, temperature, BOD, COD, dissolved oxygen, coliform numbers, the value of the standard is often related to the measurement done on the receiving waters, also on the basis of the quality criteria, which practically form the basis of most environmental legal methods of pollution control. This is a norm according to *mixed criteria*, which in many cases may be justified objectively. The difficulties met in the definition and realization of such norms are partly the same as those indicated in the case of *conformity criteria*. However, additional difficulties arise concerning the so-called territorial policy.

Owing to the correlation between the attainment of standards and treatment costs, it is obvious that policy may be based on the use of differentiate standards. As already mentioned, regulations and standards in the case of lakes generally are more restrictive than those concerning rivers and less restrictive than those concerning effluents into the sea. Absolute standards arouse considerable difficulties, both practical and economic, for the industrial concerns that have long operated according to permissible criteria and that must tackle the problem of effluent purification almost *de novo*. Hence, a realistic solution would be that of allowing a gradual attainment of such standards and of differentiating economic burdens between old and new establishments.

When the criterion of absolute standards is adopted, the succession of operations is simplified by eliminating the need to model the conditions (which is always valid to define the environmental situation) and that of partition of polluting loads. As already mentioned, the choice of the regional policy being developed plays a non negligible role in the choice of differential standards. A present trend, at least in some industrialized Mediterranean countries, e.g. Italy, is to start resettlement of populations away from the coastal region. This may be attained by not adopting lenient standards for effluent disposal in coastal waters. Furthermore, by unifying standards, unfair economic burdens on hinterland and coastal settlements are avoided. If such a planning scheme is adopted, it is impossible to make different standards for municipal discharges and industrial effluents. Vice-versa, the principle of forbidding dilution before discharging wastes, if carried out in order to reach standards, minimizes the error that may be made by adopting the *absolute criteria* of standards of effluents.

Clearly, the concept is to reduce the polluting load to be disposed into the sea.

3.4. Quality criteria

All laws concerned with marine pollution more or less state that waste disposal must be carried out without causing hazards to public health, damage to natural resources, and aesthetics. Some national laws give more specific indications by fixing quality criteria for receiving waters, such as pH, oxygen demand, coliform numbers, and by criteria such as the absence of gross floatables, visible oils, noxious compounds and undesirable materials conveyed by wastewaters. Some criteria also refer to the distance of the outfall from the coast, at which some characteristics, e.g. temperature must be measured.

For example, in Poland laws request that at a 1 km distance from the coast, sea waters must exhibit their natural features.

On considering carefully such laws, it appears that their observance is possible only by adopting adequate inland purification plants, the quality of their effluents practically corresponding to the levels given by several regulations adopting *absolute criteria*.

In this regard, it must be borne in mind that it is not always possible to carry out purification to obtain infinite gradualness of effluent quality; in general it is operated in steps and purification is achieved in stages.

3.5. Technical norms

The technical and management norms adopted by most countries are on a regional level.

A paper presented by Pearson at Sanremo, 1972, reported some management principles adopted in California, among which we quote the last: "Submarine outfall dispersion systems must be designed to achieve the most rapid initial dilution and the maximum practicable dispersion".

Once the norm system is chosen, the technical solutions that must be adopted are the responsibility of designers; it might even be undesirable to give too specific suggestions that might be construed as preferential indications for one purification technique or another.

Vice-versa, with regional programmes, design guides must be assumed, the validity of which is not axiomatic but is necessarily connected with the general socio-economic and geographic conditions of the environment where remedial work must be done, taking into account the available technical organization and facilities.

4. A SURVEY OF THE ITALIAN LAWS AND NORMS

The protection of the marine environment is presently entrusted to the law on sea fishing (No. 963, July 14, 1965) substantially enacted in the interests of fisheries. According to such law, there is forbidden "the direct or indirect discharge of pollutants into waters". Therefore, control concerns both direct and indirect sources (rivers, canals); it is further specified that "polluting substances are the substances — either extraneous or belonging to the normal composition of natural waters — that constitute direct damage to ichthyic fauna or effect such chemical or physical alterations of the environment as to unfavourably influence the life of aquatic organisms". Hence there are general principles and, as it has been lately observed, it is substantially left to the experts' judgement to establish whether a given discharge causes — or not — pollution, i.e. a damage to natural resources. From a legal point of view, extensive discussions arise when — due to judicial reasons — it is necessary to establish the noxiousness of a given discharge in a long since polluted environment. Since the immediate detrimental consequences cannot be detected *de facto*, the environmental damages being already chronic, one must rather refer to a potential dangerousness for the marine environment of the discharge under examination; this fact certainly arouses subtle legal debates and practical discussions on the acceptability limits (standards) of effluents.

With regard to public health protection, a first general reference consists in the Act of the Sanitary Laws promulgated in 1934; in some items, i.e. Nos. 217, 226, 227, it is established that the sanitary authority must exert a precautionary control of municipal and industrial waste disposal in order to prevent damages to the health. Such provisions may indirectly concern marine pollution. A more specific reference to the connection between public health protection and marine pollution is found in circulars of the Italian Ministry of Health concerning the criteria of water quality for bathing (average 100 *E. coli*/100 ml) and regulation of the sale of edible mussels (wider quality criteria for both types of activity will be the object of later circulars, since a special Commission has been appointed). Therefore, it may be realized how important is the Ministry Circular No. 105 giving the acceptability limits of effluents. Other specific laws for the protection of the marine environment may be referred to: according to item 71 of the navigation code, waste dumping in harbours is forbidden; according to items 22 and 30 of the law of July 21, 1967, No. 613: research on and exploitation of hydrocarbons in territorial waters must be developed in line with the directions of maritime authorities entrusted with the safeguard and preservation of biological resources and of the marine environment.

As may be seen, all laws, even the most up-to-date ones, refer to damage that may result to the marine environment and to public health as a consequence of pollution, but do not specifically rule, the "matter of discharges"; in practice, only by authorizations is it possible to take preventive action and safeguard receiving waters from municipal and industrial pollutions.

Since regulations in this matter are lacking, for a long time the whole subject has lacked control except in particular cases where pollution actually occurred. The Bench has lately obliged — or anyhow stimulated — industries and public administrations to actually tackle the problem of the treatment of waste waters. In the absence of adequate norms and by passing from an extremely lenient situation to a quite rigorous one, pollution control was initiated with much confusion and fear rather than through scientific and civic action.

In the wake of an obscurantism that lasted for many decades, which actually caused the decay of our environment, industrial concerns (with very few exceptions) tried to delay whenever possible all action aimed at purification under the pretext of the lack of laws, regulations and standards for effluents.

At a FAST meeting held in Milan in 1970, the author — with a working group belonging to IRSA, a branch of the Italian National Council for Research — indicated the quality limits for effluents and explicitly stated that such values were a reference for further discussions on the problem. But no debate or polemics followed.

L. Mendia

Table 5. Italian standards for final effluents to be discharged into the sea.

Parameter	Circular of Ministry of Health	Bill to Parliament Provisional situation	Bill to Parliament Definitive situation
pH	$5 - 9.5$*	$5.5 - 9.5$	$5.5 - 9.5$
Temperature	35°C**	$\leqslant 3^\circ$C/basic T[c]	$\leqslant 3^\circ$C/basic T[h]
Colour	1:40/10 cm	1:40/10 cm	1:20/10 cm
Odour	no nuisance	no nuisance	no nuisance
Gross materials	absent	absent	absent
Settleable materials ml/l	2.0	2	0.5
Suspended materials mg/l	150*	Min: def. table[d] Max: 200	80
BOD$_5$ mg/l	80*	Min: def. table[e] Max: 250	40
COD mg/l	250*	Min: def. table[f] Max: 500	160
Total toxic metals and non metals mg/l (As–Cd–Cr–Cu–Hg– Ni–Pb–Se–Zn)	$\dfrac{C_1}{L_1} + \dfrac{C_2}{L_2} + \dfrac{C_n}{L_n}^{(a)} \leqslant 3$	$\leqslant 3$	$\leqslant 3^{(g)}$
Aluminium mg Al/l	–	2	1
Arsenic mg As/l	0.5	0.5	0.5
Barium mg Ba/l	40	–	20
Boron mg B/l	20	4	2
Cadmium mg Cd/l	0.1	0.02	0.02
Chromium III mg Cr/l	2	4	2
Chromium VI mg Cr/l	0.5	0.2	0.2
Iron mg Fe/l	2[b]	4	2
Manganese mg Mn/l	2[b]	4	2
Mercury mg Hg/l	0.01	0.005	0.005
Nickel mg Ni/l	4	4	2
Lead mg Pb/l	1	0.3	0.2
Copper mg Cu/l	0.05	0.4	0.10
Selenium mg Se/l	0.1	0.03	0.03
Tin mg Sn/l	–	–	10
Zinc mg Zn/l	1	1	0.5
Free cyanides mg CN^-/l	1	–	–
Total cyanides mg CN^-/l	–	1	0.5
Active chlorine mg Cl_2/l	2	0.3	0.2
Sulfides mg H_2S/l	2	2	1
Sulfites mg SO_3/l	10	2	1
Sulfates mg SO_4/l	–	–	–
Chlorides mg Cl^-/l	–	–	–

Parameter	Circular of Ministry of Health	Concentrations Bill to Parliament Provisional situation	Definitive situation
Fluorides mg F^-/l	20	12	6
Total phosphorus mg P/l	20 (PO_4^{-3})	10	10
Ammonia mg NH_4^+/l	5	30	15
Nitrous nitrogen (NO_2^-) mg/l	2	0.6 (=N)	0.6 (=N)
Nitric nitrogen (NO_3^-) mg/l	50	30 (=N)	20 (=N)
Greases and animal and vegetable oils	20	40	20
Mineral oils mg/l	3	10	5
Phenols mg C_6H_5OH/l	0.5	1	0.5
Aldehydes mg H CHO/l	4	2	1
Organic aromatic solvents mg/l	1	0.4	0.2
Organic nitrogenous solvents mg/l	1	0.2	0.1
Chlorinated solvents		2	0.05
Surface active agents mg MBAS/l	6	4	2
Chlorinated pesticides mg/l	0.1	0.05	0.05
Phosphorous pesticides mg/l	0.2	0.1	0.1
Toxicity test	*Carrassius auratus*[i]	*Carrassius auratus*[l]	*Salmo gairdnerrii* Rich[m]
Total coliforms MPN/100 ml	Ref. quality criteria	20 000	20 000
Fecal coliforms MPN/100 ml	Ref. quality criteria	12 000 [n]	12 000 [n]
Fecal streptococci MPN/100 ml	Ref. quality criteria	12 000	2 000
		(o)	(o)

Notes:

(*) Except variations when dilution conditions are ascertained, allowing the uses for which the discharge area is intended.

(**) If, by the action of discharges, a region of the receiving waters exists where temperature variation takes place, some modifications occur that negatively affect the ecosystem, e.g. an obstacle to the migration of ichthyc fauna. In order to limit alterations, for example, it must be shown that only one part of the area be affected by the thermal phenomenon and in the case of sea, such a phenomenon must not occur close to a mouth.

(a) Being established that the limit fixed for each element must not be exceeded, the sum of the ratios between the concentration of each element and the related limit concentration must not exceed 3.

$$\frac{C_1}{L_1} + \frac{C_2}{L_2} + \frac{C_n}{L_n} \leqslant 3$$

(b)   Iron as Fe + Manganese as Mn = 4 mg/l.

(c)   As to the sea, the discharge temperature must not exceed 35°C and temperature increase of the receiving waters must never exceed 3°C beyond a distance of 1000 m from the discharge point.

(d)   No more than 40% of the value before the purification plant; the minimum limit that may be imposed is that of the final table, max 200 ml/l.

(e)   No more than 70% of the value before the plant; minimum that may be imposed; max limit 250 ml/l.

(f)   No more than 70% of the value before the plant: minimum that may be imposed: max limit 500 ml/l.

(g)   As (a).

(h)   As (c). Furthermore, the formation of thermal barriers at the river mouth must be avoided.

(i)   Use is made of *Carrassius Auratus* placed in the effluent from which free residual chlorine has been removed and a 1:1 dilution made with standard water at a temperature of 20°C $\pm$ 1. After 6 hours, survival must not be less than 50%.

After such an initiative, in 1971, the "Consiglio Superiore della Sanita" appointed a special Commission that, after the author's communication, issued a circular (No. 105) stating the acceptability limits of wastewaters in sewer systems, rivers, lakes and sea: hence directions were given to peripheral bodies, when called upon, for granting discharge authorizations or anyhow for controlling wastewater discharges.

The values concerning sea dumping are reported in Table 5. As to some properties (pH, colour, suspended solids, BOD, COD) limits are less restrictive than those foreseen in the same circular for disposal into rivers; derogations are foreseen when "dilution conditions are ascertained that allow the area involved to be used for purposes for which it is intended".

The different values mostly refer to the polluting features that may make use of the dilution and dispersion capacity of the marine environment, the damage caused or the inconvenience being limited within time and space. Therefore, the criterion adopted is of the mixed type.

With regard to biological indices of pollution, reference is made to the quality requirement for bathing in the receiving waters (100 *E. coli*/100 ml). It is also required that the chlorine residue must be further controlled in the receiving waters: within 50 m from the outfall, it must not exceed 0.2 mg/l.

The lack of laws on water protection from pollution has been discussed for more than 15 years. After a preliminary bill drawn by ANDIS in the early sixties, on the initiative of various Government commissions several other bills have been proposed, but none of them has received Parliament's approval. The most recent bill, which, as it seems, will be brought to a successful issue, was drawn by a Parliamentary Committee presided by Hon. Merli. The bill also contains Tables concerning acceptability limits for effluents, the values of which are reported in Table 5. The data were determined by the Commission after technical meetings with representatives of regional bodies and of industrial concerns.

The most striking features of the bill under consideration are summarized here:

uniqueness of the acceptability limits in respect of receiving waters (with some exceptions for lakes) on the basis of absolute standards for effluents;

gradualness in reaching the limits: Table A (more restrictive) for new production units, Table C for existing establishments; as to civil community developments, the gradualness of adjustment to the limits of Table A is determined by regional reclamation plans;

prohibition of dilution in order to reach limits;

reference to quality criteria of receiving waters for some parameters such as:

possibility of making the limits proposed more restrictive, at the decision of regional authorities, after considering particular local situations;

forecast of a series of economic measures in order to reduce the burdens incurred in building purification plants.

Obviously, such tough regulations and the inevitably rigorous standards have aroused widespread and manifold criticism. The most frequent arguments are: the non-reference to the "sacrificed" hydrographic body unit in the name of the administrative regional system, the system of fixed standards (though flexible and mixed) and the severity of some limits.

In particular cases, autochthonous species may be used, for which the experimental methods must be defined every time. The standard dilution was prepared as follows:

$CaCO_3$ (8.4 g) was dissolved by bubbling with CO_2 in deionized water (40 l). This amount is later brought to 120 l by additional 80 l deionized water. Then a solution (120 l) of NaCl (0.4 mole) + $MgSO_4$ (0.3 mole) + K_2SO_4 (0.025 mole) was added. Then vigorous aeration is performed until the pH is above 7 and dissolved oxygen saturation is reached. Total hardness must be further controlled, which must be about 100 mg/l of $CaCO_3$ (tentatively 100 ± 5 mg/l $CaCO_3$).

(1) The 1:1 diluted sample with standard water must allow — under aeration conditions — the survival of at least 50% test animals, employed for the test over a 24-h period at $20°C$. The species used for the tests must be *Carrassius auratus*.

(2) As in (1) at $15°C$. The species used for the test must be *Salmo gairdnerii* Rich.

(3) The limit is applied when requested by Control Authorities for joint uses of the receiving waters (bathing, molluscs culture; note of the author).

(4) Analyses must be carried out on an average sample, drawn at minimum intervals of 3 h. The analytical sampling methods to be adopted in the determination of parameters are those described in the "Analytical Methods for Waters" published by the "Istituto di Ricerca sulle Acque" (C.N.R.), Rome, and subsequent issues.

It is clear that, in addition to valid reasons, other reasons are raised to delay law enactment in the hope that the present state of ambiguity will continue, which is less burdensome to the polluter from the economic standpoint. Everyone realizes that economic and managerial burdens are involved in the application of the "Merli Law" but obviously, no environmental policy can be developed without considerable economic sacrifice by the community. It is not even worth checking costs/benefits, as the answer is so obvious. Efforts to protect our water bodies and our coasts cannot be delayed any longer; perhaps it is already impossible to make up for lost time: in any case, whatever delay occurs obviously increases the remedial costs later. As a matter of fact, owing to increased deterioration of receiving waters, purification plants will now have to guarantee a higher efficiency; furthermore as it is well known, prices will continue to increase.

In considering the present deterioration of Italian waters and coasts, the lack of economic and territorial planning concerned with the environmental protection, the lack of a convenient quality requirements for water bodies and coasts, the lack of control services of environmental quality, overcrowding in urban and industrial dwellings along most national coasts, the need to protect natural coastal scenery, which are a source of tourist attraction and economic development, the administrative reality and policy of Regions, the need to avoid economic imbalance on a regional level (with possible lack of uniformity in fixing standards), the opportunity of limiting the polluting load to be disposed of in the receiving waters, we believe that the bill, even with the limitations that are necessarily involved in rigid regulations (though theoretical in the case of Italy) — constitutes one of the best possible compromises as regards the protection of waters and of the sea coasts from pollution, after the 15-year long period of "ecological games".

REFERENCES

Mendia, L., Aspetti della lotta contro l'inquinamento marino. *Ing. San.* No. 4-5-6, July—Dec (1971).

Mendia, L., Inquinamento industriale ed urbano: fonti, effetti, prevenzione, rimedi. Proceedings of the Interparliamentary Conference of Coastal Countries on Pollution Abatement in the Mediterranean Sea. Chamber of Deputies, Rome, April 29, 1974, "Servizio Studi Legilazione ed Inchieste Parlamentary".

Marchetti, R., Gerletti, M., Colamari, D. and Chiandani, G., Elementi e Criteri per la Definizione del livello di Accettabilita della acque di scarico. *Quaderni dell'IRSA* — CNR No. 24, Roma 1973.

Pearson, E.A., Criteria for marine waste disposal in California, *Marine Pollution and Marine Waste Disposal*, Ed. E. de F. Frangipane and E.A. Pearson. Pergamon Press, Oxford, 1975.

AUTHOR INDEX

Aubert, M. 33, 41

Bascom, W. N. 99

Emmett, R. 230

Frangipane, E. de F. 1, 11
Fried, A. 191

Giaccone, G. 51
Gilad, A. 1, 113

Jenkins, S. H. 237

Klapow, L. A. 77

Lacy, W. J. 213
Lewis, R. H. 77
Lucas, R. S. 231
Ludwig, H. F. 1, 223

Markantonatos, G. 131
Mearns, A. J. 19
Mendia, L. 271

Oakley, H. R. 167

Paoletti, A. 149
Pearson, E. A. 1
Portmann, J. E. 59

Sheinberg, Y. 191
Shelef, G. 191
Snook, W. G. G. 179, 197
Storrs, P. N. 87

Yoshpe-Purer, Y. 191

SUBJECT INDEX